Phytochemistry of Fruit and Vegetables

PROCEEDINGS OF THE PHYTOCHEMICAL SOCIETY OF EUROPE

23. **The Genetic Manipulation of Plants and its Application to Agriculture**
    Edited by P.J. Lea and G.R. Stewart
24. **Membranes and Compartmentation in the Regulation of Plant Functions**
    Edited by A.M. Boudet and others
25. **The Biochemistry of Plant Phenolics**
    Edited by C.P. Van Sumere and P.J. Lea
26. **Plant Products and the New Technology**
    Edited by K.W. Fuller and J.R. Gallon
27. **Biologically Active Natural Products**
    Edited by K. Hostettmann and P.J. Lea
28. **Biochemistry of the Algae and Cyanobacteria**
    Edited by L.J. Rogers and J.R. Gallon
29. **Bryophytes: Their Chemistry and Chemical Taxonomy**
    Edited by H.D. Zinsmeister and R. Mues
30. **Secondary Plant Products from Tissue Culture**
    Edited by B.V. Charlwood and M.J.C. Rhodes
31. **Ecological Chemistry and Biochemistry of Plant Terpenoids**
    Edited by J.B. Harborne and F.A. Tomás-Barberán
32. **Biochemistry and Molecular Biology of Plant–Pathogen Interactions**
    Edited by C.J. Smith
33. **Nitrogen Metabolism of Plants**
    Edited by K. Mengel and D.J. Pilbeam
34. **Phytochemistry and Agriculture**
    Edited by T.A. Van Beek and H. Breteler
35. **Seed Storage Compounds: Biosynthesis, Interactions, and Manipulation**
    Edited by Peter R. Shewry and Keith Stobart
36. **Biochemical Mechanisms Involved in Plant Growth Regulation**
    Edited by C.J. Smith, J. Gallon, D. Chiatante, and G. Zocchi
37. **Phytochemistry of Plants Used in Traditional Medicine**
    Edited by K. Hostettmann, A. Marston, M. Maillard and M. Hamburger
38. **Plant Membrane Biology**
    Edited by I.M. Møller and P. Brodelius
39. **Protein Phosphorylation in Plants**
    Edited by P.R. Shewry, N.G. Halford and R. Hooley
40. **Applications of Modern Mass Spectrometry in Plant Science Research**
    Edited by R.P. Newton and T.J. Walton

PROCEEDINGS OF THE
PHYTOCHEMICAL SOCIETY OF EUROPE

Phytochemistry of Fruit and Vegetables

Edited by

F.A. TOMÁS-BARBERÁN
Laboratorio de Fitoquímica, Departamento de Ciencia y Tecnología de Alimentos, CEBAS (CSIC), Murcia, Spain

and

R.J. ROBINS
Laboratoire de d'Analyse Isotopique et Électrochimique de Métabolismes, URA-CNRS 472, Faculté des Sciences, Nantes, France

CLARENDON PRESS · OXFORD
1997

This book has been printed digitally in order to ensure its continuing availability

OXFORD
UNIVERSITY PRESS

Great Clarendon Street, Oxford OX2 6DP

Oxford University Press is a department of the University of Oxford.
It furthers the University's objective of excellence in research, scholarship,
and education by publishing worldwide in

Oxford New York

Auckland Bangkok Buenos Aires Cape Town Chennai
Dar es Salaam Delhi Hong Kong Istanbul Karachi Kolkata
Kuala Lumpur Madrid Melbourne Mexico City Mumbai Nairobi
São Paulo Shanghai Singapore Taipei Tokyo Toronto

with an associated company in Berlin

Oxford is a registered trade mark of Oxford University Press
in the UK and in certain other countries

Published in the United States
by Oxford University Press Inc., New York

Reprinted 2001

A catalogue record for this book is available from the British Library

Library of Congress Cataloging in Publication Data
Phytochemistry of fruits and vegetables / edited by F. A. Tomás-Barberán and R. J. Robins. - (1st ed.)
(Proceedings of the Phytochemical Society of Europe; 41)
Revised versions of papers presented at the International
Symposium on Phytochemistry of Fruit and Vegetables
held in Murcia, Spain, Sept. 1995.
1. Fruit-Physiology-Congresses. 2. Vegetables-Physiology-Congresses. 3. Fruit-composition-Congresses. 4. Vegetables-Composition-Congresses. 5. Botanical chemistry-Congresses.
I. Tomás-Barberán, F. A. (Francisco A.) II. Robins, Richard J.
III. International Symposium of Phytochemistry of Fruit and Vegetables (1995: Murcia, Spain) IV. Series.
SB357.28.P48 1997 581.19'2—dc20 96-9372

ISBN 0-19-857790-7 (Hbk)

Preface

This book constitutes the Proceedings of the International Symposium on Phytochemistry of Fruit and Vegetables, which was held in Murcia, Spain, in September 1995. This meeting was organized by the Phytochemical Society of Europe and the Spanish Consejo Superior de Investigaciones Científicas. This was the second meeting of the Phytochemical Society to take place in Spain.

The choice of subject enabled us to invite speakers from the United States and Canada as well as Europe to survey some of the exciting progress made in this field in the last 15 years. Some 150 participants from 15 different countries attended the meeting and more than 90 poster communications were presented.

The speakers have prepared review articles which constitute revised versions of their original presentations, and these make up the present volume.

The editors, on behalf of the Phytochemical Society of Europe, wish to thank the local Organizing Committee (F. Ferreres, C. Garcia-Viguera, M.I. Gil-Muñoz, M. Castañer, F. Artés, M.E. Candela, D. Rivera and C. Obón) for the excellent organization of the meeting.

In addition to this symposium a parallel international workshop on EU food-related RTD projects in the field of fruit and vegetables was also organized, in which the results of selected EU-funded research projects in the field of fruit and vegetables were presented.

Murcia and Nantes F.A.T.-B.
February 1996 R.J.R.

Acknowledgements

The Organizing Committee gratefully acknowledges the generous support of the following Institutions:

European Commission, Directorate General XII
Spanish Comisión Interministerial de Ciencia y Tecnología
Spanish Dirección General de Investigación Científica y Técnica
Consejería de Agricultura de la Región de Murcia
Caja Murcia
Dirección General de Turismo de la Región de Murcia
Consejo Superior de Investigaciones Científicas
Universidad de Murcia
Centro de Edafología y Biología Aplicada del Segura
Phytochemical Society of Europe

Contents

List of contributors ix

1. Introduction 1
F.A. Tomás-Barberán and R.J. Robins

2. Carotenoids and colour in fruit and vegetables 11
George Britton and Dámaso Hornero-Méndez

3. Molecular interactions of phenolic compounds in relation to the colour of fruit and vegetables 29
Raymond Brouillard, Paulo Figueiredo, Mourad Elhabiri and Olivier Dangles

4. Phenolic compounds and oxidative mechanisms in fruit and vegetables 51
Marie J. Amiot, Annie Fleuriet, Veronique Cheynier and Jacques Nicolas

5. Astringency 87
Michael N. Clifford

6. Fruit aroma precursors with special reference to phenolics 109
J. Crouzet, M. Sakho and D. Chassagne

7. Aroma biochemistry of fruits and vegetables 125
Carlos Sanz, José M. Olías and Ana G. Pérez

8. The psychology of food choice: the influence of attitudes and flavour perception 157
Richard Shepherd

9. Coumarins in fruit and vegetables 173
Alicja M. Zobel

10. Physical and physiological changes in minimally processed fruits and vegetables 205
Mikal E. Saltveit

11. Role of phytoalexins and other antimicrobial compounds from fruits and vegetables in postharvest disease resistance 221
Julien Mercier

12. ACC oxidase gene family: characterization and down-regulation by genetic manipulation 243
M. Bouzayen, M. Ferrer, P. Guillen, R. Ayub, S. Bidonde, M. Ben Amor, M. Guis, S. Ramassamy, H. Zegzouti, J.C. Pech and A. Latché

13. Options for the genetic manipulation of astringent and antinutritional metabolites in fruit and vegetables 251
Mark P. Robbins, Adrian D. Bavage and Phillip Morris

14. Gibberellins and fruit development 263
José L. García-Martínez and Peter Hedden

15. The possible role of antioxidants in fruit and vegetables in protecting against coronary heart disease 287
David S. Leake

16. Potentially anticarcinogenic secondary metabolites from fruit and vegetables 313
Michaël G.L. Hertog, Geert van Poppel and Dorette Verhoeven

17. Toxic compounds from fruit and vegetables 331
J.P. Felix D'Mello

18. Phytochemistry of fruit and vegetables: an ecological overview 353
Jeffrey B. Harborne

Index 369

Contributors

Amiot, Marie-J.: Institut National de la Recherche Agronomique, Station de Technologie des Produits Végétaux, 84914 Avignon Cedex 9, France.

Ayub, R.: Ecole Nationale Supérieure Agronomique, UA INRA, 145, Avenue de Muret, 31076 Toulouse Cedex, France.

Bavage, Adrian D.: Cell Manipulation Group, Institute of Grassland and Environmental Research, Aberystwyth Research Centre, Plas Gogerddan, Aberystwyth, Dyfed, Wales. SY23 3EB, UK.

Ben Amor, M.: Ecole Nationale Supérieure Agronomique, UA INRA, 145, Avenue de Muret, 31076 Toulouse Cedex, France.

Bidonde, S.: Ecole Nationale Supérieure Agronomique, UA INRA, 145, Avenue de Muret, 31076 Toulouse Cedex, France.

Bouzayen, M.: Ecole Nationale Supérieure Agronomique, UA INRA, 145, Avenue de Muret, 31076 Toulouse Cedex, France.

Britton, George: Department of Biochemistry, University of Liverpool, PO Box 147, Liverpool L69 3BX, UK.

Brouillard, Raymond: Laboratoire de Chimie des Polyphénols, associé au CNRS, Université Louis Pasteur, Institut de Chimie, 4, rue Blaise Pascal, 67008 Strasbourg, France.

Chassagne, D.: Laboratoire de Genie Biologique et Sciences des Aliments, Equipe de Microbiologie et Biochimie Industrielles, associée à l INRA Université de Montpellier II, 34095 Montpellier Cedex 05, France.

Cheynier, Veronique: Institut National de la Recherche Agronomique, Institut des Produits de la Vigne, 34060 Montpellier Cedex, France.

Clifford, Michael N.: Food Safety Research Group, School of Biological Sciences, University of Surrey, Guildford, Surrey GU2 5XH, UK.

Crouzet, J.: Laboratoire de Genie Biologique et Sciences des Aliments, Equipe de Microbiologie et Biochimie Industrielles, associée à l INRA, Université de Montpellier II, 34095 Montpellier Cedex 05, France.

Dangles, Olivier: Laboratoire de Chimie des Polyphénols, associé au CNRS, Université Louis Pasteur, Institut de Chimie, 4, rue Blaise Pascal, 67008 Strasbourg, France.

D'Mello, J.P. Felix: The Scottish Agricultural College, Department of Crop Science and Technology, West Mains Road, Edinburgh EH9 3JG, UK.

Elhabiri, Mourad: Laboratoire de Chimie des Polyphénols, associé au CNRS, Université Louis Pasteur, Institut de Chimie, 4, rue Blaise Pascal, 67008 Strasbourg, France

Ferrer, M.: Ecole Nationale Supérieure Agronomique, UA INRA, 145, Avenue de Muret, 31076 Toulouse Cedex, France.

Figueiredo, Paulo: Laboratoire de Chimie des Polyphénols, associé au CNRS, Université Louis Pasteur, Institut de Chimie, 4, rue Blaise Pascal, 67008 Strasbourg, France.

Fleuriet, Annie: Université Montpellier II, Laboratoire de Biotechnologie et Physiologie Végétale Appliquée, 34095 Montpellier Cedex 5, France.

García-Martínez, José L.: Instituto de Biología Molecular y Celular de Plantas, Universidad Politécnica de Valencia-CSIC, Camino de Vera 14, 46022 Valencia, Spain.

Guillén, P.: Ecole Nationale Supérieure Agronomique, UA INRA, 145, Avenue de Muret, 31076 Toulouse Cedex, France.

Guis, M.: Ecole Nationale Supérieure Agronomique, UA INRA, 145, Avenue de Muret, 31076 Toulouse Cedex, France.

Harborne, Jeffrey B.: Department of Botany, School of Plant Sciences, The University of Reading, Whiteknights, Reading, RG6 6AJ, UK.

Hedden, Peter: IACR, Long Ashton Research Station, Department of Agricultural Sciences, University of Bristol, Long Ashton, Bristol BS18 9AF, UK.

Hertog, Michaël G.L.: Department of Chronic Diseases and Environmental Epidemiology, National Institute of Public Health and Environmental Protection, PO Box 1, 3720 BA Bilthoven, The Netherlands.

Hornero-Méndez, Dámaso: Department of Biochemistry, University of Liverpool, PO Box 147, Liverpool L69 3BX, UK.

Latché, A.: Ecole Nationale Supérieure Agronomique, UA INRA, 145, Avenue de Muret, 31076 Toulouse Cedex, France.

Leake, David S.: School of Animal and Microbial Sciences, University of Reading, Whiteknights, PO Box 228, Reading RG6 6AJ, UK.

Mercier, Julien: Department of Plant and Microbial Biology, 111 Koshland Hall, University of California, Berkeley, CA 94720-3102, USA.

Morris, Phillip: Cell Manipulation Group, Institute of Grassland and Environmental Research, Aberystwyth Research Centre, Plas Gogerddan, Aberystwyth, Dyfed SY23 3EB, UK.

Nicolas, Jacques: Conservatoire National des Arts et Métiers, Chaire de Biochimie Industrielle et Agroalimentaire, 75141 Paris Cedex 03, France.

Olías, José M.: Department of Physiology and Technology of Plant Products, Instituto de la Grasa, CSIC, Padre García Tejero 4, 41012 Seville, Spain.

Pech, J.C.: Ecole Nationale Supérieure Agronomique, UA INRA, 145, Avenue de Muret, 31076 Toulouse Cedex, France.

Pérez, Ana G.: Department of Physiology and Technology of Plant Products, Instituto de la Grasa, CSIC, Padre García Tejero 4, 41012 Seville, Spain.

Poppel, Geert van: TNO Nutrition and Food Research Institute, PO Box 360, 3700 AJ Zeist, The Netherlands.

Ramassamy, S.: Ecole Nationale Supérieure Agronomique, UA INRA, 145, Avenue de Muret, 31076 Toulouse Cedex, France.

Robbins, Mark P.: Cell Manipulation Group, Institute of Grassland and Environmental Research, Aberystwyth Research Centre, Plas Gogerddan, Aberystwyth, Dyfed SY23 3EB, UK.

Robins, R.J.: Laboratoire d'AIEM, URA-CNRS 472, Faculté des Sciences, 2, rue de la Houssinière, 44072 Nantes Cedex 03, France.

Sakho, M.: Laboratoire de Genie Biologique et Sciences des Aliments, Equipe de Microbiologie et Biochimie Industrielles associée à l'INRA, Université de Montpellier II, 34095 Montpellier Cedex 05, France.

Saltveit, Mikal E.: Mann Laboratory, Department of Vegetable Crops, University of California, Davis, CA 95616-8631, USA.

Sanz, Carlos: Department of Physiology and Technology of Plant Products, Instituto de la Grasa, CSIC, Padre García Tejero 4, 41012 Seville, Spain.

Shepherd, Richard: Institute of Food Research, Earley Gate, Whiteknights Road, Reading RG6 6BZ, UK.

Tomás-Barberán, F.A.: Laboratorio de Fitoquímica, Departamento de Ciencia y Tecnología de Alimentos, CEBAS (CSIC), PO Box 4195, 30080 Murcia, Spain.

Verhoeven, Dorette: TNO Nutrition and Food Research Institute, PO Box 360, 3700 AJ Zeist, The Netherlands.

Zegzouti, H.: Ecole Nationale Supérieure Agronomique, UA INRA, 145, Avenue de Muret, 31076 Toulouse Cedex, France.

Zobel, Alicja M.: Department of Chemistry, Trent University, Peterborough, Ontario, Canada K9J 7B8.

1. Introduction

FRANCISCO A. TOMÁS-BARBERÁN[1] and RICHARD J. ROBINS[2]

[1]Laboratorio de Fitoquímica, Departamento de Ciencia y Tecnología de Alimentos, CEBAS (CSIC), PO Box 4195, Murcia 30080, Spain and [2]Laboratoire de RMN-RC, URA-CNRS 472, Faculté des Sciences, 2, rue de la Houssinière, 44072 Nantes Cedex 03, France

Secondary metabolites play a number of crucial roles in fruit and vegetables, being responsible for colour, appearance, flavour and taste. Thus, it is primarily the content of these metabolites within the product that is also responsible both for consumer selection of the commodity and for the enjoyment derived from its consumption. While it has long been recognized that 'fruit and vegetables are good for you', studies of fruit and vegetables over the last few years have demonstrated that these products contain constituents—notably vitamins and fibre—essential to a healthy, well-balanced diet. Furthermore, it has also been shown that some of the secondary metabolites of fruit and vegetables are beneficial to health in directly combating the onset of cancer and coronary diseases.

The content of secondary metabolites in these plant organs is affected by different aspects of their pre- and postharvest treatment. Thus, much effort is devoted to understanding the mechanisms underlying their production, storage and degradation. The complex metabolism involved in fruit ripening has been shown to involve genetic induction of biosynthetic pathways involved in the formation of secondary metabolites—notably for flavour and colour in fruit—and of enzymes involved in the alteration of cell-wall structure. Genetic manipulation of fruit is now well advanced; for example, there now exist commercially available varieties of tomato which have extended shelf-lives.

The biochemistry of fruit and vegetables was the topic of a previous Symposium of the Phytochemical Society of Europe (PSE) held in Norwich (UK) in 1980. The lectures presented at that meeting were published by the PSE in the book *Recent advances in the biochemistry of fruit and vegetables* (Friend and Rhodes, 1982). The book focuses on the primary metabolism of fruit and vegetables and contains reviews on ripening and respiration (Biale, Young and Vance), carbohydrate metabolism (Ap Rees *et al.*; Hobson, Knee and Bartley), the biosynthesis of ethylene and its action (Yang and Beyer), the molecular biology of ripening (Grierson *et al.*) and the biochemistry of polyphenol oxidases and tyrosinases—important enzymes in fruit and vegetables (Mayer and Harel; Robb). In addition, two chapters were devoted to secondary metabolism; one describing the changes in phenolic metabolism with stress (Rhodes *et al.*), the other focusing on anthocyanins (Timberlake).

Fifteen years later, the Phytochemistry of Fruit and Vegetables was the topic of a further PSE Meeting, which was held in Murcia (Spain) in September 1995. The programme aimed to review the advances made in this field since the previous meeting in Norwich with secondary metabolites considered as the central topic of the Symposium. Several questions were addressed:

- How do secondary metabolites contribute to fruit and vegetable quality?
- What are the justifications for studying secondary metabolism in fruit and vegetables?
- Is this purely a consumer-driven interest or are there fundamental aspects of secondary metabolism unique to these plant organs?

There is overwhelming evidence that secondary metabolites are responsible for many characteristics of the quality of fruit and vegetables:

- **colour and appearance:** many changes in colour and pigmentation occur during the postharvest life of these products
- **taste:** pungency of peppers, bitterness of grapefruits and astringency of persimmons, etc.
- **flavour and aroma:** terpenoids of citrus fruits, phenolics of banana, volatile aldehydes etc. of apples.

During the postharvest life of fruit and vegetables and the different technological treatments to which these tissues are submitted in order to extend their storage life, some changes in secondary metabolism occur. Postharvest fruit and vegetables are still living, metabolically active material and ongoing metabolism will continue to have a direct influence on their quality. In some instances—such as the postharvest colour and flavour development of tomatoes—this is beneficial. In many cases, however, fruit needs to be fully matured on the plant to obtain maximum quality, and postharvest tissue is susceptible to over-ripening with associated discolourations, loss of nutrients, loss of texture and pathogenic degradation.

The important role of secondary metabolites from the nutritional point of view is not directly associated with their 'natural' functions in the plant. Nevertheless, it is now recognized that some antioxidant and free-radical scavenging properties can be related to the possible prevention of cancer and cardiovascular diseases. In contrast, however, many secondary metabolites typical of the fruit or vegetable parts of wild species have toxic effects. In the main, this problem has been overcome either by breeding commercial cultivars in which the levels of toxic secondary metabolites are very reduced—as in the potato—or by the development of complex postharvest treatment—as with cassava and manioc. Cooking, a common postharvest treatment, is not only valuable in aiding the softening of fruit or vegetable by partial degradation of the cell wall but can also be important in helping to degrade or remove toxic metabolites.

It is not the intention of this book to provide a balanced coverage of all aspects of secondary metabolites in fruit and vegetables: significant areas of research on the phytochemistry of fruit and have not been included. We have shown a bias towards the importance of secondary metabolites in providing colour (Brouillard *et al.*; Britton; Amiot *et al.*), taste and astringency (Clifford; Shepherd), and flavour and aroma (Crouzet *et al.*; Sanz *et al.*). Processing and its effects on the chemical composition of fruit and vegetables has not been considered, since our aim was to concentrate on the metabolic changes occurring in living tissues. Minimal processing involving cutting, sorting, coring and peeling, however, leads to products which are still living tissue and subject to important metabolic changes—notably oxidations—that affect secondary metabolite composition. Therefore the physiological and biochemical changes in the secondary metabolites of minimally processed products have also been reviewed (Saltveit). Other chapters cover the effect of plant hormones on the development and postharvest life of fruit and vegetables (García-Martínez and Hedden; Bouzayen *et al.*), the use of gene technology in obtaining cultivars with better quality characteristics (Robbins) and secondary metabolites in the prevention of postharvest diseases (Mercier; Zobel).

Secondary metabolites present in fruit contribute to making them attractive to a consumer, although humans are seldom the primary target. Vegetables normally represent parts of the plant not 'naturally' produced for consumption (e.g. storage organs) but which humans have adapted to their omnivorous diet. Thus, it is to be expected that the secondary metabolites encountered within the human diet play many roles of ecological importance within the plant, both attracting and repelling potential consumers. The chapter by Harborne reviews the ecological biochemistry of secondary metabolites within this context. Protective compounds are frequently toxic as well as unpleasant to the taste. The well-documented adverse effects of such compounds in animal models are considered in relation to human health (D'Mello). While some compounds are antinutritional, others have been recognized more recently to have beneficial effects, notably in the prevention of chronic diseases (Leake) and cancers (Hertog).

Plant colour is mainly due to three groups of compounds: the flavonoids (anthocyanins), the carotenoids and the chlorophylls. The first two groups of compounds are covered in detail in this book. Chlorophylls are also essential in the pigmentation of many vegetables. For instance, the degradation of chlorophylls in broccoli leads to quality loss and make the product unmarketable. Research on the preservation of chlorophylls during the post-harvest life of some vegetables is required. In other cases, the degradation of chlorophylls is an indispensable requirement to lead to quality products. For instance, chlorophylls have to be degraded to lead to fruits with characteristic orange colours. In the case of bell peppers, for example, they have to show homogeneous colour in order to have better consumer acceptance. Therefore, knowledge of the biochemical changes in bell pepper pigmentation from green (chlorophyll) to intense red (carotenoid) is fundamental. In this field it is important to remark the recent advances in the knowledge of chlorophyll degradation by chlorophyllases, and how the hormone ethylene induces the synthesis of chlorophyllases leading to the loss of green pigments during maturation and ripening (Trebitsh *et al.*, 1993).

Another relevant topic is the study of the changes in secondary metabolism which occur during the physical treatment of fruit and vegetables for extending postharvest life or enhancing quality characteristics (pigmentation, texture, etc.). Physical treatments are interesting from the technological point of view since they do not leave residues and therefore have some advantages over traditional chemical treatments. Among these physical treatments we should include ultraviolet and gamma irradiation, heat shock and storage under controlled atmospheres (CA) and modified atmospheres (MA).

Ultraviolet irradiation (254 nm) can be used for inducing anthocyanin biosynthesis and increase fruit peel colour as shown for apples (Dong *et al.*, 1995). This is important, since the development of postharvest treatment regimes that will modulate and generate a desirable fruit skin colour, without affecting other characteristics of the fruit cultivars, are vital considerations in capturing and developing competitive markets. Ultraviolet or gamma irradiation is also used to prevent postharvest fungal growth. These treatments have significant effects on secondary metabolism and the biosynthesis of some secondary metabolites, including some notably antifungal compounds, is substantially increased (Lu *et al.*, 1993; Ben-Yehoshua *et al.*, 1992). These irradiations can also retard fruit ripening.

On the other hand, treatment at high temperatures (high-temperature stress at 36–40°C) induces metabolic changes in chill-sensitive plants which prevent chilling injury. The changes induced in secondary metabolism by these heat treatments have not been studied so far and could probably help to understand how the heat-shock treatment contributes to alleviate the damage produced by chilling injury (Lurie and Klein, 1991).

The effects of controlled or modified atmospheres on secondary metabolism are also interesting topics of study. These treatments involve the storage of fruit and vegetables under atmospheres rich in carbon dioxide and sometimes poor in oxygen. Such conditions have significant effects on secondary metabolism as reflected in the changes observed in pigmentation (anthocyanins and other phenolics) of plant tissues stored under these atmospheres. Other secondary metabolites, like glucosinolates, are also affected by these storage conditions. Thus, the glucosinolate content of broccoli florets stored in controlled atmospheres with high carbon dioxide changed significantly during seven days of storage (Hansen *et al.*, 1995).

Preharvest factors also have a strong influence on secondary metabolism and affect the composition of fruit and vegetables. The presence of nutrients in the soil where the plants were grown impacts on the secondary metabolism of fruit and vegetables and on their postharvest life. The levels of calcium and boron in soil influence the pigmentation of tomatoes (Oyewole and Aduayi, 1992) and apples. The calcium content has essential effects on phenolic metabolism (browning) and the firmness of fruits. Randle *et al.* (1995) have shown that the content in sulphur of the soil has a direct effect in the content of sulphur-containing compounds in onions, since sulphur increases the levels of alk(en)yl cysteine sulphoxides, the biosynthetic intermediates of onion flavour constituents. In addition, some environmental factors such as drought (water stress) and salt stress can have major effects on secondary metabolism and fruit and vegetable composition.

Considering the nutritional value of fruit and vegetables, it is basic to understand the fate of secondary metabolites once they reach the gastro-intestinal tract. The metabolic changes of these compounds are important since the products would be responsible for the physiological effects in vivo. Most of the current knowledge about the physiological activity of secondary metabolites in animals have been acquired in vitro and therefore the relevance of these results to the efficacity of compounds obtained from the diet must be established. For example, extracts of apple skin and isolated flavonoids such as naringenin and rutin have been shown in vitro to prevent UV-B-induced DNA damage (Kootstra, 1994). Are these compounds effective when entering the body via the digestive tract? Much work remains to be done in establishing the active forms of antinutritional and antidisease agents and the metabolism of fruit- and vegetable-derived compounds by the intestinal flora.

Since the beginning of agriculture, the human race has selected and developed varieties of fruit and vegetables. Initially, perhaps, better yield was a priority. Then, palatability must have been an important factor. Increasingly, breeding has led to varieties with specific properties, notably in the range of climatic conditions in which the product could be produced. In the modern market, appearance and reproducibility of colour and size are important factors. Indeed, there was a phase when flavour seemed less important than appearance! There now exists a vast range of varieties in every fruit and vegetable crop species and numerous cross-species hybrids providing 'new' fruits for the marketplace. What, then, can genetic engineering offer to improve further the quality of fruit and vegetables?

In the simplest most general agronomic sense, genetic engineering can be applied to these crops, as to any other, to generate plants having greater resistance to pathogen attack. This should provide both a better yield and, potentially, commodities with a better appearance. One approach is to express genes for foreign phytoalexins under organ-specific promoters (Fischer and Hain, 1994). Another, to introduce more specific disease resistance genes (Lamb *et al.*, 1992). Maize plants with a high resistance to stem-borer beetle have been generated by introducing the modified toxin from *Bacillus thuringiensis*, leading to greatly enhanced yields.

There also exists, however, the possibility of genetically engineering specific aspects of the unique properties of fruit or vegetables outlined above and discussed in detail in the following chapters. Colour modification has been achieved by selection: both the orange carrot and the yellow bell pepper are natural mutants. Pigmentation of flowers by anthocyanins is already being altered by genetic manipulation (Courtney-Gutterson, 1994) and similar approaches to anthocyanins in fruit could be taken. As our understanding of the biosynthesis of carotenoids improves (see Britton, this volume) similar approaches to produce novel colours in carotenoid-pigmented products could be taken. Might the blue hues of crustacean carotenoids be transferred to fruit?

While the value of altering colour by such techniques might be questionable, direct benefits can be seen from engineering other aspects of fruit and vegetable physiology. Flavour and aroma compounds are just as susceptible to manipulation as are pigments. The pathways of some aromatic and terpenoid aromas are steadily being unravelled, and genes isolated. The option to create novel flavours within fruits and vegetables lies in the near future. The intensity of the flavour compounds and their continued biosynthesis in the postharvest product could be enhanced. Simple transformation of lemon geranium (*Pelargonium* sp.) provided plants with increased production of their essential oils (Pellegrinischi *et al.*, 1994). Our understanding of both the enzymology and the regulation of the biosynthesis of *Mentha* monoterpenes is rapidly progressing (Croteau and Gershenzon, 1994) and genetic engineering of these products is in progress. While mint oils are not strictly within the domain of fruit and vegetables, they nevertheless represent a class of compounds important in providing fruit aromas.

The processing characteristics of fruit and vegetables have received much attention. As stated above, processing *per se* is not a topic being considered in this volume but mention must be made of the progress made in genetically manipulating fruit and vegetables to alter their postharvest properties. The tomato fruit has been the subject of intensive study and varieties ('Flavr Savr' and 'Calgene') are now on the market with extended shelf-life as a result of decreasing polygalacturonidase activity (Gray *et al.*, 1994; Sheey *et al.*, 1988) or inhibiting ethylene production (Gray *et al.*, 1994; Oeller *et al.*, 1991). Tomatoes with altered solids content are not far behind.

Perhaps the most valuable results to be obtained by genetically engineering fruit and vegetables, however, will be to alter their nutritional and antinutritional properties. Take a crop like cassava, the staple diet for over 200 million people. The tubers are rich in starch but have a very low protein content (1–2% fresh weight) and this is low in sulphur amino acids. They also contain cyanogenic glucosides, which cause chronic poisoning, goitre and spastic paraparesis in children when ingested. These toxins are largely removed by storage and washing of the ground meal. As, however, the crop is particularly intensively eaten in times of drought, the washing is frequently only partially effective and large numbers of people suffer the subsequent debilitating effects. All three of these aspects of the crop are potentially susceptible to genetic manipulation. Both starch content and quality can be altered (Stitt and Sonnewald, 1995), potentially producing a better and more readily digested diet. Already, the starch content of potato tubers has been both increased (Stark *et al.*, 1992) and changed in composition (Visser *et al.*, 1991). The protein content can be improved by such approaches as enriching the tubers with sulphur-rich seed albumins (Rafiqul *et al.*, 1996) or stimulated lysine biosynthesis (Lea and Forde, 1994). These objectives are realistic.

To delete the cyanogenic glycosides from cassava could be a problem, as these compounds are present in the plant to protect the tissues. There are, however, varieties of cassava which naturally contain low levels of these toxins and so it should be possible to diminish or even delete the production of these compounds in some varieties. The ways in which this might be achieved are discussed by Ibrahim *et al.* (1994) in relation to the glucosinolates of *Brassica*, which present similar problems of cumulative toxicity. The work discussed in this volume (Robbins), relating to the manipulation of tannin quality, follows a parallel line of thinking. Tannins play an important role in astringency (see Clifford, this volume) and some act as antinutritional factors. In sheep, however, they can cause bloat. To alter the concentration or the nature of the tannins present in a fruit or vegetable could have major advantages in the diet.

Regrettably, problems such as toxicity from ingesting cassava cyanogenic glycosides and the equally serious problem of lathyrism, caused by amino acid analogues in *Lathyrus sativus*, relate to the diet of the less well developed countries and, as a result, receive inadequate attention (Bunders *et al.*, 1989). Much of the application of genetic engineering in crops is directed towards providing more marketable products for consumers in developed countries. While many of these goals are admirable, they seems to be providing the 'icing on the cake' rather than the fundamental raw materials with which to do the baking.

All the developments in the genetic engineering of fruit and vegetables, however, rest on a sound basis of biochemical and physiological research.

The 'Flavr Savr' and 'Calgene' tomatoes represent the results of over twenty years of painstaking work by such groups as that of Grierson into the understanding of the biochemistry and genetic regulation of fruit development. We hope that this volume will serve to stimulate readers, whatever their discipline, to consider the biochemical and phytochemical problems specifically associated with fruit and vegetables and to apply their own skills to solving the many questions that remain to be answered.

References

Ben-Yehoshua, S., Rodov, V., Kim, J.J. and Carmeli, S. (1992). *J. Agric. Food Chem.* **40**, 1217–21.

Bunders, J., Broerse, J. and Stolp, A. (1989). *TIBTECH* **7**, S16–S24.

Courtney-Gutterson, N. (1994). In *Genetic engineering of plant seccndary metabolism*, (ed. B.E. Ellis, G.W. Kuroki and H.A. Stafford), pp. 93–124. Plenum Press, New York.

Croteau, R. and Gershenzon, J. (1994). In *Genetic engineering of plant secondary metabolism*, (ed. B.E. Ellis, G.W. Kuroki and H.A. Stafford), pp. 193–229. Plenum Press, New York.

Dong, Y. H., Mitra, D., Kootstra, A., Lister, C. and Lancaster, J. (1995). *J. Amer. Soc. Hort. Sci.* **120**, 95–100.

Fischer, R. and Hain, R. (1994). *Curr. Opinion Biotech.* **5**, 125–30.

Friend, J. and Rhodes, M.J.C (1982). *Recent advances in the biochemistry of fruit and vegetables*, Vol. 19. Proceedings of the Phytochemical Society of Europe. Cambridge University Press, Cambridge, UK.

Gray, J.E., Picton, S., Giovannoni J.J. and Grierson, D. (1994). *Plant Cell Environ.* **17**, 557–72.

Hansen, M., Moller, P., Sorensen, H. and Cantwell de Trejo, M. (1995). *J. Amer. Soc. Hort. Sci.* **120**, 1069–74.

Ibrahim, R.K., Chavedej, S. and De Luca, V. (1994). In *Genetic engineering of plant secondary metabolism,* (ed. B.E. Ellis, G.W. Kuroki and H.A. Stafford), pp. 125–52. Plenum Press, New York.

Khan, M.R.I., Ceriotti, A., Tabe, L., Aryan, A., McNabb, W., Moore, A., Craig, S., Spencer, D. and Higgins, T.J.V. (1996). *Trans. Res.* **5**, in press.

Kootstra, A. (1994). *Plant. Mol. Biol.* **26**, 771–74.

Lea, P.J. and Forde, B. G. (1994). *Plant Cell Environ.* **17**, 541–57.

Lamb, C.J., Ryals, J.A., Ward, E.R. and Dixon, R.A. (1992). *Bio/Technology* **10**, 1436–45.

Lu, J.Y., Lukombo, S.M., Stevens, C., Khan, V.A., Wilson, C.L., Pusey, P.L. and Chaultz, E. (1993). *J. Food Qual.* **16**, 301–309.

Lurie, S. and Klein, J.D. (1991). *J. Amer. Soc. Hort. Sci.* **116**, 1007–12.

Oeller, P.W., Wong, L.M., Taylor, L.P., Pike, D.A. and Theologis, A. (1991). *Science* **254**, 437–9.

Oyewole, O.I. and Aduayi, E.A. (1992). *J. Plant Nutr.* **15**, 199–209.

Pellegrinischi, A., Damon, J.P., Valtorta, N., Paillard, N. and Tepfer, D. (1994). *Bio/Technology* **12**, 64–8.

Randle, W.M., Lancaster, J.E., Shaw, M.L., Sutton, K. H., Hay, R. L. and Bussard. M.L. (1995). *J. Amer. Soc. Hort. Sci.* **120**, 1075–81.

Sheey, R.E., Kramer, M. and Hiatt, W.R. (1988). *Proc. Natl. Acad. Sci. USA* **85**, 8805–9.

Stark, D.M., Timmerman, K.P., Barry, G.F., Preiss, J. and Kishore G.M. (1992). *Science* **258**, 287–92.

Stitt, M. and Sonnewald, U. (1995). *Annu. Rev. Plant Physiol. Plant Mol. Biol.* **46**, 341–68.

Trebitsh, T., Goldschmidt, E.E. and Riov, J. (1993). *Proc. Natl. Acad. Sci. USA* **90**, 9441–5

Visser, R.G.F., Somhorst, I., Kuipers, G.J., Ruys, N.J., Feenstra, W.J. and Jacobsen, E. (1991). *Mol. Gen. Genet.* **225**, 289–96.

2. Carotenoids and colour in fruit and vegetables

GEORGE BRITTON and DÁMASO HORNERO-MÉNDEZ

Department of Biochemistry, University of Liverpool, PO Box 147, Liverpool L69 3BX, UK

Introduction

Carotenoids and anthocyanins are the most important pigments for providing attractive colours in fruit and vegetables. In contrast to the water-soluble anthocyanins, the carotenoids are water-insoluble, isoprenoid lipid molecules, and are usually located in membrane structures in discrete organelles, namely chloroplasts (in green tissues) or chromoplasts (in yellow, orange or red tissues). Carotenoids may be found in, and may provide or contribute to the colour of, all parts of plants: roots, leaves, flowers, fruit and seeds.

Structures

The carotenoids are tetraterpenes, and are derived from the symmetrical C_{40} skeleton illustrated in Fig. 2.1 by lycopene (**1**, acyclic) and β,β-carotene (**2**, dicyclic). Carotenoid hydrocarbons are known as carotenes; carotenoids bearing oxygen substituents as xanthophylls. More than 600 different carotenoids have been isolated from natural sources and identified; many of these (more than 100) have been found in fruit and vegetables. Structural variations include allenic groups and five-membered rings as well as various oxygen functions. The most common oxygen substituents are hydroxy groups and these are often esterified with fatty acids, especially in fruit and flowers. Usually, carotenoids are found as the all-*trans* (all-*E*) form but small amounts of *cis* (*Z*) isomers, which are now readily detectable by HPLC, are also present. In some cases, the naturally occurring form of the carotenoid may be a *Z* isomer, notable examples being prolycopene ((7*Z*, 9*Z*, 7′*Z*, 9′*Z*)-lycopene, **3**) from the tangerine tomato strain and bixin (methyl hydrogen (9′*Z*)-6,6′-diapocarotene-6,6′-dioate, **4**), the colouring principle of annatto (*Bixa orellana*).

A detailed survey of carotenoid structure and nomenclature is given (Weedon and Moss, 1995) in the first volume of a new book series, which also describes methods for the isolation and analysis of carotenoids, and

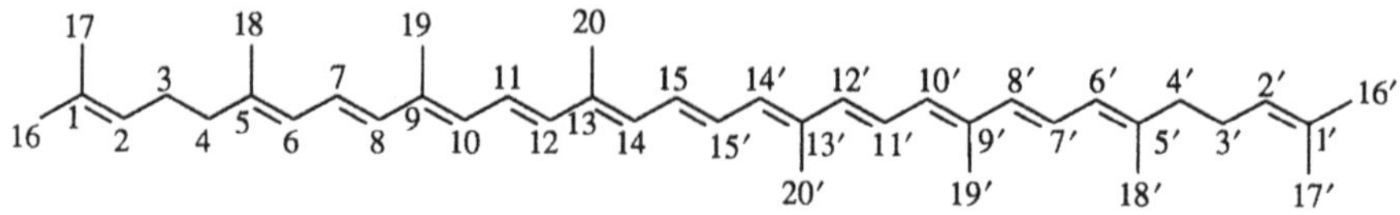

Lycopene **1**

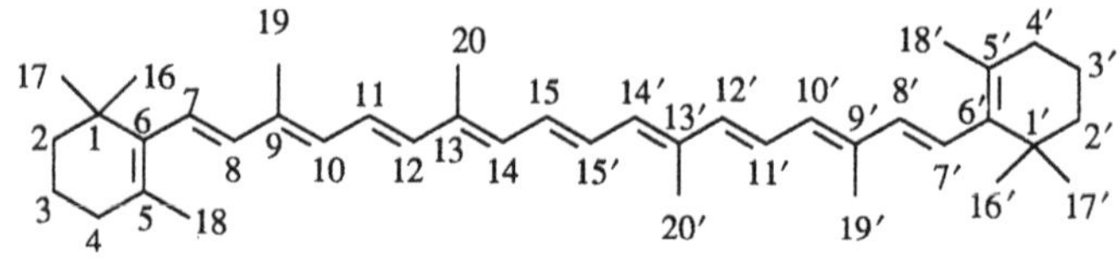

β,β-Carotene **2**

Fig. 2.1. The C_{40} skeleton and numbering scheme of carotenoids, illustrated by the acyclic hydrocarbon, lycopene (ψ,ψ-carotene, **1**) and the dicyclic hydrocarbon, β,β-carotene (**2**).

provides extensive tabulated spectroscopic date (Britton *et al.*, 1995). A 'Key to Carotenoids' (Straub, 1987; Kull and Pfander, 1995) lists the formulae and semi-systematic names of all known naturally occurring carotenoids.

Carotenoids in green fruit and vegetables

Carotenoids are important constituents of chloroplasts, in which they play essential roles in photosynthesis (Young and Britton, 1993) as accessory light-harvesting and photoprotective pigments in the pigment–protein complexes (Siefermann-Harms, 1985; Young, 1993). They therefore occur universally in leaves and other green plant tissues. There is remarkable uniformity in the carotenoid composition of higher plant chloroplasts. β,β-Carotene (**2**), which is located in reaction centre complexes and sometimes accompanied by much lesser amounts of β,ε-carotene (α-carotene) normally constitutes 25–30% of the total carotenoid. The light-harvesting antenna complexes contain a set of xanthophylls, with lutein (**6**) (40–45%), violaxanthin (**7**) (10–15%) and neoxanthin (**8**) (10–15%) as the major components (Goodwin and Britton, 1988). Zeaxanthin (**9**) and antheraxanthin (**10**) are also present in significant amounts under appropriate conditions as components of the photoprotective xanthophyll cycle (Demmig-Adams and Adams, 1993). Lettuce (*Lactuca sativa*) is exceptional in that about one-third of the lutein is replaced by lactucaxanthin (ε,ε-carotene-3,3′-diol, **11**) (Siefermann-Harms *et al.*, 1981). All the xanthophylls are present in the free (i.e. unesterified) form. In leaves that have been maintained in the dark, including the inner leaves of vegetables such as cabbage and lettuce, normal chloroplasts are not formed, but etioplasts are present which contain no chlorophyll and a modified carotenoid composition, usually with no carotenes but moderate amounts of xanthophylls, especially lutein (**6**) and antheraxanthin (**10**) (Barry *et al.*, 1991). In these etiolated tissues, the yellow colour of the carotenoids is no longer masked by the intense green of chlorophylls *a* and *b* and is clearly seen.

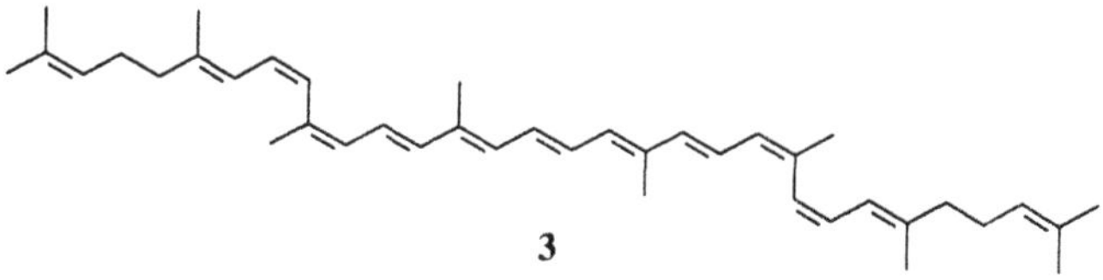

3

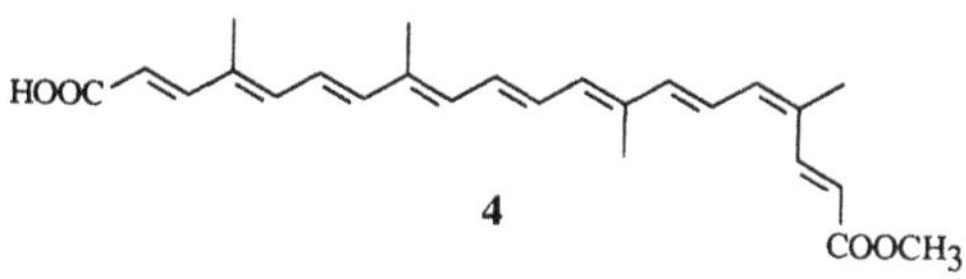

4

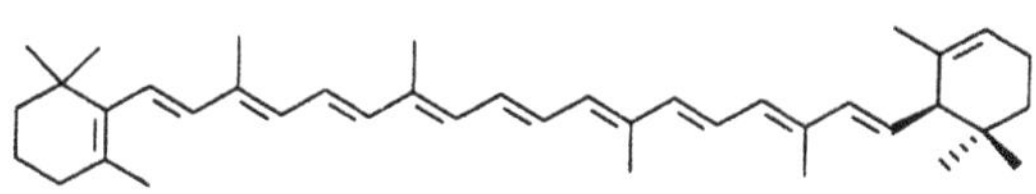

5

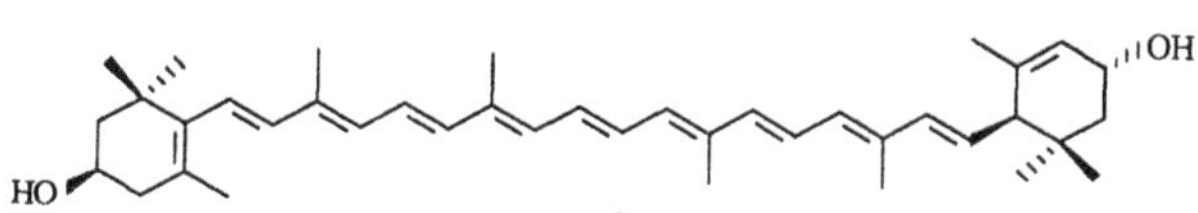

6

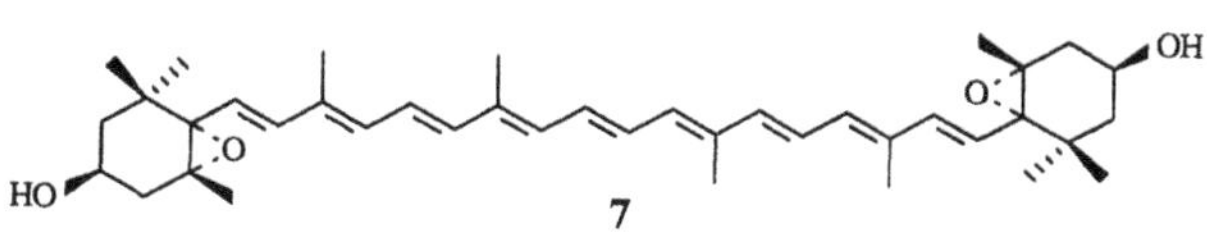

7

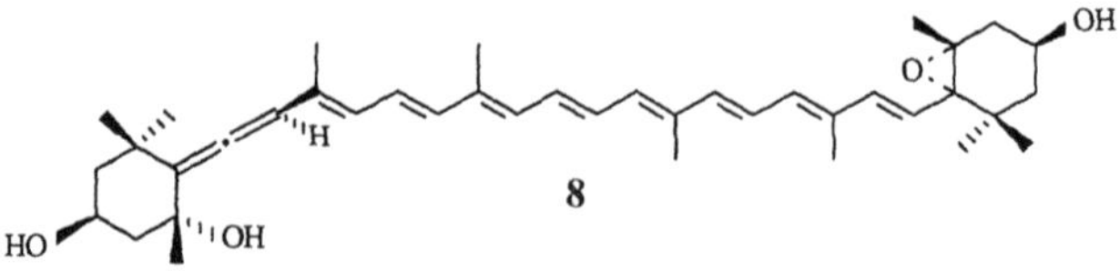

8

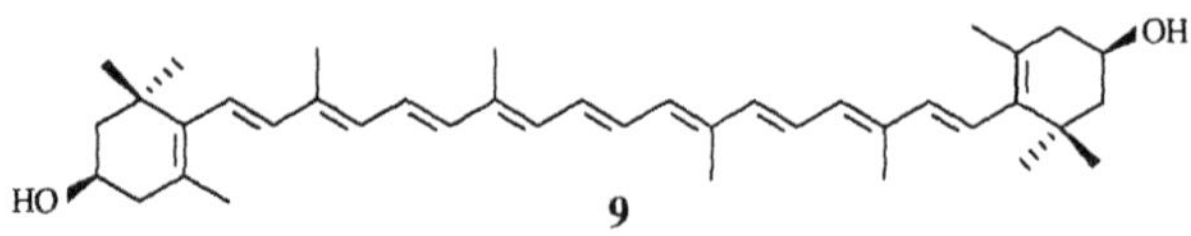

9

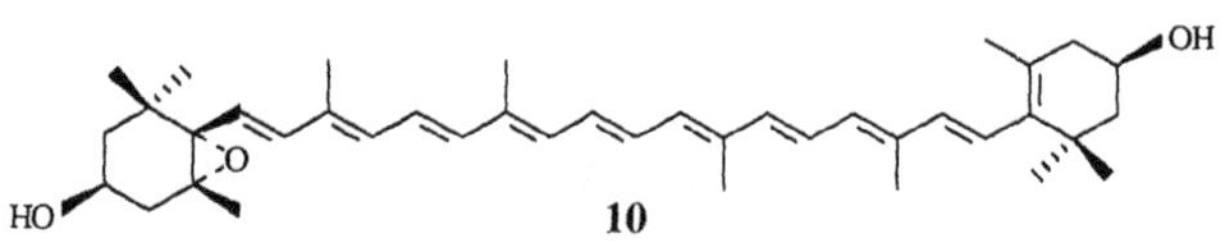

10

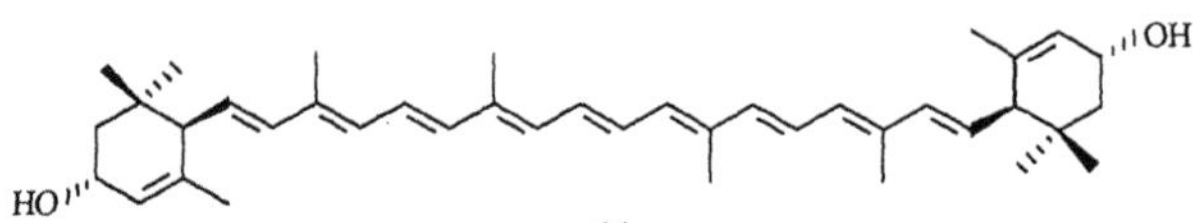

11

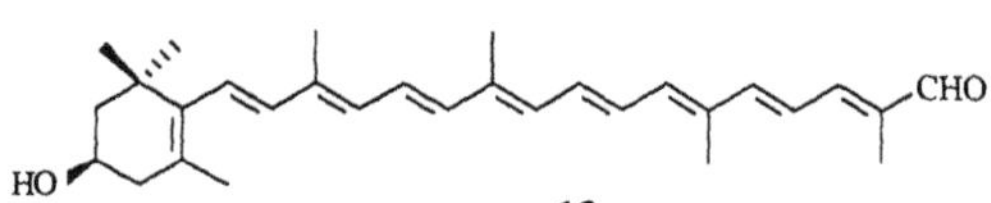

12

Not only leaves but all green parts of plants, including green fruit, contain chloroplasts and all have a more-or-less similar carotenoid composition. The total amount of carotenoid present can vary greatly. As a general rule, the darker the green colour, the higher will be the carotenoid content, reflecting the greater number of chloroplasts. In pale green cabbage, the carotene content may be less than 1% of that in dark green cabbage or spinach.

Carotenoids in non-green fruit and vegetables

As well as their universal occurrence in green tissues, carotenoids are also widespread in other parts of plants, where they may be present in large amounts. Well-known and striking examples of this include yellow-orange flowers, orange and red fruit such as orange, tomato and pepper (*Capsicum*), roots, notably carrots, and seeds like maize and *Bixa orellana*, source of the extensively used colorant annatto. Carrots and other yellow root vegetables, including swede and sweet potato, characteristically accumulate carotenes and only small amounts of xanthophylls (carrots, depending on variety, may contain up to 10–15 mg/100 g β,β-carotene).

The carotenoids of fruits, however, show much greater diversity, and exotic structures may be encountered. Apocarotenoids, in which one end group has been removed, are common in fruit, e.g. β-citraurin (**12**) in *Citrus* fruit. The carotenoids in ripe fruit are usually different from those that are present in the unripe, green fruit. The carotenoid compositions can be very complex; in *Capsicum annuum* almost 60 different carotenoids have been detected by HPLC and many of these identified (Matus *et al.*, 1991). When acyl esters and geometrical isomers are included, the complexity can be daunting. As a simplification, the carotenoid compositions of fruit have been classified into several groups (Table 2.1) (Goodwin and Britton, 1988). Although the distinctions are not always clear, this broad classification remains useful and can now be rationalized on the basis of the presence or absence of particular genes for carotenoid biosynthesis (see below).

Table 2.1. Eight general groups into which fruit can be classified on the basis of their carotenoid compositions

Group	Pigment characteristics
I	Insignificant amounts
II	Small amounts generally of chloroplast carotenoids
III	Relatively large amounts of lycopene (**1**) and its hydroxy derivatives
IV	Relatively large amounts of β,β-carotene (**2**) and its hydroxy derivatives
V	Large amounts of epoxides, particularly furanoid epoxides
VI	Unusual carotenoids, e.g. capsanthin (**19**)
VII	Poly-*Z* carotenoids, e.g. prolycopene (**3**)
VIII	Apocarotenoids e.g. β-citraurin (**12**)

Books by Goodwin (1980) and Gross (1987, 1991) give detailed accounts of the occurrence of carotenoids in fruit and vegetables and present extensive tabulated data on carotenoid compositions.

In these non-green tissues, the carotenoids are located in chromoplasts. Details of chromoplast ultrastructure and the location of the carotenoids can be found in several books and reviews (e.g. Gross 1987; Kirk and Tilney-Bassett, 1978; Sitte *et al.*, 1980). The carotenoids are located in various internal structures that may be derived from the plastid envelope membrane. Four major types of structure have been identified by electron microscopy, namely globulous, tubulous, membranous and crystallous (Sitte *et al.*, 1980), but combinations of these structural types can occur in the same plant and even within the same plastid. In the globulous structures, the carotenoids are found in the stroma, located in plastoglobuli associated with the inner membrane. Tubulous chromoplasts contain tube-shaped internal elements where the carotenoids are located. Membranous plastids are most characteristic of flowers. The carotenoids are located in spherical bodies surrounded by concentric layers of membrane. In the crystallous plastids, the carotenoids are present as crystals surrounded by a thin membrane. Lycopene in tomatoes and β,β-carotene in carrots occur as small crystals in this kind of structure.

Carotenoid biosynthesis occurs within the chromoplast as a part of the ripening process.

Carotenoid biosynthesis

The carotenoids are isoprenoid compounds and are biosynthesized by a branch of the great isoprenoid pathway (Fig. 2.2). In the carotenoids, the isoprenoid chain is built up by prenyl transferases to the C_{20} level, as geranylgeranyl diphosphate (**13**), and two molecules of this are joined tail-to-tail to give phytoene (**14**) as the first product with the C_{40} carotenoid skeleton (Fig. 2.3). This reaction is proposed to occur via an intermediate, prephytoene diphosphate, which is analogous to presqualene diphosphate in sterol biosynthesis. Phytoene is colourless, but undergoes a series of desaturation reactions, each of which creates a new double bond and extends the chromophore by two conjugated double bonds. The end product is lycopene (**1**), produced via the intermediates phytofluene (**15**), ζ-carotene (**16**) and neurosporene (**17**) (Fig. 2.3). The light absorption maximum shifts progressively to longer wavelengths as the chromophore is extended, and lycopene, with 11 conjugated double bonds, absorbs maximally at 470–500 nm and is strongly coloured orange-red. The phytoene in higher plants appears to be formed as the 15*Z* isomer, though lycopene and the other coloured carotenoids are generally in the all-*E* form. Isomerization from *Z* to *E* must therefore take place during the desaturation sequence, but the stage at which this occurs has not been established unequivocally.

The lycopene molecule may then undergo cyclization to form six-membered rings at one end or both ends of the molecule, e.g. β-rings, as illustrated in Fig. 2.4. Cyclization reduces the λ_{max} and hence alters the colour; β,β-carotene, with λ_{max} 450 nm, is yellow.

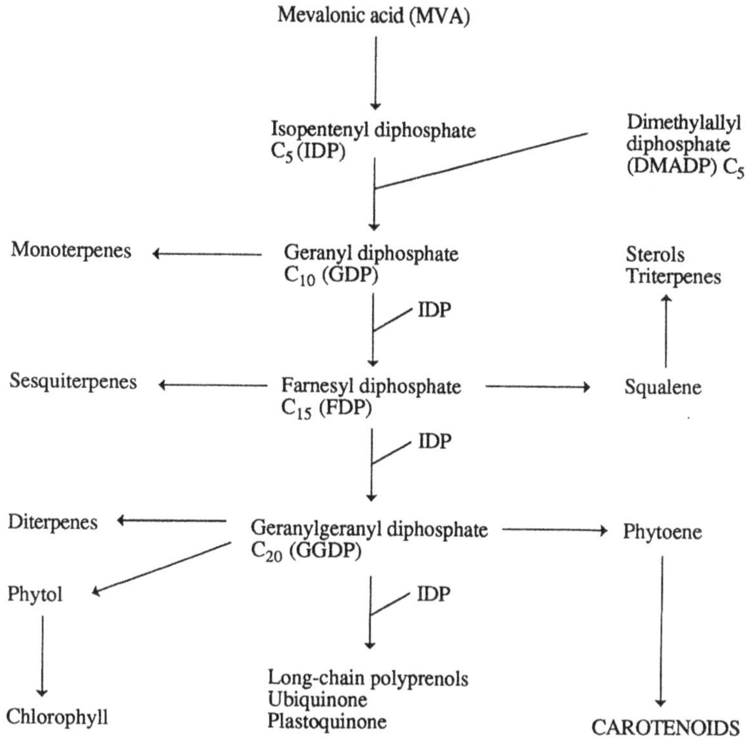

Fig. 2.2. The isoprenoid biosynthetic pathway.

The introduction of oxygen functions and other structural modifications of end groups, including esterification, then follow as the final stages of biosynthesis. Thus zeaxanthin (**9**) and lutein (**6**) are formed by the introduction of two hydroxy groups at C-3 and C-3′ of β,β-carotene (**2**) and β,ε-carotene (**5**), respectively. Schemes have been proposed for the formation of a variety of other end groups by rearrangement of a 5,6-epoxy-β ring end group (Fig. 2.5).

Because of the number of combinations of reactions that are possible in the two ends of the molecule, the conventional pathways that can be constructed can look very complicated, but the picture is greatly simplified when considered in terms of sequences of reactions that can occur in one end of the molecule or the other. Thus the formation of zeaxanthin from lycopene involves only two reactions, namely β-cyclization and hydroxylation, at each end group. The exotic looking cyclopentanone end group of capsanthin (**19**) and capsorubin (**20**) requires only two additional reactions, namely an epoxidation and a rearrangement (see Fig. 2.5). Each reaction, whether it occurs in only one or in both end groups, is catalysed by a particular enzyme

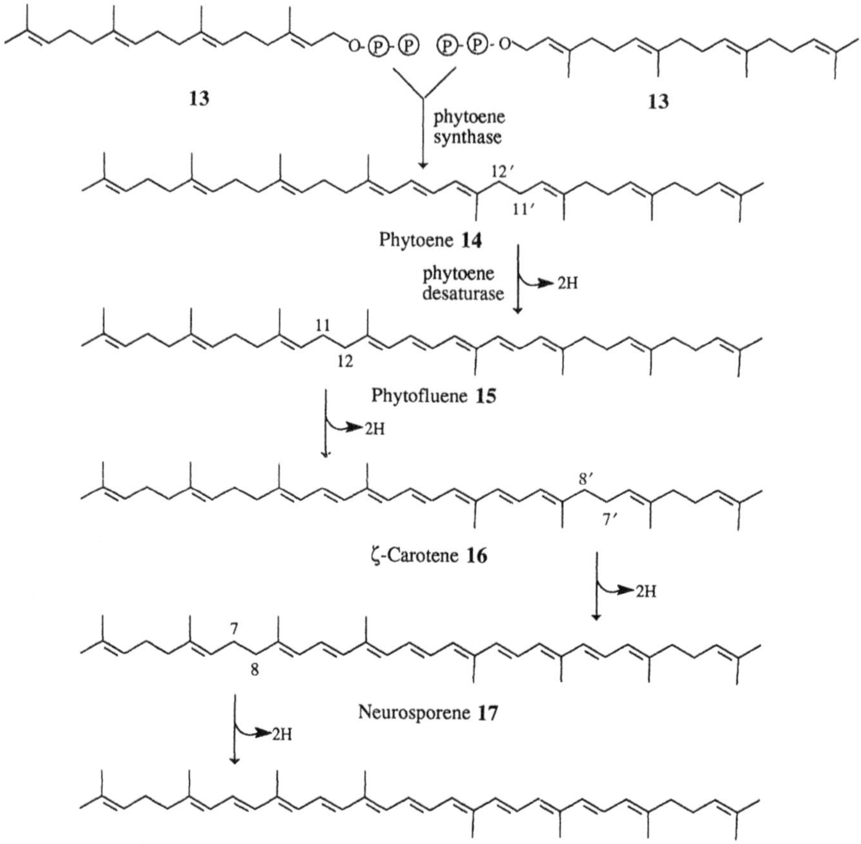

Fig. 2.3. Biosynthesis of phytoene (**14**) from two molecules of geranylgeranyl diphosphate (**13**) catalysed by the enzyme phytoene synthase, and the sequence of desaturation reactions (enzyme phytoene desaturase) by which phytoene (**14**) is converted into lycopene (**1**).

and each enzyme is the product of expression of a particular gene. The pathways and rections of carotenoid biosynthesis in plants are discussed in detail in many reviews (e.g. Britton, 1988; 1993).

Even complex carotenoid compositions containing a large number of individual carotenoids can be rationalized in terms of a relatively small number of genes. The differences between the groups of fruit identified in the broad classification given in Table 2.1 can, in most cases, be accounted for by differences in only one or two genes. Differently coloured strains or varieties of a species can be explained in a similar way.

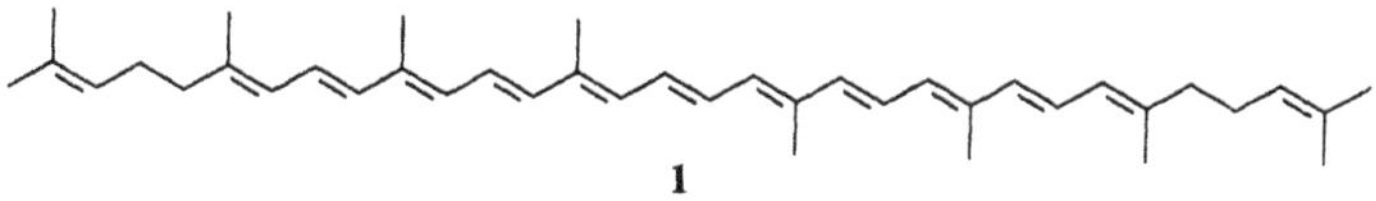

1

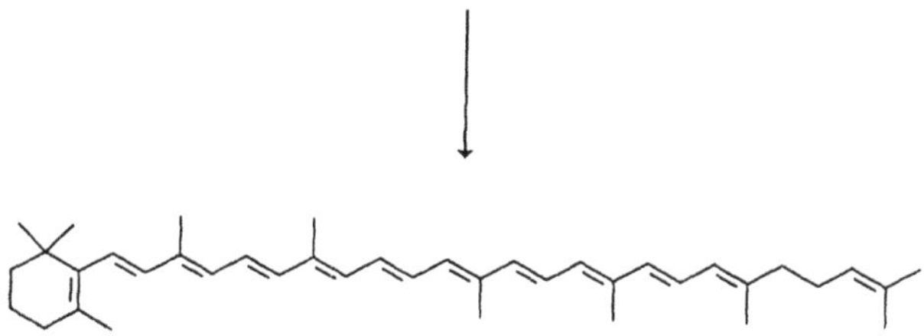

18

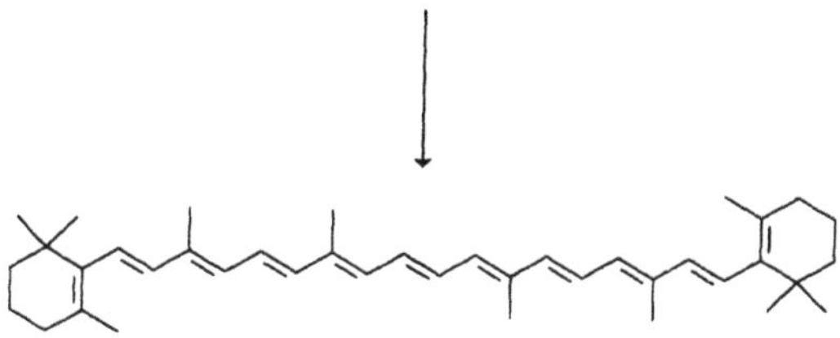

2

Fig. 2.4. The cyclization of lycopene (**1**) to form, successively, the monocyclic β,ψ-carotene (γ-carotene, **18**) and the dicyclic β,β-carotene (**2**).

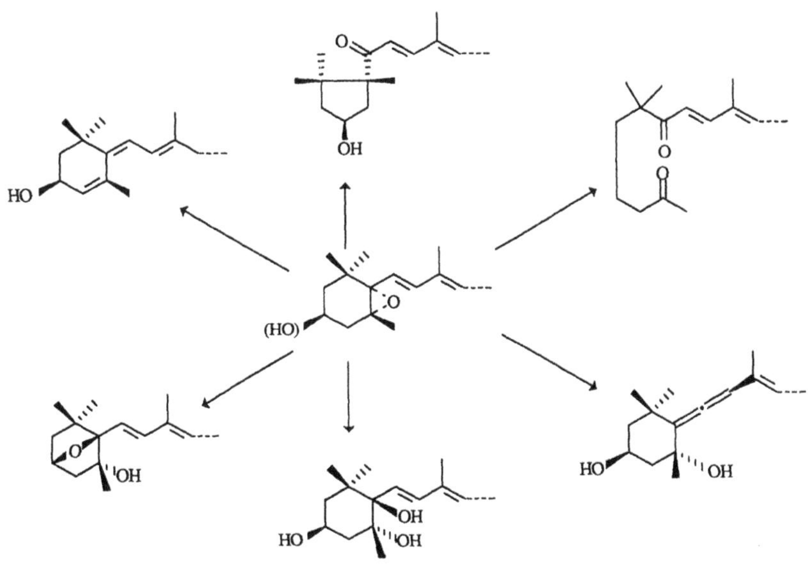

Fig. 2.5. Proposed reactions for the formation of diverse carotenoid end groups by rearrangements of the 5,6-epoxy-β ring.

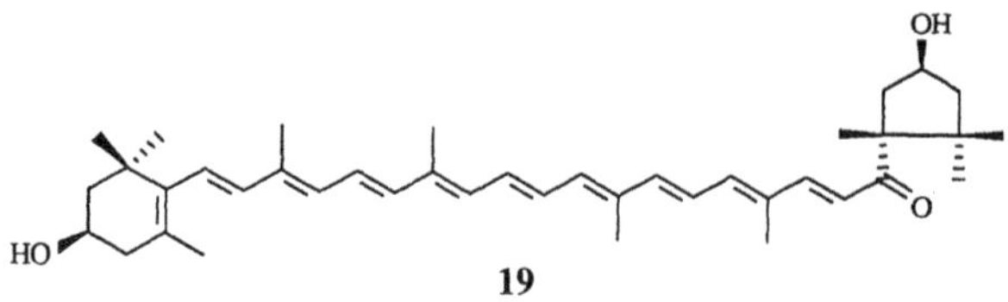

19

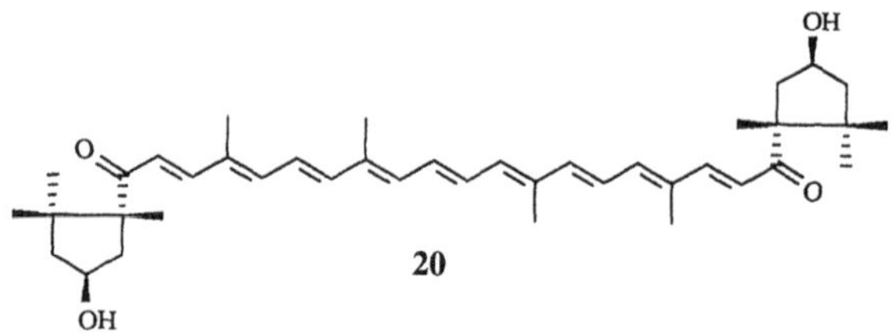

20

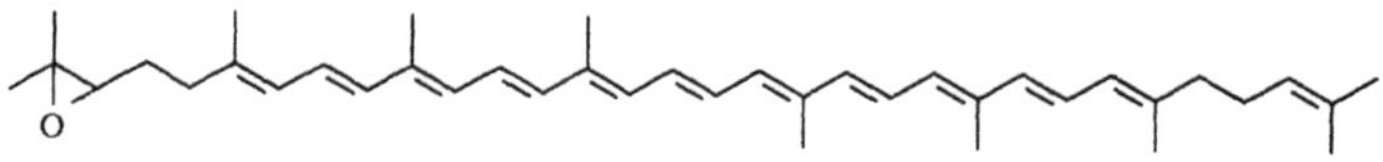

21

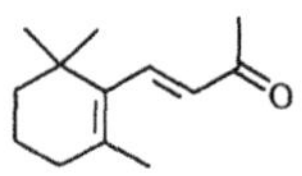

22

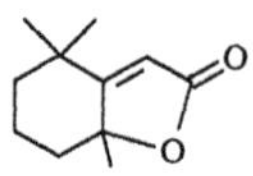

23

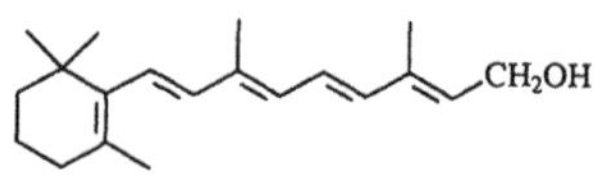

24

Factors that determine carotenoid content and composition of fruit and vegetables

In general, green, unripe fruit contain the normal collection of chloroplast carotenoids. When the fruit ripen, chromoplasts develop and carotenoids are synthesized on a large scale. The main carotenoids that accumulate are usually not the same ones as in the chloroplast. Carotenoid biosynthesis is, therefore, an important part of the overall ripening process. At the onset of ripening, many genes are 'switched on', including those for the ripening-specific carotenoid biosynthesis, and this leads to increased amounts of the biosynthetic enzymes, including new ones to catalyse the reactions that are needed to produce the carotenoids that accumulate specifically in the ripe fruit (Price *et al.*, 1993). Thus, in tomato, a ripening-specific phytoene synthase can be detected (Fray and Grierson, 1993).

Progress in the molecular genetics of carotenoid biosynthesis in tomatoes now makes possible genetic manipulations to modify the carotenoid content and composition. For example, transgenic tomato plants have been developed which contain increased amounts of phytoene synthase through the introduction of multiple copies of the phytoene synthase gene and have substantially increased lycopene levels (Fray and Grierson, 1993).

Many environmental factors also affect the biosynthesis and accumulation of carotenoids (Goodwin and Goad, 1970; Goodwin, 1980). Factors such as temperature, light regime, availability of water, soil quality, and environmental pollution, all of which have a great influence on growth, will affect carotenoid content and composition of fruit and vegetables. The effects of variations in any of these environmental factors can be different for different species and even varieties, though in most cases the carotenoid content will be greater under conditions that favour growth.

Molecular genetics

In recent years, the most exciting developments in carotenoid biosynthesis have been the advances made with molecular genetics, first in bacteria and later in higher plants. Several of the genes for carotenoid biosynthesis have now been isolated from plants and cloned, notably those coding for geranylgeranyl diphosphate synthase, phytoene synthase, phytoene desaturase and lycopene cyclase. Most of the work with higher plants has been done with tomato, *Capsicum*, soyabean and *Arabidopsis;* this work is reviewed by Bartley and Scolnik (1994).

In terms of carotenoid content and colour, the most interesting of these genes are *PSY* and *PDS*, which code for the enzymes phytoene synthase and phytoene desaturase, respectively (see Fig. 2.3). A cDNA, pTOM5, for a gene that is expressed during the ripening of tomato fruit, was found to code for phytoene synthase (Ray *et al.*, 1992). Transgenic tomato plants that were constructed to express a segment of this gene in an antisense direction showed a significant decrease in carotenoid (lycopene) content in the fruit but leaf carotenoids were not affected (Bramley *et al.*, 1992). This gene, now referred to as *PSY1*, appears to be expressed only during fruit ripening, and directs the synthesis of a 47 kDa protein which is imported into the plastids and processed into an active 42 kDa form which largely associates with membranes (Bartley *et al.*, 1992). The gene *PSY1* is the key factor in determining the massive synthesis of lycopene during the ripening process; during fruit development, expression of *PSY1* increases more than 20-fold. Its expression is also induced by ethylene (Maunders *et al.*, 1987). The expression of *PDS* is also increased during ripening but only by about 3-fold. A second phytoene synthase gene, termed *PSY2*, is active in the biosynthesis of the chloroplast carotenoids in the leaves and other tissues (Bartley and Scolnik, 1993).

The pace of progress in the molecular genetics of carotenoid biosynthesis in plants is now very rapid and many exciting advances are likely in the next few years. Already transgenic tomato plants with increased or decreased levels of lycopene in the fruit have been produced.

Effects of storage and processing on carotenoids in fruit and vegetables

Carotenoids, when isolated, are unstable towards heat, light and especially oxygen. In plant tissues *in situ* they are greatly stabilized by their associations with proteins, membranes etc, or by aggregation into microcrystals. Disruption of the tissues, e.g. by chopping or homogenizing, can, in some cases, lead to substantial rapid destruction of carotenoids through the action of lipoxygenases or other oxidation processes (Simpson *et al.*, 1976). Thus, for example, up to 30% of β,β-carotene can be lost within a few minutes when green leaves are macerated.

There can be great variation in the stability of carotenoids in stored fruit and vegetables. Usually, in mature fruit and vegetables that can be stored for long periods, losses or changes of carotenoids are small and occur slowly. Carotenoid biosynthesis and accumulation may even continue *post* harvest, e.g. during the ripening of fruit after they have been detached from the plant (Goodwin, 1980). In soft fruit, however, which deteriorate rapidly in storage (especially when not refrigerated), rapid losses of carotenoid and changes in carotenoid composition can occur as the tissues collapse and the carotenoids are exposed to oxidizing conditions or degradative enzymes. A good example of this is the tomato where, in over-ripe fruit, significant amounts of lycopene epoxides are found, especially lycopene-1,2-epoxide (**21**), in optically active form, indicating its formation by enzymic processes, possibly by the action of squalene epoxidase which comes into contact with the lycopene as the cells are disrupted (Ben-Aziz *et al.*, 1973). Generally carotenoid esters are more stable than the free carotenols.

Although the pathways of degradation of carotenoids have not been established, the main changes that occur are isomerization, leading to increased amounts of *Z* isomers, 5,6-epoxide to 5,8-furanoid oxide rearrangement, structural modification by released enzymes, and oxidative cleavage of the polyene chain by free-radical reactions to give shorter-chain apocarotenals and ketones as products (Simpson *et al.*, 1976). If unchecked, this last process will lead to bleaching of the tissues. It should be noted that some products of oxidative breakdown of carotenoids, e.g. β-ionone (**22**) and dihydroactinidiolide (**23**), make a significant contribution to flavour and aroma (Weedon, 1971). Carotenoids would be expected to be stable to freezing or freeze-drying, but losses can occur (bleaching) if oxygen is not rigorously excluded, even with deep-frozen material when this is stored for long periods.

The carotenoids are usually reasonably stable during heat processing and cooking of vegetables and fruit (Khachik *et al.*, 1992). Increased amounts of *Z*-isomers are usually found, however. Since the λ_{max} of the *Z*-isomers is usually only 2–5 nm lower than that of the all-*E* form, and the *Z*-isomers are formed only as a small percentage of the total, differences in colour due to isomerization are unlikely to be detectable.

Subtle changes in hue may be seen, however, as a result of changes in the subcellular structure in which the carotenoid is located, and of altered molecular interactions between the carotenoid and proteins, lipids, etc. within these subcellular structures. Dispersal or partial solubilization of microcrystalline deposits may also be a significant factor.

Carotenoids in nutrition and health

In view of the vital importance of fruit and vegetables as food, a brief consideration of the significance of carotenoids in nutrition and health is appropriate. For reviews on these topics, see Canfield *et al.* (1993). It is well known that carotenoids in fruit and vegetables provide the main dietary source of vitamin A for most people, especially in the poorer countries of the world, where vitamin A deficiency remains a major problem (Bauernfeind, 1986). β,β-Carotene (**2**) is the main provitamin, and is cleaved by an intestinal dioxygenase enzyme to provide two molecules of vitamin A (retinol, **24**) (Olson, 1986). β,β-Carotene is not the only provitamin A carotenoid, however. Any carotenoid with an unsubstituted β-ring can undergo similar cleavage and provide one vitamin A molecule. Carotenoids such as β,ε-carotene (α-carotene) (**5**), β,β-carotene-5,6-epoxide (**25**) and β-cryptoxanthin (**26**) can therefore make a substantial contribution to the vitamin A nutritional value of fruit and vegetables.

In recent years, increasing evidence has been obtained to show that this may not be the only health value of carotenoids. Many epidemiological surveys have indicated that diets rich in carotenoids are correlated with the lowest risks of cancer, heart disease and other major human disorders (Peto *et al.*, 1981; Gaziano and Hennekens, 1993; Ziegler, 1993). This in itself does not show that carotenoids are the active factor in this beneficial effect, but subsequent human intervention trials and experiments with animals have shown that a direct effect of carotenoids, usually β,β-carotene, is likely. Currently, attention is mainly focused on the proposed action of β,β-carotene as an antioxidant that may interfere with free-radical oxidations, such as lipid peroxidation, that are a feature of many degenerative diseases (Palozza and Krinsky, 1992). Although it can readily be demonstrated that β,β-carotene can have a significant protective antioxidant effect *in vitro* (Burton and Ingold, 1984), there is still no real proof of such a role, *in vivo*, at the low concentrations of carotenoids found, and under physiological conditions.

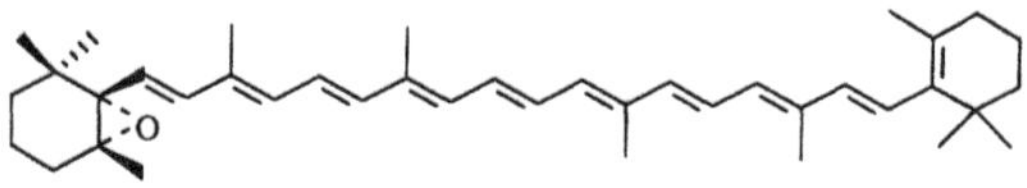

25

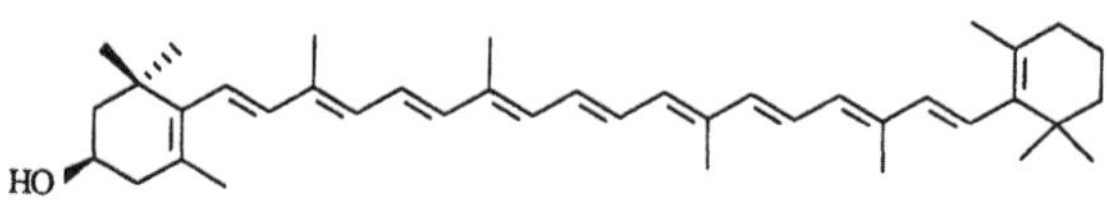

26

Recent work showing that carotenoids can modify the structure and properties of membranes (Subczynski *et al.*, 1992), and that they can have substantial effects on the human immune response system (Bendich, 1991) and on processes such as cell gap junction communication (Bertram, 1994), opens up exciting new possibilities for investigation.

To have any beneficial action, the carotenoids must be absorbed, transported and deposited in the required tissues (Erdman *et al.*, 1993). Any of the carotenes and many of the xanthophylls can be utilized, but it appears that the epoxycarotenoids such as violaxanthin and neoxanthin, although they occur abundantly in plants, cannot be used by the body. The bioavailability of dietary carotenoids is dependent on other dietary factors, especially the availability of fat. Carotenoids may be absorbed more efficiently from cooked than from raw vegetables, probably because cooking releases the carotenoids from their normal molecular environment and thus renders them more accessible for solubilization. Because of their reduced tendency to form microcrystalline aggregates and their consequent easier solubilization, carotenoid *Z*-isomers may show better bioavailability than the all-*E* form.

The carotenoids are of great nutritional importance, as intact molecules and not simply as vitamin A precursors, so carotenoid content and composition are clearly important factors in the nutritional value of fruit and vegetables. The newer work in this area shows that all carotenoids, not only β,β-carotene and other provitamins A, could have important beneficial actions and must be included whenever the nutritional value of fruit and vegetables is assessed.

Future perspectives

With the wide availability of HPLC, the analysis of carotenoid compositions of fruit and vegetables has now become routine. A new era has begun for biosynthetic studies following recent advances in molecular genetics. The way is now open to tackle hitherto unapproachable questions about the organization of enzyme complexes and the regulation of carotenoid biosynthesis, especially in relation to plastid development. Genetic manipulations to increase or modify carotenoid content and composition are already a reality with the tomato and the development of other transgenic plants can be expected.

For many years the nutritional value of carotenoids has been equated with provitamin A activity. Now it is realized that not only β,β-carotene but also other carotenoids that are abundant in fruit and vegetables, notably lycopene, lutein and zeaxanthin, may have important effects in protection against major disorders including cancer and heart disease. The general view that such effects involve antioxidant action is not proved and other possibilities such as effects on the immune response system and on membrane-associated processes show greater promise. Any biological activity of carotenoids in humans and other animals clearly depends on the efficiency of absorption from the food. New developments in the area of bioavailability can be expected, especially in relation to the release of carotenoids from the structural matrix of the food and the absorption, transport and deposition of the carotenoids within the body. Systematic studies of the ultrastructure of chromoplasts and the physical state and molecular environment of the carotenoids within the chromoplast are highly relevant to this topic as well as being of intrinsic plant physiological interest.

References

Barry, P., Young, A.J. and Britton, G. (1991). *J. Exp. Bot.* **42**, 229–34.

Bartley, G.E. and Scolnik, P.A. (1993). *J. Biol. Chem.* **268**, 25718–21.

Bartley, G.E. and Scolnik, P.A. (1994). *Ann. Rev. Plant Physiol. Plant Mol. Biol.* **45**, 287–301.

Bartley, G.E., Viitanen, P.V., Bacot, K.O. and Scolnik, P.A. (1992). *J. Biol. Chem.* **267**, 5036–9.

Bauernfeind, J.C. (ed.) (1986). *Vitamin A deficiency and its control.* Academic Press, Orlando, Florida.

Ben-Aziz, A., Britton, G. and Goodwin, T.W. (1973). *Phytochemistry* **12**, 2759–64.

Bendich, A. (1991). *Proc. Nutr. Soc.* **50**, 263–74.

Bertram, J.S. (1994). *Pure Appl. Chem.* **66**, 1025–32.

Bramley, P., Teulieres, C., Blain, I., Bird, C. and Schuch, W. (1992). *Plant J.* **2**, 343–9.

Britton, G. (1988). In *Plant pigments* (ed. T.W. Goodwin), pp. 133–82. Academic Press, London.

Britton, G. (1993). In *Carotenoids in photosynthesis* (ed. A.J. Young and G. Britton), pp. 96–126. Chapman & Hall, London.

Britton, G., Liaaen-Jensen, S. and Pfander, H. (1995). *Carotenoids Vol. 1A: Isolation and analysis; Vol. 1B: Spectroscopy.* Birkhäuser Verlag, Basel.

Burton, G.W. and Ingold, K.U. (1984). *Science* **224**, 569–73.

Canfield, L.M., Krinsky, N.I. and Olson, J.A. (eds) (1993). *Carotenoids in human health.* (*Ann. N.Y. Acad. Sci.* Vol. **691**). New York Academy of Sciences, New York.

Demmig-Adams, B. and Adams, W.W. III (1993). In *Carotenoids in photosynthesis* (ed. A.J. Young and G. Britton), pp. 206–51. Chapman & Hall, London.

Erdman, J.W., Bierer, T.L. and Gugger, E.T. (1993). *Ann. N.Y. Acad. Sci.* **691**, 76–85.

Fray, R.G. and Grierson, D. (1993). *Plant Mol. Biol.* **22**, 589–602.

Gaziano, J.M. and Hennekens, C.H. (1993). *Ann. New York Acad. Sci.* **691**, 148–55.

Goodwin, T.W. (1980). *The biochemistry of carotenoids, 2nd edition, Vol. 1 (Plants).* Chapman & Hall, London.

Goodwin, T.W. and Britton, G. (1988). In *Plant pigments* (ed. T.W. Goodwin), pp. 61–134, Academic Press, London.

Goodwin, T.W. and Goad, L.J. (1970) In *The biochemistry of fruits and their products, Vol. 1* (ed. A.C. Hulme), pp. 305–68. Academic Press, London.

Gross, J. (1987). *Pigments in fruits.* Academic Press, London.

Gross, J. (1991). *Pigments in vegetables. Chlorophylls and carotenoids.* Van Nostrand Reinhold, New York.

Khachik, F., Goli, M.B., Beecher, G.R., Holden, J., Lusby, W.R., Tenorio, M.D. and Barrera, M.R. (1992). *J. Agr. Food Chem.* **40**, 390–8.

Kirk, J.T.O. and Tilney-Bassett, R.A.E. (1978). *The plastids: their chemistry, structure, growth and inheritance,* 2nd edn. Elsevier-North Holland, Amsterdam.

Kull, D. and Pfander, H. (1995). In *Carotenoids Vol. 1A: Isolation and analysis* (ed. G. Britton, S. Liaaen-Jensen and H. Pfander), pp. 295–317. Birkhäuser Verlag, Basel.

Matus, Z., Deli, J. and Szabolcs, J. (1991). *J. Agric. Food Chem.* **39**, 1907–14.

Maunders, M.J., Holdsworth, M.J., Slater, A., Knapp, J.E. and Bird, C.R. (1987). *Plant Cell Environ.* **10**, 177–84.

Olson, J.A. (1986) In *Vitamin A deficiency and its control* (ed. J.C. Bauernfeind), pp. 19–67. Academic Press, Orlando, Florida.

Palozza, P. and Krinsky, N.I. (1992). *Meth. Enzymol.* **213**, 403–20.

Peto, R., Doll, R.J., Buckley, J.D. and Sporn, M.B. (1981). *Nature* **290**, 201–8.

Price, C.A., Cervantes-Cervantes, M., Hadjeb, N., Newman, L.A. and Oren-Shamir, M. (1993) In *Pigment–protein complexes in plastids: synthesis and assembly* (ed. C. Sundqvist and M. Ryberg), pp. 485–505. Academic Press, San Diego.

Ray, J., Moureau, P., Bird, C., Bird, A., Grierson, D., Maunders, M., Truesdale, M., Bramley, P. and Schuch, W. (1992). *Plant Mol. Biol.* **19**, 401–4.

Siefermann-Harms, D. (1985). *Biochim. Biophys. Acta* **811**, 325–55.

Siefermann-Harms, D., Hertzberg, S., Borch, G. and Liaaen-Jenen, S. (1981). *Phytochemistry* **20**, 85–8.

Simpson, K.L., Lee, T.-C., Rodriguez, D.B. and Chichester, C.O. (1976). In: *Chemistry and biochemistry of plant pigments*, 2nd edn (ed. T.W. Goodwin), pp. 780–842. Academic Press, London.

Sitte, P., Falk, H. and Liedvogel, B. (1980) In *Pigments in plants,* 2nd edn (ed. F.-C. Czygan), pp. 117–48. Gustav Fischer, Stuttgart.

Straub, O. (1987). *Key to carotenoids,* 2nd edn (ed. H. Pfander). Birkhäuser Verlag, Basel.

Subczynski, W.K., Markowska, E., Gruszecki, W.I. and Sielewiesiuk, J. (1992). *Biochim. Biophys. Acta* **1105**, 97–108.

Weedon, B.C.L. (1971). In *Carotenoids* (ed. O. Isler), pp. 29–59. Birkhäuser Verlag, Basel.

Weedon, B.C.L. and Moss, G.P. (1995). In *Carotenoids Vol. 1A: Isolation and analysis* (ed. G. Britton, S. Liaaen-Jensen and H. Pfander), pp. 27–70. Birkhäuser Verlag, Basel.

Young, A.J. (1993). In *Carotenoids in photosynthesis* (ed. A.J. Young and G. Britton), pp. 72–95. Chapman & Hall, London.

Young, A.J. and Britton, G. (eds) (1993). *Carotenoids in photosynthesis.* Chapman & Hall, London.

Ziegler, R.G. (1993). *Ann. New York Acad. Sci.* **691**, 110–19.

3. Molecular interactions of phenolic compounds in relation to the colour of fruit and vegetables

RAYMOND BROUILLARD, PAULO FIGUEIREDO, MOURAD ELHABIRI and OLIVIER DANGLES

Laboratoire de Chimie des Polyphénols, associé au CNRS, Université Louis Pasteur, Institut de Chimie, 4, rue Blaise Pascal, 67008 Strasbourg, France

---

Introduction

Three large families of pigments give colour to plants: chlorophylls, carotenoids and flavonoids (Britton, 1983). The two latter are seen by humans on the green background provided by the chlorophylls. Carotenoids and flavonoids are widespread in fruit, flower and vegetable tissues. Flavonoids belong to a larger group of natural organic plant compounds, the polyphenols. Good recent introductions to the biochemistry of polyphenols may be found in Scalbert (1993) and in Brouillard *et al.* (1995).

About ten thousand or so natural plant polyphenols have been presently identified, half of them probably being of the flavonoid type (Harborne, 1993). Coloured flavonoids are best seen in flower and fruit epidermal tissues. Such pigments are synthesized by almost all flowering vegetables and can be found in all plant organs. They are also found both unaltered or chemically or enzymatically modified in food products of vegetable origin and they constitute a regular component in human and animal diets. Every day about 10 g are consumed by an adult person. Among their many biological activities (Cody *et al.*, 1986; 1988), only their prominent role in producing the most vivid colours to be found in plants will be taken into consideration.

Almost all polyphenols strongly absorb light in the ultraviolet range but only a few members of this huge family do so in the visible range. In fact, the anthocyanins are the only meaningful subgroup of polyphenols visible to the human eye. Anthocyanins gradually appear in fruit, flower and other plant tissues when chlorophylls are fading away. Anthocyanins are stored, if not biosynthesized, within the vacuoles of mature epidermal cells. Extraction from the vacuoles of their coloured liquid occurs in fruit processing for industrial or home-made fruit juice production. The vacuolar aqueous liquid is always slightly acidic, the characteristic acidity of grape juice for instance, ranges from pH 3 to 3.5, although other values may be found with other juices. In the food industry there is a great concern about anthocyanin colour stability and variation. *In vivo,* colour stability can last from a few hours to several weeks or even months, and may be limited only by the lifetime of a given fruit or flower. This is usually not the case with plant preparations, where colour may vary, or even worse, be lost more or less rapidly, a problem which is incompatible with the often long storage periods until the product reaches the consumer. The natural physicochemical environment of anthocyanins is well defined: it is a slightly acidic aqueous solution kept at room temperature under nearly atmospheric pressure. Nevertheless, it remains to be said that most naturally occurring anthocyanins, when kept in slightly acidic aqueous model solutions, tend to decolourize quickly and sometimes almost completely (Brouillard, 1983). We will see that this initial fading is in fact vital to the production of many different colours arising from a single anthocyanin and the colour loss is extremely useful in revealing the many interactions, covalent and molecular, that are a feature of this type of natural organic pigment (Brouillard *et al.,* 1990). Previous reviews on molecular interactions of phenolic compounds, with a special focus on anthocyanins, have been written by Timberlake (1981), Brouillard (1988) and Brouillard and Dangles (1993).

Access to pure natural anthocyanin pigments is not so easy as one would expect on the basis of their extremely large and widespread production in the plant kingdom. For instance, to obtain just a few milligrammes of pure anthocyanin requires skill in many purification techniques as well as powerful spectroscopic and chemical identification methods (see Johansen *et al.,* 1991; Markham and Geiger, 1993; Strack and Wray 1993; Santos *et al.,* 1993; Saito *et al.,* 1994; 1995). In addition to ultraviolet-visible, NMR and mass spectroscopies, vibrational spectroscopy as well as computer-assisted theoretical methods can be helpful in the elucidation of anthocyanin chemical structures (see Merlin *et al.,* 1994*a,b*; Yoshida *et al.,* 1992). In this context, chemical synthesis can be an alternative to providing large amounts of pure synthetic analogues where the 3-glucosyl residue, common to most natural anthocyanins, may be replaced by a 3-methyl group (Wigand *et al.,* 1992; Dangles and Elhajji 1994; Dangles *et al.,* 1994). Surveys of the routes of chemical synthesis leading to anthocyanins have been published recently (Elhabiri *et al.,* 1995*a, b*).

Anthocyanins of fruits and vegetables

Anthocyanins are secondary metabolites of higher plants. From a chemical point of view their structure is made of a C_{15} heterocyclic nucleus bearing at least one, and often several, sugar residues. The sugar hydroxyls may be esterified by aliphatic and/or aromatic organic acids. The most common sugar residues are glucosyl, galactosyl, rutinosyl and sophorosyl. A typical anthocyanin, found in petals of a *Pharbitis nil* cultivar is shown in Fig. 3.1 (Dangles *et al.*, 1993*b*). It features the structural elements of many natural anthocyanins, i.e. a 2-phenylbenzopyrylium skeleton, two sugars (3-sophorosyl and 5-glucosyl) and a caffeic acid moiety. In strongly acidic aqueous media it exists predominantly as the flavylium cation form represented in the figure, with the positive charge delocalized through the pyrylium aromatic system, although the carbons 2 and 4 were found to be the two more positively charged positions of the chromophore (Amić *et al.*, 1990).

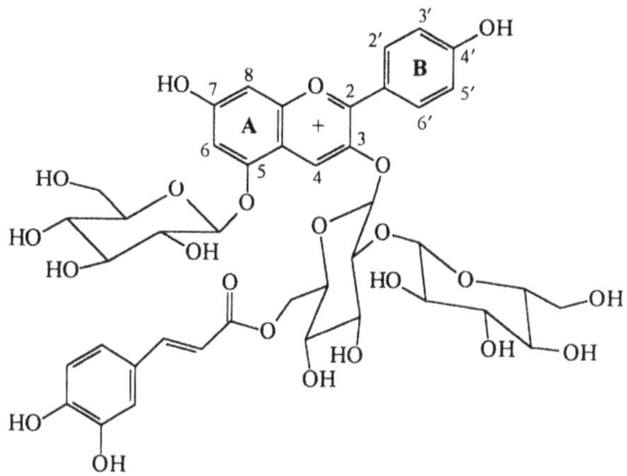

Fig. 3.1. A typical anthocyanin with an aglycone (pelargonidin in this case), two different sugars (glucosyl and sophorosyl residues) with one of the sugars bearing a caffeoyl group. For a list of natural anthocyanin structures see Harborne and Grayer (1988).

Furthermore, the two acidic hydroxyl groups present at positions 4′ and 7 of the aglycone contribute to its changes in colour. One deprotonates around pH 4 and the second close to neutrality producing the quinonoidal bases, which are also coloured and exhibit absorbances at longer wavelengths than the flavylium cation. Of these species, the one formed first is neutral (A, Fig. 3.2) while that arising at higher pH is negatively charged (A^-, Fig. 3.2). Therefore, the flavylium cation remains stable only in sufficiently acidic media. Moreover, being positively charged, the flavylium species acts as an electrophile, nucleophilic covalent addition taking place at positions 2 and/or 4. In an aqueous environment, the water molecule, albeit not a very strong nucleophile but present in large amounts, binds to the flavylium system more or less rapidly, starting at pH values near 2, with the disappointing result of colour loss, essentially by formation of the colourless hemiacetalic adduct (B_2, Fig. 3.2). This corresponds to a pH-dependent

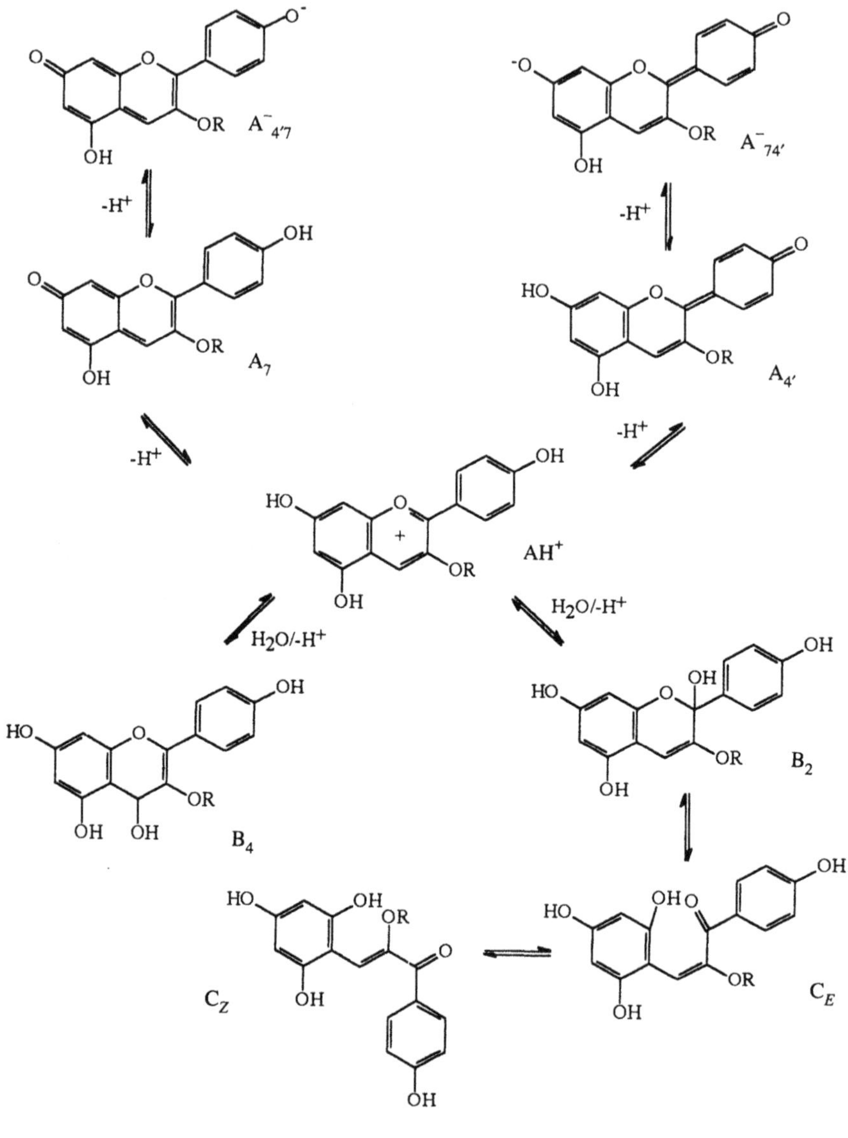

Fig. 3.2. Structural transformations of callistephin, the strawberry anthocyanin. R stands for a glucosyl group (see Brouillard 1988).

chemical equilibrium, allowing the flavylium ion to be restored merely by subsequent acidification (see Brouillard and Dangles, 1993). For anthocyanins the situation is therefore more complex than encountered with simple acid and base systems. Here we have two fast acid–base systems accompanied by a much slower pseudo-acid–base equilibrium (covalent hydration), eventually followed by a reversible central-ring opening leading to the formation of two isomeric chalcones (C_E and C_Z, Fig. 3.2). The general mechanism is exemplified in Fig. 3.2, for the case of callistephin, the orange-red pigment of strawberries.

These reactions are characteristic of the aglycone moiety and do not directly involve the participation of the sugar and/or acyl moieties, although such groups could have a profound effect on the hydration, ring-opening and isomerization steps. In the natural anthocyanin series, structural diversity, hence colour differences, comes from the varied number and positions of the hydroxyl, methoxyl, glucosyl and acyl substituents leading to about 200 anthocyanins identified at present (Harborne, 1993). There are not so many aglycones, however, and, if one deletes the rare ones from the list, only a few well-known types remain, viz. pelargonidin, peonidin, malvidin, cyanidin, delphinidin and petunidin, ranked according to their substitution pattern in the B-ring (Harborne, 1967). Those anthocyanins possessing a catechol group are expected to react further with metallic cations (Dangles *et al.*, 1994).

In natural or processed food anthocyanin-containing media, water is always present and hence the colour-damaging nucleophilic hydration reaction. Therefore, depending upon the acidity of a given medium, the total amount of pigment is shared between coloured chromophores (flavylium and quinonoidal species) and colourless ultraviolet-light-absorbing forms. In the chalcone form, an anthocyanin is much more vulnerable to physicochemical factors which may even lead to its irreversible degradation. In cell vacuoles, colourless anthocyanin species have seldom been found: this indicates that, *in vivo*, there exist stabilizing mechanisms strongly favouring the coloured structures. Of these mechanisms, copigmentation is probably the most efficient. It has been demonstrated to occur in apples (Lancaster *et al.*, 1994) although it has been much more studied in flowers (Osawa, 1982; Goto and Kondo, 1991). When compared with living tissues, food products derived from edible coloured plant organs (fruits, leaves, tubers, etc.), show a fairly poor colouration. Crushing tissues leads, on the one hand, to the mixing of biochemically active molecules with quite favourable conditions for pigment bleaching (water, oxygen, temperature, light and acidity) and, on the other hand, to the partial or total disappearance of the mechanisms that protect anthocyanin coloured species (destruction of the natural supramolecular edifices).

Comprehensive reports on anthocyanin distribution in fruit and vegetables have been issued recently (Macheix *et al.*, 1990; Mazza and Miniati, 1993). Emphasis has been placed on their usefulness as food colourants (Francis, 1993). Fruit and vegetable anthocyanin structures fit into the general structural scheme outlined above. In some instances several different anthocyanins, sometimes in large amounts, can be present, as with red cabbage leaves, red fruit berries or yam tubers (see Hong and Wrolstad, 1990; Yoshida *et al.*, 1991b; Baublis *et al.*, 1994; Gao and Mazza, 1994). Grapes constitute a very special case because of their widespread occurrence as a food crop and the numerous scientific investigations of their natural products, including polyphenols, in relation to wine elaboration. Oenin (malvidin 3-glucoside), a simple anthocyanin, is the main pigment of *Vitis vinifera* fruits, and the bright red colour of young red wines comes from its diffusion from the skins of the mature red grapes into the must (Brouillard and Dangles, 1994; Dallas and Laureano, 1994).

At this stage it should be pointed out that the colour of anthocyanins is deeply affected by the nature of their physicochemical environment. This is really a distinctive feature of these pigments, leading to the following remarks.

- The tint or hue of each fruit is not necessarily associated with a chemically distinct chromophore (aglycone) responsible for that particular colour only. Indeed, as there exists only a very small number of natural anthocyanin aglycones, it is by combining different substitution patterns, in different physicochemical environments and loose molecular associations that anthocyanins are capable of producing an infinite variety of hues.
- Anthocyanins are large organic molecules, although much smaller than most biomolecules. Nevertheless, they display binding properties usually featured only by large natural macromolecules such as proteins, nucleic acids and polysaccharides. Especially interesting are the numerous ways they interact with water molecules in both the liquid and solid states, including the establishment of hydrogen bonds between the phenolic hydroxyls and water molecules, giving proton transfer reactions (several such sites are available on any natural anthocyanin); nucleophilic addition of water to the electrophilic flavylium cation leading to covalent bond formation; good solvation of the hydrophilic glycosyl residues which are believed to adapt to the tetrahedral hydrogen bonded network of liquid water (David, 1995); solvation of the hydrophobic aromatic rings of the aglycones.

Intramolecular and intermolecular copigmentations: effects of pH, temperature and medium

A copigment can be defined as a molecule that has no colour by itself, but when put into an anthocyanin (pigment)-containing solution, it usually intensifies and modifies the initial colour given by this pigment to the solution. The corresponding colour change reveals the existence of a molecular complex, taking place between the pigment and the copigment, which, a long time ago, was called a copigmentation complex (Robinson and Robinson, 1931). It seems that almost all polyphenols have the property to behave as copigments although with more or less efficiency depending upon their chemical structure (Brouillard and Dangles, 1993; Davies and Mazza, 1993). Copigments are, however, not limited to polyphenols and copigmentation complexes are also formed with purines and alkaloids (Asen *et al.*, 1972; Osawa, 1982). In fact, the list of potential copigments cannot be restricted to any chemical family and many more compounds possessing the ability of anthocyanin colour-enhancement might be uncovered. To produce colour intensification and variation a copigment has to associate, in a loose manner, with the anthocyanin coloured structures, especially the flavylium cation species. Such associations give, in electronic absorption spectroscopy terms, an hyperchromic shift, which can be very spectacular, and which is generally accompanied by a bathochromic displacement of the visible absorption band, formerly called the bluing effect. In the formation of the copigmentation molecular complex, partial desolvation of the pigment and copigment molecules occurs with solvent molecules rearranging around the complex, allowing a closer contact between both molecules (copigmentation complexes are usually of the 1 : 1 type). This is the mechanism at the origin of the hyperchromic shift in the anthocyanin visible absorption band (more chromophores are formed due to reduced access of water molecules to the C-2 and/or C-4 electrophilic sites in the flavylium structure), and of the bathochromic shift (there is a concomitant change in polarity on the immediate neighbourhood of the chromophore, some water molecules having been displaced by an organic, less polar molecule). These effects are illustrated by malvin associated with chlorogenic acid, a polyphenolic species often found in fruits (Fig. 3.3).

As shown in Fig. 3.3 (upper part), the association is barely detectable to the human eye when strongly acidic solutions are considered, but for a pH 3.50 solution (lower part), spectral changes are remarkable, making this phenomenon clearly visible. In solutions containing an organic buffer, as is the case in fruit juices, there is a large reservoir of anthocyanins in the colourless hemiacetalic forms which may be turned into the flavylium cation/quinonoidal base system by copigmentation; hence the colour change and increase. In strongly acidified solutions the reservoir of colourless forms is negligible and copigmentation comes down to an ordinary molecular association, with a small hypochromic shift accompanying the always present bathochromic shift (Dangles and Brouillard, 1992*b*). Copigmentation consists of the formation of 1 : 1 vertical stacking complexes between a π-electron rich copigment moiety and a coloured planar anthocyanin form (Brouillard, 1983). Interestingly, a special and rare type of association (copigmentation with self-association) has been postulated to explain the strong colouring power of alatanin C extracted from tubers of the purple yam (Yoshida *et al.*, 1991a). Another new mechanism of association was described by Nerdal and Andersen (1991; 1992), for the case of petanin, a monoacylated petunidin derivative found in tubers of *Solanum tuberosum*. This pigment was shown, by means of NOESY NMR data, to self-associate, the aglycones being head-to-tail along the long and the short aglycone axis.

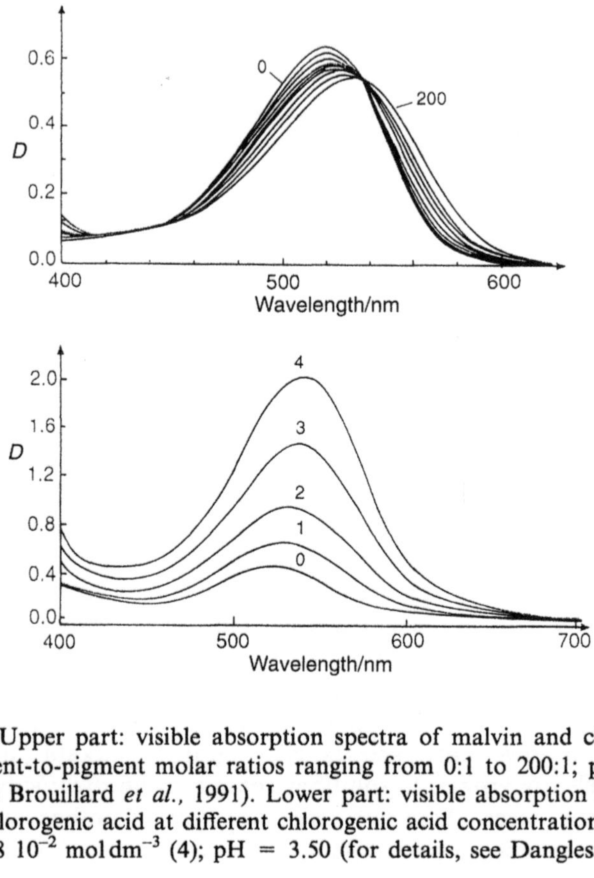

Fig. 3.3. Upper part: visible absorption spectra of malvin and chlorogenic acid at copigment-to-pigment molar ratios ranging from 0:1 to 200:1; pH = 0.65 (for details, see Brouillard *et al.*, 1991). Lower part: visible absorption spectra of malvin and chlorogenic acid at different chlorogenic acid concentrations ranging from 0 (0) to 1.8 10^{-2} mol dm^{-3} (4); pH = 3.50 (for details, see Dangles and Brouillard 1992*b*).

Colour stabilization and variation is expected to occur to different extents with different copigments (Brouillard *et al.*, 1991; Mistry *et al.*, 1991). Copigmentation complexes of a 1 : 2 ratio—two copigment molecules providing the frame within which the pigment inserts (sandwich-type copigmentation)—have been observed using synthetic sugar free anthocyanin analogues (Dangles and Elhajji, 1994). Glucosylated anthocyanins appear to be capable of only forming 1 : 1 associations, a good explanation for this being that access to one of the aglycone almost planar faces is obstructed by the sugar residue(s) folded over that face (steric hindrance), as can be demonstrated through computer-assisted molecular modelling. In the special case of caffeine, it has been demonstrated that copigmentation can also occur with colourless forms, in particular the chalcone forms, although the interaction was not strong enough to cause colour loss in the glycosylated series (Dangles and Elhajji, 1994). Apart from electronic spectroscopy, NMR methods are good tools in the qualitative and quantitative evaluation of the association between an anthocyanin and a copigment (Mistry *et al.*, 1991). In our group we used ^{1}H NMR spectroscopy to study the association between the synthetic 7-hydroxy-3,4′-dimethoxyflavylium chloride and gallic acid (Wigand *et al.*, 1992). In the ^{1}H NMR experiments, gallic acid features an unique singlet which, fortunately, does not interfere with the spectrum of the 7-hydroxy-3,4′-dimethoxyflavylium cation. In the presence of gallic acid all anthocyanin protons are shifted upfield which means that they experience the gallic acid ring-current effect. Such a result provides good evidence for the vertical stacking of copigment and pigment molecules and adequately explains the reduction in the extent of the hydration reaction on addition of a copigment to a slightly acidic aqueous anthocyanin solution. To sum up the effect of pH on intermolecular copigmentation, one can say that:

- at pH $\leqslant$ 1 complexation is meaningless from the point of view of colour changes;
- from pH 1–2 to neutrality, copigmentation produces astonishing colour enhancements and variations, depending on pigment and copigment structures and concentrations, temperature and the nature of the medium where it takes place, aqueous solutions producing the more spectacular results (Brouillard *et al.*, 1991; Dangles and Brouillard, 1992*a*).

Intramolecular copigmentation can only occur with those anthocyanins possessing at least one copigment residue in their structure—usually a cinnamic ester derivative—that can interact through its π-orbitals with the benzopyrylium nucleus. The existence of intramolecular copigmentation was first demonstrated in the case of zebrinin, the *Zebrina pendula* anthocyanin (Brouillard, 1981). In this case, colourless forms could not be detected whatever the pH of the solution. A thorough examination of intramolecular copigmentation could be made possible only recently when a structurally homogeneous anthocyanin series of five pigments was extracted, purified and identified from red-purple corolla limbs of *Pharbitis nil* cultivars (Lu *et al.*, 1992).

Figure 3.4 shows this series of five derivatives of pelargonidin. They possess in common the aglycone (pelargonidin), the glucosyl at C-5 and the sophorosyl at C-3. They share the same chromophore and classical chemical views would predict them to feature identical absorption spectra in the visible range. Their visible absorption spectra are, however, quite different (Dangles *et al.*, 1993*a*). Given this observation, it is assumed that the caffeoyl residues linked to the sophorosyl unit interact with the chromophore through space and not through chemical bonds. This hypothesis is further confirmed by NOE ^{1}H NMR data that demonstrate the existence of long-range interactions between the caffeoyl protons and those of the aglycone (Lu *et al.*, 1992). Such a kind of remote effect is usually observed with large biomolecules but rarely seen with smaller organic molecules whose conformations do not allow long distance interactions to occur. The number of 'internal' copigments in the case of the pharbitins shown in Fig. 3.4 ranges from zero (reference pigment **1**) to three (pigment **5**). At a given pH, colour enhancement occurs gradually along the series **1–5**. By means of kinetic, spectroscopic and thermodynamic measurements, it was possible to conclude that intramolecular stacking complexes were formed, as well as sandwich-type complexes for compounds *4* and *5*, the latter structures possessing the required number of 'internal' copigments to demonstrate this behaviour (Dangles *et al.*, 1993*b*). Intramolecular complexation has an important entropic advantage over intermolecular copigmentation, since the former does not need to bring together molecules initially separated in solution.

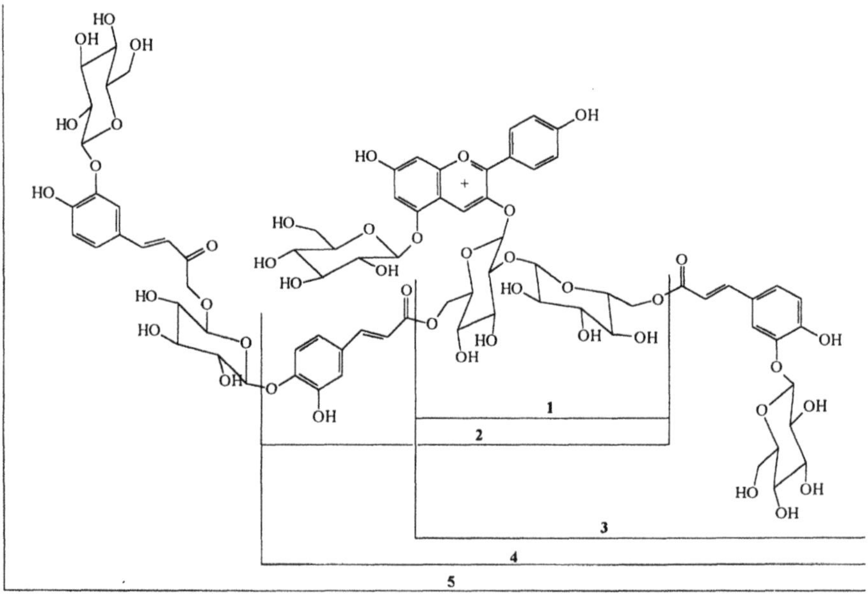

Fig. 3.4. Chemical structures of pharbitins *1–5* when in the flavylium form (see Lu *et al.*, 1992).

Preorganization is also provided, in the case of the pharbitins, by the sophorosyl spacer which may help in the formation of the stacking between the pelargonidin aglycone and the differently situated caffeoyl groups. Acidity constant values characteristic of the 7-OH and 4′-OH do not appreciably change within the series (pK_a close to 4.1), suggesting that the positive charge of the flavylium cation remains distributed in the same manner in all five pharbitins. This also means that the partial positive charge at C-2 remains the same with this carbon keeping its intrinsic electrophilic character intact (Dangles *et al.*, 1993*a*;*b*). The almost constant value of the deprotonation constant also constitutes a proof that the two hydroxyl groups remain free through the series **1–5**, thus not participating in any hydrogen bonding with the copigment moiety. The value of the hydration rate constant steadily decreases when going from **1** to **5**, however, strongly suggesting that the new chemical environment of the aglycone is responsible for the reduction of the colour-damaging hydration reaction which takes place on the pyrylium system. At the same time, the thermodynamic hydration constant diminishes when going from *1* to *5* (pK_h ranges from 2.30 for **1** to 4.01 for **5**). The last value is very close to the mean pK_a value (4.1, see above), which translates into a predominance of coloured forms (flavylium and quinonoidal bases) through all the acidic to neutral pH range. Good colour stabilization is therefore provided to pigment **5** by its caffeoyl residues. Similar properties were very recently demonstrated with two series of delphinidin derivatives, extracted from blue-purple flowers of *Eichhornia crassipes* and the blue flowers of *Evolvulus pilosus* cv Blue Daze, possessing different glycosylation and acylation patterns (Figueiredo *et al.*, 1995). Particularly interesting, in this new series, is pigment **6**, a dimalonyl ester substituted on one side by a 3-gentiobiosyldelphinidin moiety and, on the other side by a 7-glucosylapigenin moiety (Fig. 3.5). This is the only known natural pigment that covalently links an anthocyanin to a flavone (Toki *et al.*, 1994) and it gives rise to a conformation where the apigenin and delphinidin aglycones are brought together by the intramolecular copigment effect, efficiently preventing the pigment from nucleophilic attack ($pK_h = 2.9$).

Another pigment, delphinidin 3-gentiobioside yielded an unexpected result, its hydration constant ($pK_h = 1.30$), being probably the largest ever observed for natural anthocyanins (Figueiredo *et al.*, 1995). With the help of molecular mechanics calculations, it was concluded that the free primary OH group of the gentiobiosyl terminal glucosyl residue folds back over the pyran ring allowing the formation of an hydrogen bond with the C-2 hemiacetalic hydroxyl thus displacing the equilibria depicted in Fig. 3.2 towards the formation of the colourless hemiacetal, hence the strengthening of the hydration process. Finally, it was shown that the building capability of the intramolecular copigmentation effect in producing sandwich complexes culminates in the *Zebrina* anthocyanin which possesses four unsubstituted caffeoyl residues well situated around the aglycone and which can even give a more or less globular shape to this molecule (Brouillard, 1981; Goto and Kondo, 1991).

To our knowledge, intramolecular copigmentation has only been demonstrated to occur in flowers and not in fruits or vegetables. Nevertheless, some of the red cabbage anthocyanins fit the structural requirements to show this effect (for structures see Mazza and Miniati, 1993). In contrast, it has been shown that intermolecular copigmentation is a minor factor in producing red shades in apple skin (Lancaster *et al.*, 1994). The question remains open whether or not copigmentation is important in the stabilization and variation of fruit and vegetable colours.

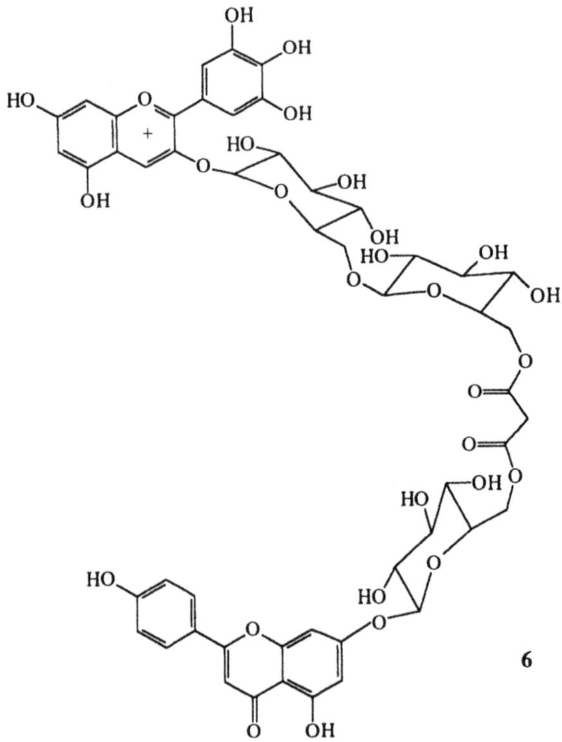

6

Fig. 3.5. Delphinidin gentiobiosyl-apigenin glucosyl malonate (see Toki *et al.*, 1994).

Structures of pigments and copigments are important parameters in the copigmentation phenomenon. One of the most determining factors for the formation of vertical stacking complexes, however, is the nature of the medium within which pigment and copigment can come close enough to interact. In particular, it was shown that in water copigmentation increases when temperature diminishes and that the phenomenon vanishes close to the boiling point of water (Brouillard *et al.*, 1989). The temperature effect occurs with all copigments and is best seen by recording absorption spectra in the visible range for solutions with suitable copigment to pigment molar ratios (Dangles and Brouillard, 1992*a*). It is apparent that copigmentation is enthalpically driven (exothermic process), with usually unfavourable entropy changes. Only small values in the range 100–300 M^{-1} were measured for the association constants (25 °C, in water). Such values are in line with weak molecular interactions allowing permanent formation and dissociation of the copigmentation complexes (chemical equilibrium). Among the copigments tested are the popular polyphenols chlorogenic acid, (+)-catechin and rutin. Caffeine was also thoroughly investigated for its ability to react with anthocyanins. Similar experiments in water–cosolvent binary mixtures showed a drastic weakening of copigmentation on cosolvent addition (Brouillard *et al.*, 1991; Dangles and Brouillard, 1992*a*). An interesting conclusion has been reached: no solvent promotes copigmentation better than water, a fact which indicates that the strength of the copigment effect parallels the cohesion of the hydrogen-bonded tetrahedral network of liquid water molecules. Ionic salts in molar quantities enhance the colour of anthocyanin-containing solutions (Goto *et al.*, 1976). The phenomenon has been interpreted as a decrease of water availability in the hydration of the pyrylium ring by competitive solvation of the inorganic salt. The effect of ionic salts on copigmentation leads to a small but significant decrease in the strength of the association between pigment and copigment (Dangles and Brouillard, 1992a). Another interesting interpretation for the salt effect suggests that ion-pair formation might take place between the mineral anion and the organic flavylium cation (Figueiredo and Pina, 1994). Evidence for ion-pairing mainly rests on the upfield shift in the ^{1}H NMR spectrum of the peak attributed to the 2′ and 6′ protons of the flavylium cation, as opposed to the downfield shifts observed for the remaining flavylium protons, when molar quantities of simple salt anions are added to a mildly acidic anthocyanin solution. The proximity of the anion may be at the origin of the special behaviour of the 2′ and 6′ protons.

Theoretical methods have been used to determine solvent effects on the stability of the various tautomeric forms of apigeninidin quinonoidal bases (Rastelli *et al.*, 1994). Preferential solvation for an *in vacuo* unstable tautomer was demonstrated (Rastelli *et al.*, 1993; Costantino *et al.*, 1995).

The stability of anthocyanins is also affected by their exposure to light and heat. Investigations reporting these effects usually consider pigments in a liquid aqueous environment exposed to light of different intensities and wavelengths going from summer sunlight to monochromatic irradiation. It is known that anthocyanidins degrade quite rapidly at room temperature in moderately acidic solutions, even when kept away from light sources (Furtado *et al.*, 1993). In this case, it is the formation of the chalcone form that results in the splitting of the pigment into two parts, one coming from the A-ring, the other from the B-ring each being characterized by the corresponding substitution pattern. Exposure of delphinidin and malvidin to monochromatic light sources (254 nm, 313 nm and 546 nm) not only accelerates the reaction of thermal decomposition through the formation of more chalcone but also leads to the direct formation of the same degradation products from a photodissociative flavylium excited state (Furtado *et al.*, 1993; Figueiredo, 1994). Anthocyanins are much more stable with respect to light and temperature effects; they can be kept in the dark for several hours, if not days, in acidic solutions their slow degradation (fading) begins to occur through a similar process to the one described above for anthocyanidins. For instance, oenin, the main pigment of *Vitis vinifera* red grapes, remains stable for one day at pH 1 and room temperature. It appears that the presence of a 3-glucosyl residue makes the pigment more resistant toward the degradation process (in this context, compare the values obtained for the degradation constants of oenin and malvidin, its aglycone, by Maccarone *et al.*, 1987 and Furtado *et al.*, 1993). Glucosylation also provides a protective effect against photodegradation, which was shown to increase in the order 3,5-diglucoside < 3-glucoside < aglycone for the malvidin series (Figueiredo, 1994). Formation of intermolecular copigmentation complexes and ion-pairs provides, to some extent, a means to diminish these degradative reactions (Figueiredo, 1994; Figueiredo and Pina, 1994). With some synthetic flavylium salts, photoisomerization between the *Z* and *E* chalcones can occur (note that the relative abundances in the ground state are reversed relative to the more common anthocyanins where the *E* form is predominant). The 4′,7-dihydroxyflavylium chloride provides a good example of such behaviour. In the dark, at mildly acidic pH, there exists a thermal equilibrium between the flavylium cation and the very stable *Z*-chalcone, resulting in a faint yellow colour. When exposed to light, which provides the necessary energy, the *Z*-chalcone isomerizes to the *E* form allowing the recyclization and an almost quantitative reformation of the flavylium form (Figueiredo *et al.*, 1994).

These pigments exhibit fluorescence emission arising from all equilibrium forms, whose intensity and quantum yield strongly depend on the substitution pattern of the aglycone (Figueiredo *et al.*, 1990; Lima *et al.*, 1994). As is found with simple phenolics (Becker, 1969), the polyphenolic anthocyanins are characterized by a much larger flavylium acidity constant in the excited state than in the ground state (i.e. the excited state pK_a is lower than the ground state pK_a).

When the anthocyanin concentration reaches high enough values, self-association between two flavylium cations, two quinonoidal bases or even between a quinonoidal base and a flavylium cation may occur (Hoshino,

1992; Dangles *et al.*, 1994). It is unlikely that this type of molecular association happens on a large scale *in vivo*, essentially due to the high concentrations required for its appearance. So far it has not been reported to occur in fruits or vegetables.

At this stage one may pose the following question: how can one prevent anthocyanins from losing their brilliant colours? What to do and what not to do? Firstly, have plant organs dried as quickly and as gently as possible. Secondly, store the dried material in the open air at room temperature sheltering it from light. As far as possible use fresh plants which have not been pretreated in any way. From our own work, we came to the conclusion that in order to keep natural anthocyanin preparations for a long time the best thing to do is to quickly remove water from samples. Undoubtedly, an initial extraction of the pigments with an aqueous mixture is the worst thing to do. Indeed, this procedure readily allows reactions which, whether biochemically or chemically initiated, will result in pigment loss. This is because, on the one hand, molecules can now interact, while in a solid state this is not so likely to happen, and on the other hand, water promotes anthocyanin reactivity and hence decomposition by forming hydrated colourless species, especially chalcones vulnerable to degradation by physical and chemical factors. Water also permits enzyme activity to continue, at least for some time. Our partial conclusion may be of the utmost importance to industries who want to colour food using natural additives. A good process for the retention of natural anthocyanin colours could be the same as that used in the preparation of herbal tea bags.

Inclusion complexes of polyphenols with cyclodextrins and interactions of anthocyanins bearing a catechol group with metallic cations

The formation of inclusion complexes of natural anthocyanins with natural cyclodextrin macrocycles was demonstrated as early as 1980 by Yamada *et al.* (1980). Twelve years later it was reported by Dangles and Brouillard (1992*c*) and Dangles *et al.* (1992*a*) that simple natural and synthetic anthocyanins can also give inclusion complexes with cyclodextrins. Interestingly, this interaction leads, with small anthocyanins, to a strong decrease in the anthocyanin visible absorbance, β-cyclodextrin being the macrocycle producing by far the strongest hypochromic shift in the pigment's visible absorption band. Two good candidates for inclusion into the β-cyclodextrin cavity, among natural anthocyanins, are callistephin and chrysanthemin. This phenomenon (anticopigmentation) has a strong impact on colour intensity since almost complete loss of colour may occur in the presence of large amounts of the macrocyclic compound (Dangles *et al.*, 1992*a*,*b*). It has been shown that selectivity of the inclusion process strongly favours the colourless forms of anthocyanins, especially the extremely flexible *Z*-chalcones. For instance, in the case of the synthetic 3,4′-dimethoxy-7-hydroxyflavylium chloride, the open chain *Z*-chalcone gives the best fit into the β-cyclodextrin cavity (Dangles *et al.*, 1992*a*). The computerized structure obtained for the corresponding inclusion complex is shown in Fig. 3.6.

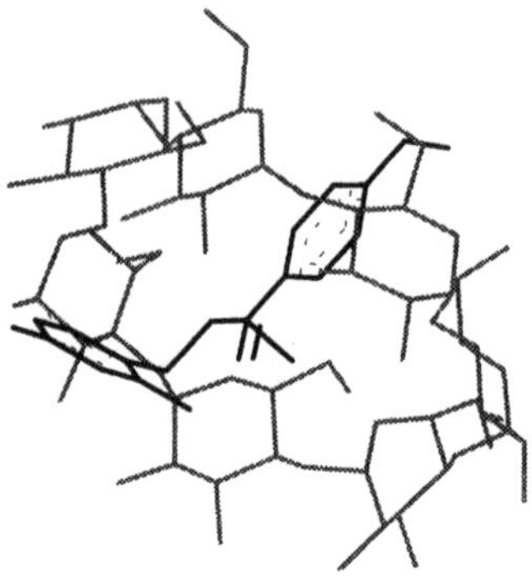

Fig. 3.6. Computer modelled (MM+ parametrization) spatial structure for the 3,4′-dimethoxy-7-hydroxyflavylium chalcone form–β-cyclodextrin inclusion complex. The pigment is represented by dark lines and the macrocycle is indicated in gray.

The anthocyanin–β-cyclodextrin system has been used to investigate competitive inclusion complex formation between the anthocyanin colourless structures and different types of molecules which fit into the macrocyclic cavity (Dangles and Brouillard, 1992*b*). The method remains valid whether the guest molecule (polyphenol) is a copigment or not. In this case, anthocyanins play the role of very sensitive indicators. Malvin copigmentation with chlorogenic acid and chlorogenic acid inclusion complex formation have been studied in this way, copigmentation occurring between malvin and chlorogenic acid and inclusion taking place with chlorogenic acid only. This is a straightforward application of the copigmentation molecular association to the detection of another type of molecular association (Dangles and Brouillard, 1992*b*).

It has been demonstrated that aluminium, magnesium or iron chelation to anthocyanins may be at the origin of blue petal colour in a number of cases (Goto and Kondo, 1991; Brouillard and Dangles, 1993; Kondo *et al.*, 1994). To our knowledge, similar results concerning the impact of metallic ions on fruit or vegetable colours have not so far been reported. It remain to be said that those anthocyanins with a catechol structure in their B-ring have the potential to give more or less stable metallic complexes with many metals, not just aluminium, magnesium or iron ionic species. One example of a recent thorough study on a model system between a synthetic anthocyanin and

Al^{3+} is to be found in Dangles *et al.* (1994). Using the synthetic 3′,4′,7-trihydroxy-3-methoxyflavylium chloride, the authors were able to show that the aluminium cation strongly binds to the coloured forms of the pigment. At pH 3.75, very large bathochromic and hyperchromic shifts accompany aluminium complex formation. The mechanism suggests that two protons are removed from the flavylium cation by Al^{3+} turning the pigment into the strong visible light-absorbing ionized quinonoidal base, which normally is only formed in alkaline medium (see Fig. 3.7).

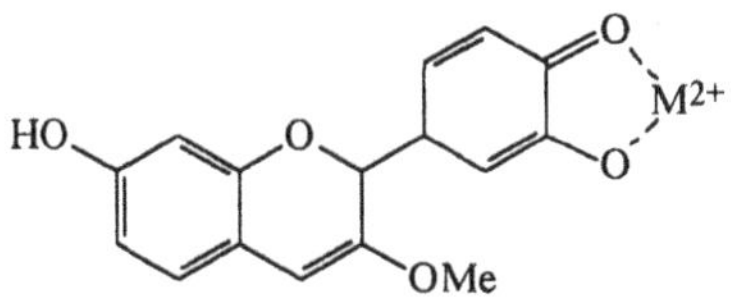

Fig. 3.7. The aluminium complex of 3′,4′,7-trihydroxy-3-methoxyflavylium chloride (adapted from Dangles *et al.*, 1994).

From ultraviolet-visible spectroscopic data, the binding constants of complexation in aqueous solution have been calculated and the percentage of the different free and complexed pigment forms plotted as a function of pH. Such results are important to our understanding of plant pigmentation. For instance, a narrow pH domain, in which colour amplification due to complexation is at a maximum, has been found. Relaxation kinetics measurements (pH-jump) have provided information on the mechanism of metallic complexation and ^1H NMR analysis in CD_3OD (in which complexation is much stronger than in water) has confirmed the conversion of the anthocyanin from a red flavylium form to a deep-purple quinonoidal form upon coordination to Al^{3+} (Dangles *et al.*, 1994). High stoichiometry molecular complexes could appear when the coordination ability of the metallic ion strengthens a pigment–copigment interaction (Takeda *et al.*, 1990; Goto and Kondo 1991; Kondo *et al.*, 1994). Here again, studies of molecular edifices which involve metals in their structures and are responsible for plant colours have received much more attention in the case of flowers than in the case of any other plant organs including fruits, leaves, stems, tubers and roots.

Perspectives

A deepening of our knowledge on copigmentation could be obtained in the near future if answers to the following questions could be given.

- Do the anthocyanin sugar residues play a crucial role in the strength of the association between pigments and copigments?

- What are the structural requirements for a molecule to behave as a good copigment?
- What is the extent of complexation which could occur between a copigment and a given anthocyanin colourless form?
- Which metals are capable of increasing the copigment effect? What are the compounds accompanying the anthocyanin pool in vacuoles, fruit juices, etc?

Another kind of approach rests on *in vivo* microspectrometric investigations. This could reveal detailed results cell by cell and even vacuole by vacuole. Two different spectrometric techniques are easily applicable to water-containing media such as vacuoles or fruit juices, ultraviolet-visible absorption and resonance Raman spectroscopy. The equipment is commercially available, and with the technical evolution which has taken place over the past decade (powerful PCs, Fourier transform analysis, diode-array detection), much more valuable results can now be obtained. Comparison with results from model systems should improve the efficiency of this strategy.

A very exciting new field of research might be explored without delay: why does copigmentation only take place in aqueous media and how could it be turned into an efficient tool in the investigation of the moleular structure of liquid water? Remember that liquid water is the most important medium in biochemical investigations, and although many data and theories have been accumulated on its unique properties, a definitive model taking into account all those physicochemical preperties remains to be established.

Finally, copigmentation is probably an important factor in the elaboration of colour of fruit and vegetables. More work has been done, however, on the influence of the copigmentation effect on the colour of flowers than on the colour of edible fruit and vegetables. It is therefore not surprising that the more complex anthocyanins, possessing efficient colour-protecting and colour-varying mechanisms, have been identified almost exclusively within flowers. Since intramolecular copigmentation is better at stabilizing the coloured forms than the intermolecular effect appears to be, could it be concluded that the former effect is to be observed more often with flowers and the latter essentially encountered in fruit and vegetables? To return to phytochemical preoccupations, could that mean that from the point of view of evolution, flowers are in advance of fruits and other parts of a plant, in the development of plant pigmentation processes due to the presence of anthocyanins?

References

Amić, D., Baranac, J. and Vukadinović, V. (1990). *J. Agric. Food Chem.* **38**, 936–40.
Asen, S., Stewart, R. N. and Norris, K. H. (1972). *Phytochemistry* **11**, 1139–44.

Baublis, A., Spomer, A. and Berber-Jiménez, M. D. (1994). *J. Food Sci.* **59**, 1219–21.

Becker, R. S. (1969). In *Theory and interpretation of fluorescence and phosphorescence.* Wiley Interscience, New York.

Britton, G. (1983). *The biochemistry of natural pigments.* Cambridge University Press, Cambridge.

Brouillard, R. (1981). *Phytochemistry* **20**, 143–5.

Brouillard, R. (1983). *Phytochemistry* **22**, 1311–23.

Brouillard, R. (1988). In *The flavonoids. Advances in research since 1980* (ed. J.B. Harborne), pp. 525–38. Chapman & Hall, London.

Brouillard, R. and Dangles, O. (1993). In *The flavonoids. Advances in research since 1986* (ed. J. B. Harborne), pp. 565–88, Chapman & Hall, London.

Brouillard, R. and Dangles, O. (1994). *Food Chem.* **51**, 365–71.

Brouillard, R., Mazza, G., Saad, Z., Albrecht-Gary, A. M. and Cheminat, A. (1989). *J. Am. Chem. Soc.* **111**, 2604–10.

Brouillard, R., Wigand, M. C. and Cheminat, A. (1990). *Phytochemistry* **29**, 3457–60.

Brouillard, R., Wigand, M. C., Dangles, O. and Cheminat, A. (1991). *J. Chem. Soc. Perkin Trans.* **2**, 1287–96.

Brouillard, R., Jay, M. and Scalbert. A. (eds) (1995). *Polyphenols 94.* INRA Editions, Paris.

Cody, V., Middleton, E. and Harborne, J. B. (eds) (1986). *Plant flavonoids in biology and medicine. Biochemical, pharmacological, and structure-activity relationships.* Alan R. Liss, New York.

Cody, V., Middleton, E., Harborne, J. B. and Beretz, A. (eds.) (1988). *Plant flavonoids in biology and medicine. II. Biochemical, cellular, and medicinal properties.* Alan R. Liss, New York.

Costantino, L., Rastelli, G., Rossi, M. C. and Albasini, A. (1995). *J. Chem. Soc. Perkin Trans.* **2**, 227–34.

Dallas, C. and Laureano, O. (1994). *J. Sci. Food Agric.* **65**, 477–85.

Dangles, O. and Brouillard, R. (1992*a*). *Can. J. Chem.* **70**, 2174–89.

Dangles, O. and Brouillard, R. (1992*b*). *J. Chem. Soc. Perkin Trans.* **2**, 247–57.

Dangles, O. and Brouillard, R. (1992*c*). In *Minutes of the sixth international symposium on cyclodextrins* (ed. A.R. Hedges), pp. 117–22. Editions de Santé, Paris.

Dangles, O. and Elhajji, H. (1994). *Helv. Chem. Acta* **77**, 1595–610.

Dangles, O., Stoeckel, C., Wigand, M. C. and Brouillard, R. (1992*a*). *Tetrahedron Lett.* **33**, 5227–30.

Dangles, O. Wigand, M. C. and Brouillard, R. (1992*b*). *Phytochemistry* **31**, 3811–2.

Dangles, O., Saito, N. and Brouillard, R. (1993*a*) *J. Am. Chem. Soc.* **115**, 3125–32.

Dangles, O., Saito, N. and Brouillard, R. (1993*b*). *Phytochemistry* **34**, 119–24.

Dangles, O., Elhabiri, M. and Brouillard, R. (1994). *J. Chem. Soc. Perkin Trans.* **2**, 2587-96.

David, S. (1995). *Chimie moléculaire et supramoléculaire des sucres. Introduction chimique aux glycosciences.* InterEditions/CNRS Editions, Paris.

Davies, A. J. and Mazza, G. (1993). *J. Agric. Food Chem.* **41**, 716–20.

Elhabiri, M., Figueiredo, P., Fougerousse, A. and Brouillard, R. (1995*a*). *Tetrahedron Lett.* **36**, 4611–14.

Elhabiri, M., Figueiredo, P., Fougerousse, A., Dangles, O. and Brouillard, R. (1995*b*). *Polyphénols Actualités* **13**, 11–13.

Figueiredo, P. (1994). *Ph. D. Thesis*, Universidade Nova de Lisboa, Portugal.

Figueiredo, P. and Pina, F. (1994). *J. Chem. Soc. Perkin Trans.* **2**, 775–8.

Figueiredo, P., Pina, F., Vilas-Boas, L. and Maçanita, A. (1990). *J. Photochem. Photobiol., A: Chem.* **52**, 411–24.

Figueiredo, P., Lima, J. C., Santos, H., Wigand, M. C., Brouillard, R., Maçanita, A. L. and Pina, F. (1994). *J. Am. Chem. Soc.* **116**, 1249–54.

Figueiredo, P., Elhabiri, M., Toki, K., Saito, N., Dangles, O. and Brouillard, R. (1995). *Phytochemistry* **41**, 301–8.

Francis, F. J. (1993). In *Polyphenolic phenomena* (ed. A. Scalbert), pp. 209–20. INRA Editions, Paris.

Furtado, P., Figueiredo, P., Neves, H. C. and Pina, F. (1993). *J. Photochem. Photobiol., A: Chem.* **75**, 113–18.

Gao, L. and Mazza, G. (1994). *J. Food Sci.* **59**, 1057–9.

Goto, T. and Kondo, T. (1991). *Angew. Chem. Int. Edn. Engl.* **30**, 17–33.

Goto, T., Hoshino, T. and Ohba, M. (1976). *Agr. Biol. Chem.* **40**, 1593–6.

Harborne, J. B. (1967). *Comparative biochemistry of the flavonoids*, Academic Press, London.

Harborne, J. B. (ed.) (1993). *The flavonoids. Advances in research since 1986*. Chapman & Hall, London.

Harborne, J. B. and Grayer, R. J. (1988). In *The flavonoids. Advances in research since 1986* (ed. J.B. Harborne), pp. 1–20. Chapman & Hall, London.

Hong, V. and Wrolstad, R. E. (1990). *J. Agric. Food Chem.* **38**, 698–708.

Hoshino, T. (1992). *Phytochemistry* **31**, 647–53.

Johansen, O. P., Andersen, O. M., Nerdal, W. and Aksnes, D. W. (1991). *Phytochemistry* **30**, 4137–41.

Kondo, T., Ueda, M., Tamura, H., Yoshida, K., Isobe, M. and Goto, T. (1994). *Angew. Chem. Int. Ed. Engl.* **33**, 978–9.

Lancaster, J. E., Grant, J. E. and Lister, C. E. (1994). *J. Amer. Soc. Hort. Sci.* **119**, 63–9.

Lima, J. C., Danesh, P., Figueiredo, P., Pina, F. S. and Maçanita, A. (1994). *Photochem. Photobiol.* **59**, 412–18.

Lu, T. S., Saito, N., Yokoi, M., Shigihara, A. and Honda, T. (1992). *Phytochemistry* **31**, 289-95.

Maccarone, E., Ferrigno, V., Longo, M. L. and Rapisarda, P. (1987). *Ann. Chim.* **77**, 499-508.

Macheix, J. J., Fleuriet, A. and Billot, J. (1990). *Fruit phenolics*. CRC Press, Boca Raton.

Mazza, G. and Miniati, E. (1993). *Anthocyanins in fruits, vegetables and grains*. CRC Press, Boca Raton.

Markham, K. R. and Geiger, H. (1993). In *The flavonoids. Advances in research since 1986* (ed. J. B. Harborne), pp. 441–97. Chapman & Hall, London.

Merlin, J. C., Cornard, J. P., Statoua, A., Saidi-Idrissi, M., Lautié, F. and Brouillard, R. (1994*a*). *Spectrochim. Acta* **50A**, 703–12.

Merlin, J. C., Statoua, A., Cornard, J. P., Saidi-Idrissi, M. and Brouillard, R. (1994*b*). *Phytochemistry* **35**, 227–32.

Mistry, T. V., Cai, Y., Lilley, T. H. and Haslam, E. (1991). *J. Chem. Soc. Perkin Trans.* **2** 1287–96.

Nerdal, W. and Andersen, O. M. (1991). *Phytochem. Anal.* **2**, 263–70.

Nerdal, W. and Andersen, O. M. (1992). *Phytochem. Anal.* **3**, 182–9.

Osawa, Y. (1982). In *Anthocyanins as food colors* (ed. P. Markakis), pp. 41–68. Academic Press, New York.

Rastelli, G., Costantino, L. and Albasini, A. (1993). *J. Mol. Struc. (THEOCHEM)* **279**, 157–66.

Rastelli, G., Costantino, L. and Albasini, A. (1994). *Tetrahedron Lett.* **35**, 9751–4.

Robinson, G.M. and Robinson, R. (1931). *Biochem. J.* **25**, 1687–705.

Saito, N., Lu, T.S., Akaizawa, M., Yokoi, M., Shigihara, A. and Honda, T. (1994). *Phytochemistry* **35**, 407–11.

Saito, N., Tatsuzawa, F., Nishiyama, A., Yokoi, M., Shigihara, A. and Honda, T. (1995). *Phytochemistry* **38**, 1027–32.

Santos, H., Turner, D. L., Lima, J. C., Figueiredo, P., Pina, F. and Maçanita, A. L. (1993). *Phytochemistry* **33**, 1227–32.

Scalbert, A. (ed), (1993). *Polyphenolic phenomena.* INRA Editions, Paris.

Strack, D. and Wray, V. (1993). In *The flavonoids. Advances in research since 1986* (ed. J. B. Harborne), pp. 1–22. Chapman & Hall, London.

Takeda, K., Yamashita, T., Takahashi, A. and Timberlake, C. F. (1990). *Phytochemistry* **29**, 1089–91.

Timberlake, C. F. (1981). In *Recent advances in the biochemistry of fruits and vegetables* (ed. J. Friend and M. J. C. Rhodes), pp. 221–47. Academic Press, London.

Toki, K., Saito, N., Imura, K., Suzuki, T. and Honda, T. (1994). *Phytochemistry* **36**, 1181–3.

Wigand, M. C., Dangles, O. and Brouillard, R. (1992). *Phytochemistry* **31**, 4317–24.

Yamada, T., Komiya, T. and Akaki, M. (1980). *Agric. Biol. Chem.* **44**, 1411–3.

Yoshida, K., Kondo, T. and Goto, T. (1991*a*). *Tetrahedron Lett.* **32**, 5579–80.

Yoshida, K., Kondo. T., Kameda, K., Kawakishi, S., Lubag, A. J. M., Mendoza, E. M. T. and Goto, T. (1991*b*). *Tetrahedron Lett.* **32**, 5575–8.

Yoshida, K., Kondo, T. and Goto, T. (1992). *Tetrahedron* **48**, 4313–26.

4. Phenolic compounds and oxidative mechanisms in fruit and vegetables

MARIE J. AMIOT[1], ANNIE FLEURIET[2], VERONIQUE CHEYNIER[3] and JACQUES NICOLAS[4]

[1]Institut National de la Recherche Agronomique, Station de Technologie des Produits Végétaux, 84914 Avignon Cedex 9, France, [2]Université Montpellier II, Laboratoire de Biotechnologie et Physiologie Végétale Appliquée, 34095 Montpellier Cedex 5, France, [3]Institut National de la Recherche Agronomique, Institut des Produits de la Vigne, 34060 Montpellier Cedex, France and [4]Conservatoire National des Arts et Métiers, Chaire de Biochimie Industrielle et Agroalimentaire, 75141 Paris Cedex 03, France

Introduction

Phenolic compounds form one of the main classes of secondary metabolites with a large range of structures (Table 4.1; Fig. 4.1), and contribute to the organoleptic and nutritional qualities of fruit and vegetables (Macheix *et al.*, 1990). Oxidative mechanisms involving phenolic compounds may modify the quality of fresh or processed fruit and vegetables. Enzymatic browning is one of the main oxidative phenomena inducing the development of undesirable colour, flavour and a loss of nutrients. In heat-processed fruit and vegetables products, non-enzymatic browning (Maillard reaction) can also occur. In some cases, such as in plums, figs or dates, enzymatic browning is required for quality.

In the large group of oxidoreductases, polyphenoloxidases and peroxidases are involved in the oxidative degradation of phenolic compounds. Although peroxidases are also believed to be responsible for deterioration in flavour, colour, texture and nutritional qualities of postharvest plant foods (Robinson, 1991), polyphenoloxidases are related in most cases to discoloration and browning of fruits and vegetables. Although peroxidases play an important role in plant physiology, their involvement in enzymatic browning is certainly limited by the avaibility of hydrogen peroxide. In some cases, however such as internal browning in pineapple, the involvement of peroxidases remains possible. Polyphenoloxidase activities have been extensively studied in fruit and vegetables during the last two decades and many reviews have been published (Mayer and Harel, 1979, 1981, 1991; Vamos-Vigyazo, 1981; Mayer, 1987; Eskin, 1990; Zawistowski *et al.*, 1991; Macheix *et al.*,

Table 4.1. The major classes of phenolic compounds

Number of carbon atoms	Basic skeleton	Class	Examples	Food plant
7	C_6–C_1	Hydroxybenzoic acids	*p*-Hydroxybenzoic acid	Spices, carrot, strawberry
9	C_6–C_3	Hydroxycinnamic acids	Chlorogenic acid	Apple, potato,
		Coumarins	Scopolin	Citrus
10	C_6-C_4	Naphthoquinones	Juglone	Walnut
13	C_6–C_1–C_6	Xanthones	Mangiferin	Mango
15	C_6–C_3–C_6	Flavonoids	Quercetin, cyanidin	Many fruits
		Isoflavonoids	Daidzein	Soybean
n	$(C_6–C_3)n$	Lignins		Stone fruits
n	$(C_{15})n$	Tannins		Grape, persimmon

1990; Marquès *et al.*, 1995*b*). Since 1992, much progress has been made concerning the structure, localization and expression of polyphenol oxidases, particularly due to the use of immunochemical and molecular biological methods.

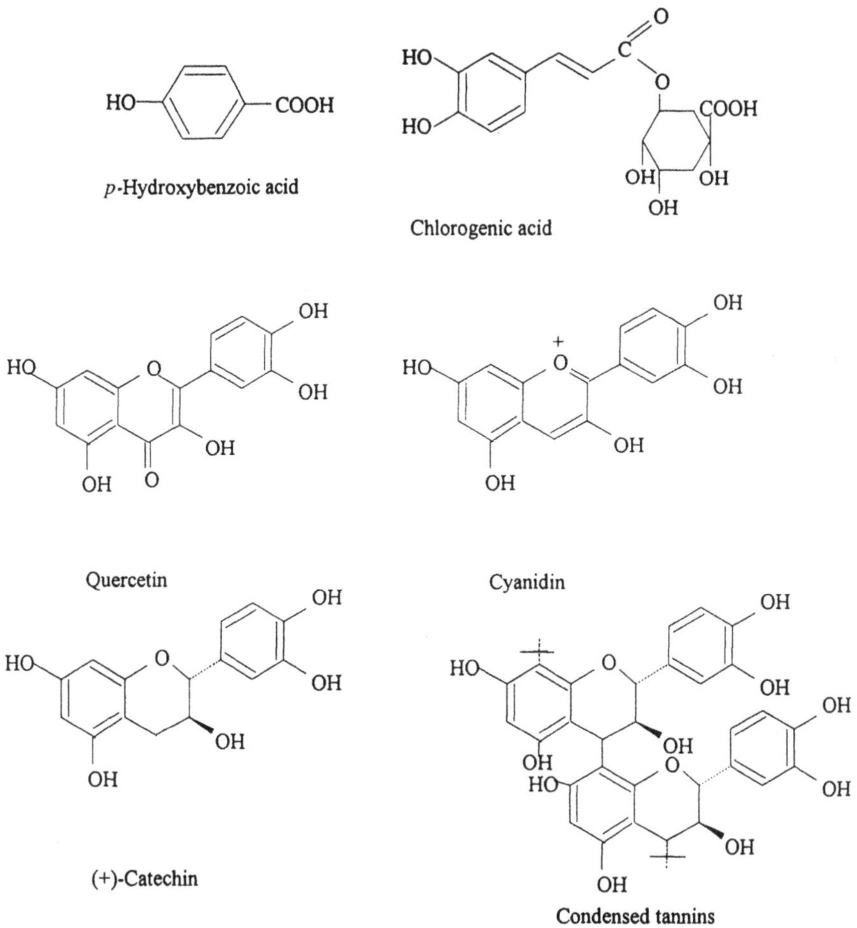

Fig. 4.1. Examples of the main phenolic compounds found in fruit and vegetables.

The initial step of browning is the enzymatic oxidation of phenols into slightly colored quinones. These quinones are very active entities which are subjected to various reactions leading to the formation of more intensely coloured pigments. Recent advances concern non-enzymatic reactions that follow the enzymatically formation of quinones, and, the colour (hue and intensity) which depends on the phenolic composition, the speed of the enzymatic primary reaction and on environmental factors.

In addition, the control of browning is a challenge in fruit and vegetable industry, especially with the development of minimally processed fruit and vegetables (Varoquaux and Wiley, 1994). Sulfite and its derivatives, powerful antibrowning chemicals are currently used in industry, but are increasingly being considered to introduce potential hazards (Taylor *et al.*, 1986). Finding non-sulfite antibrowning agents for the food industry is therefore a currently active field of research (Sapers and Hicks, 1989; McEvily *et al.*, 1991).

Although the oxidation of phenolic compounds leads to discoloration, the products have been considered as natural antioxidants by preventing oxidative deterioration of lipids. Lipid oxidation, defined as the oxidative degradation of unsaturated fatty acids, is a free radical-mediated phenomenon involving a chain reaction mechanism. The oxidation of lipids is catalysed by numerous factors such as oxygen, light, heat, metals and lipoxygenases. The use of synthetic antioxidants (butylated hydroxyanisole, butylated hydroxytoluene) is decreasing because of their suspected action as promoters of carcinogenesis. The use of plant phenolic compounds will be therefore of great interest to the food industry.

Recent developments concerning some oxidative mechanisms involving phenolic compounds will be presented in this chapter.

Enzymatic browning

Basically, enzymatic browning is related to the oxidation, in the presence of oxygen, of phenolic compounds—predominantly located in the vacuole—by polyphenol oxidases (PPOs)—located in cytoplasm. Different situations may cause browning: physiological during maturation, some disorders during storage, and technological processes involving wounding or crushing. The tendency of plants to turn brown results from the action of various factors, which are naturally involved in limiting browning and its intensity: oxidative enzyme activities, the nature or content of the oxidizable substrates. All these factors vary with the age of fruit and vegetables, physiological stage and variety, health and treatments to which they are subjected (Vamos-Vigyazo, 1981; Macheix *et al.*, 1990; Amiot *et al.*, 1995).

New data on PPOs

Polyphenoloxidases are copper proteins using dioxygen as acceptor and comprise essentially three types of enzymes depending on their substrate specificity (Fig. 4.2):

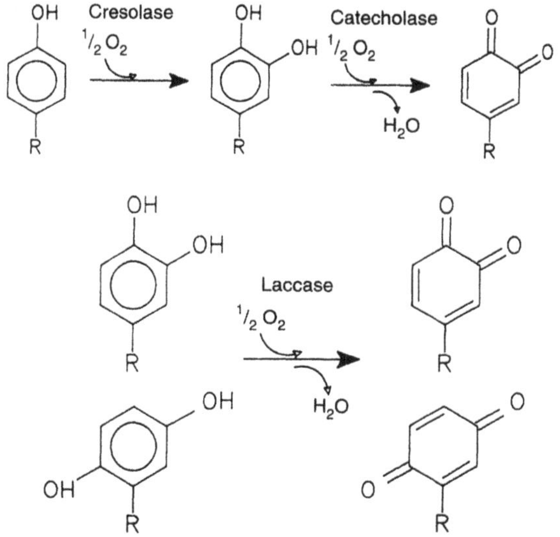

Fig. 4.2. Reactions catalysed by polyphenoloxidases (cresolase and catecholase) and by laccase.

- monophenol oxidase (E.C. 1.14.18.1) or cresolase
- *o*-diphenoloxidase (E.C. 1.10.3.1) (also named catecholase or phenolase and referred to here as PPO)
- *p*-diphenoloxidase or laccase (E.C. 1.10.3.2).

All these enzymes appear to be widely distributed in plants and animals. Although their implication in oxidation of phenolic compounds to quinones, melanines and brown polymers is certain, their physiological role in plants remains unclear (Mayer and Harel, 1991). It is thought that PPOs may be involved in plant defence mechanisms against some microbial or viral pathogens. The participation of laccase in the polymerization of lignin monomers has recently been demonstrated (O'Malley *et al.*, 1993) and possible cooperation between laccase and peroxidase in the lignification process has been suggested (Sterjades *et al.*, 1993). The catecholase has been considered to have a possible defensive role against herbivore and pathogen attacks and very recent results indicate that PPO genes are activated by wounding and methyljasmonate, suggesting their regulation through the octadecanoid signalling pathway (Constabel *et al.*, 1995).

Occurence and changes PPO activity has been detected in all higher plant organs (fruits, seeds, roots, tubers, leaves) and in most tissues as well. Immunostaining of the nitrocellulose tissue print of cross-sections of apple fruit has recently implied that chlorogenic acid oxidase is mainly localized at the core and near the skin (Murata *et al.*, 1993, 1995). PPO localization corresponds to the spreading of browning in the same parts of the apple, but differences may be observed between cultivars: PPO activity was higher in the cortex than in the peel for seven cultivars, equivalent for four others and lower for one (Nicolas *et al.*, 1994).

Many authors report the presence of soluble cytoplasmic activity, but in most cases, the use of detergents (Triton X-100, X-114, SDS, etc.), acetone powder or limited digestion with proteolytic enzymes is required to obtain maximum PPO activity since it is a membrane-bound enzyme (Vaughn and Duke, 1984). The intracellular localization of PPO has been detected in chloroplasts particularly on the inner face of the thylakoid membrane (Shomer *et al.*, 1979) and the binding strength to membranes appears to vary according to tissues, species and stages of development. PPO is a nuclear-encoded protein and it has recently been observed, in potato and tomato, that it was routed to the thylakoid lumen in two steps, consistent with the two-domain structure of the transit peptide (Sommer *et al.*, 1994). Browning always results from a subcellular decompartmentation following physical modifications to plastidic and vacuolar membranes leading to enzyme–substrate contact. Previous reports concerning the presence of PPO in mitochondria, microbodies or the cell wall have not yet been confirmed using modern cytolocalization techniques (Mayer, 1987).

PPO activity is generally maximal at early stages of development (Macheix *et al.*, 1990; Mayer and Harel, 1981; 1991). Changes in activity are often accompanied by changes in PPO solubility. For example, in unripe olives, PPO is tightly bound to chloroplast, whereas in the ripe fruit it is essentially soluble (Ben-Shalom *et al.*, 1977). Practically all of the activity was found to be soluble when olive fruit was deeply coloured by anthocyanins. It is assumed that at this stage chloroplast membranes are disrupted and their lamellar structure has desintegrated, facilitating liberation of oxidative enzyme. Similar increased solubilization has been reported in other fruits (Mayer and Harel, 1981; Macheix *et al.*, 1990) and was also observed in apple fruit tissue cultures (Volk *et al.*, 1977).

The changes that occur in PPO activity during storage are often contradictory (Macheix *et al.*, 1990): an increase is sometimes noticed, but the activity often remains almost constant. With gamma irradiation, there is a good correlation between browning and PPO activity: in banana and mango PPO activity was higher in irradiated fruit suggesting that this increase was due to conformational changes, activation of latent enzyme or *de novo* synthesis.

Biochemical, immunochemical and molecular characteristics The overall results indicate high heterogeneity in the expression of PPO concerning enzymatic activity (optimum pH, latency, specificity, etc.), the number of isoforms and the apparent molecular mass. This heterogeneity may be due to differential genomic expression depending on the species, the physiological stage or the nature of the tissues studied. Furthermore, artefacts may appear during the extraction and purification steps. PPOs have some intrinsic characteristics such as latency, resistance to SDS or proteolysis which may also explain the differences observed.

Inactive or latent PPO forms that require activation have frequently been reported. This activation can be triggered by several agents and treatments: the most effective are proteolytic attack (King and Flurkey, 1987; Sanchez-Ferrer *et al.*, 1989), changes in pH (Valero and Garcia-Carmona, 1992; Marquès *et al.*, 1995*a*) and the release of phospholipids (surfactants) which occurs *in vivo* (King and Flurkey 1987; Sanchez-Ferrer *et al.*, 1990). In many cases, PPO activity was detected fully in the presence of SDS (Flurkey, 1986; Moore and Flurkey, 1990). This is particularly interesting as most enzymes are known to be inactivated by SDS which modifies the tertiary and quaternary protein structures. Thus, PPO activity can be detected directly on SDS–PAGE using appropriate substrates, such as 3-hydroxy-L-tyrosine (L-DOPA). Latent forms have been characterized particularly in leaves of broad bean, spinach, pea (Flurkey, 1986) or potato (Sanchez-Ferrer *et al.*, 1993*a*), but not in tubers (Sanchez-Ferrer *et al.*, 1993*b*). In fruits, there is less evidence of PPO latency than in leaves. PPO latency has only been reported in mango (Robinson *et al.*, 1993), in avocado (Kahn, 1977) and in grape (Sanchez-Ferrer *et al.*, 1989). In grape, a latent PPO form has been isolated by temperature-induced phase separation in Triton X-114; the most effective activators being trypsin and cetrinium bromide (CTAB) (Sanchez-Ferrer *et al.*, 1989). However, this latency has not been reported in other grape cultivars (Valero *et al.*, 1988). These differences appear to be more related to the pH used to measure PPO activity, since it has been shown that the SDS effect was pH-dependent in apple (Fig. 4.3) and in several *Prunus* varieties (Fraignier *et al.*, 1995; Marquès *et al.*, 1995*b*).

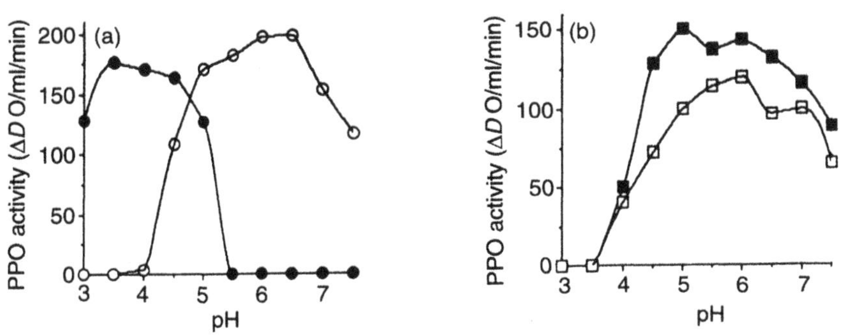

Fig. 4.3. Effect of pH and SDS on apple PPO activities of the native (a) and proteolysed (b) extracts (from Marquès *et al.*, 1995*a*). Black symbols, without SDS; light symbols, SDS 3.5 mM.

Different pH optima have been reported for PPO activity, ranging from the acidic to the neutral range (Janovitz-Klapp *et al.*, 1989; Moore and Flurkey, 1990; Robinson *et al.*, 1993; Goupy *et al.*, 1994*a*; Marquès *et al.*, 1995*a*). Sometimes two different pH optima have been noted in the same species (Sherma and Ali, 1980; Fujita and Tono, 1988), or within the same extract (Sanchez-Ferrer *et al.*, 1993*b*).

The specificity of PPO to its phenolic substrates has been extensively studied and many discrepancies in the apparent K_m values appear between species, cultivars and experimental conditions (Vamos-Vigyazo, 1981; Macheix *et al.*, 1990, 1991; Zawistowski *et al.*, 1991; Nicolas *et al.*, 1994). Only a few studies have been devoted to the effect of the second substrate, oxygen. With apple PPO, it has been shown recently that the reaction is an ordered mechanism, whereby oxygen is the first substrate to be bound by the enzyme. An equation has also been drawn up from model systems to predict initial oxygen uptake in complex multiphenolic solution based on the relative concentrations and kinetic parameters of each individual phenolic substrate (Janovitz-Klapp *et al.*, 1990*b*; Nicolas *et al.*, 1994).

There is also inter- and intraspecific diversity in the number of PPO isoforms and their molecular masses. In apple fruit, previous studies have described one form of 46 kDa (Janovitz-Klapp *et al.*, 1989) or 26 kDa (Goodenough *et al.*,1983) or several forms from 24 to 134 kDa (Demenuyk *et al.*, 1974). This variation may be related to genetic differences, the stage of development, storage at cold or freezing temperatures, or the tissues studied. Nevertheless, many of these forms are thought to be artefactual due to partial release from membranes, cross-linking reactions, covalent reactions with quinones, partial denaturation with phenols or proteolysis that appear during PPO extraction and purification (Mayer and Harel, 1991). In pear, anion exchange resins were particularly effective at removing phenolics, did not adsorb PPO and reduced the electrophoretically separable PPO bands from 11 to 3 (Smith and Montgomery, 1985). In apple (Marquès *et al.*, 1994) and in several *Prunus* fruits (Fraignier *et al.*, 1995), the *in vitro* appearance of multiple forms of PPO has been detected during mild proteolysis of an extract by proteinase-K. Mostly, these appear very rapidly and then disappear, except for one form which appears to be resistant to proteolysis as all other proteins disappeared completely during proteolysis. This unique behaviour was used to purify apple PPO and to obtain polyclonal antibodies (Marquès *et al.*, 1994). This proteolysed PPO form, resistant to further proteolysis, was an active form, also present *in vivo*, since it was detected in many plants even when antiproteases were used during extraction (Marquès *et al.*, 1994; Robinson and Dry, 1992).

The use of antibodies raised against purified PPO enzyme has led to much progress in

- the characterization of different PPO forms (active and inactive)
- the determination of molecular masses in totally denaturing conditions
- the comparison of profiles in several species and varieties.

Anti-PPO has now been obtained from broad bean (Hutcheson *et al.*, 1980), apple (Murata *et al.*, 1993; Marquès *et al.*, 1994), grape (Rathjen and Robinson, 1992) and tomato (Yu *et al.*, 1992). The use of anti-PPO has shown that some domains of PPO are highly conserved, especially around the two active copper binding sites, since the common forms were recognized by antibodies raised against PPO from very different species (Fraignier *et al.*, 1995). This has been confirmed by the high degree of homology observed in amino acid sequences in tomato (Newman *et al.*, 1993), broad bean (Cary *et al.*, 1992) and potato leaves (Hunt *et al.*, 1993).

For a long time the molecular mass of the major PPO was reported as 45 kDa, confirmed by product obtained *in vitro* by translation from mRNA (Flurkey, 1986). This value was the apparent molecular mass due to an inadequate denaturation of the enzyme sample (Cary *et al.*, 1992). This is probably related to the presence of internal disulfide bonds as shown after denaturation by guanidinium chloride and reduction by dithiothreitol (Marquès *et al.*, 1995*a*). This strengthening of the PPO structure could explain the great resistance to unfolding. Since 1992, it has seemed clear that the PPO molecular mass in all species studied was within the range 60–65 kDa (Flurkey, 1986; Steffens *et al.*, 1990; Ganesa *et al.*, 1992; Robinson and Dry, 1992; Fraignier *et al.*, 1995; Marquès *et al.*, 1995*a*).

In grapes and broad bean leaves, this 60 kDa PPO was inactive and has been described as being a precursor of active 45 kDa PPO (Rathjen and Robinson, 1992; Robinson and Dry, 1992). Amino acid sequencing revealed that the active protein was derived from the N-terminal end of the proto-PPO. In this case, the regulation of PPO activity is closely associated with the degree of proteolysis. In apple, anti-PPO has also shown the existence of an inactive PPO form, but it has approximatively the same molecular mass than the main native PPO (63 and 64 kDa respectively). It has been suggested that this inactive form was in fact a partially unfolded PPO corresponding to a translocation-competent form entering the thylakoidal membranes (Marquès *et al.*, 1995*b*). In this case the regulation of PPO activity would be dependent on the folding process.

In contrast with to the structural heterogeneity of PPO protein suggested by biochemical studies, the PPO genes known up to the preesnt are represented by approximatively 2 kb mRNA trancripts encoding 65–70 kDa precursor polypeptides. The deduced amino acid sequences of the PPO cDNA (Steffens *et al.*, 1990; Cary *et al.*, 1992; Shahar *et al.*, 1992; Hunt *et al.*, 1993; Newman *et al.*, 1993) contain N-terminal regions characteristic of plastid transit peptides. These studies have been conducted partly to elucidate the import, targeting and processing of plant PPOs. The 67 kDa precursor was first imported into the stroma; it was processed to the 62 kDa intermediate and translocation into the lumen was light dependent and involved processing of the 62 kDa to the 59 kDa mature form (Sommer *et al.*, 1994). The two-step routing may explain some multiple PPO forms observed *in vivo*.

It seems that PPO is encoded by a gene family: two (Shahar *et al.*, 1992) and seven (Newman *et al.*, 1993) tomato genes were isolated and characterized in tomato plants; three in broad bean (Cary *et al.*, 1992), two in potato leaves (Hunt *et al.*, 1993), five in potato tuber (Bachem *et al.*, 1994) and at least four in apple (Boss *et al.*, 1995). The predicted polypeptide of apple PPO shows significant homology with other plant PPO clones: 65% amino acid identity with a broad bean clone, 45–50% with a range of tomato clones and 49% with a potato clone. The CuA region within the apple clone shows 94% identity with those in the other plant PPO clones, while the CuB region shows 82% with broad bean, 62% with a potato clone and 67% with tomato. However, sequences in the 5′ and 3′ flanking regions of the genomic clones were found to be divergent in the tomato PPO gene family.

These differences in the regulatory regions may impart tissue-specific expression on otherwise similar proteins (Newman *et al.*, 1993). It has been proposed that PPO expression is regulated differently in vegetative and floral tomato organs (Shahar *et al.*, 1992).

Isolation of genes has led to one of the most successful methods of inhibiting gene expression, the introduction of an antisense gene. When this method was used for the first time with PPO, it showed the browning of potato tuber tissue from transgenic plants to be inhibited by the presence of the antisense gene PPO construction (Bachem *et al.*, 1994).

Research on these topics should be continued in order to explain the differences in expression, regulation in different tissues during development and elucidation of the function of these enzymes in plants.

Secondary reactions involving quinones

The primary products of enzymatic oxidation are quinones. These molecules have different spectral characteristics and their colour depends on both the phenolic compound from which they originate and on the pH (Taylor and Clydesdale, 1987). Quinones are very reactive entities (Metche, 1986; Finley, 1974). Different evolution pathways of *o*-quinones involving phenolic (Fig. 4.4) or non-phenolic (Fig. 4.5) molecules have been described in the literature (Rouet-Mayer *et al.*, 1993; Nicolas *et al.*, 1994):

- nucleophilic additions involving sulfites (Wedzicha, 1984; Singleton, 1987), thiols (Prota, 1988; Dudley and Hotchkiss, 1989; Molnar-Perl and Friedman, 1990*a*,*b*), benzene sulfinic acid (Piretti *et al.*, 1977; Cheynier *et al.*, 1989*b*), amino groups of aminoacids and of peptides (Mason and Peterson, 1965; Pierpoint, 1969), water (Richard-Forget *et al.*, 1992*b*) and phenol (Nicolas *et al.*, 1994)
- coupled oxidation with another phenol (Cheynier *et al.* 1989*a*; Cheynier and Moutounet, 1992; Cheynier *et al.*, 1994) or ascorbic acid , sulfites and also thiols.

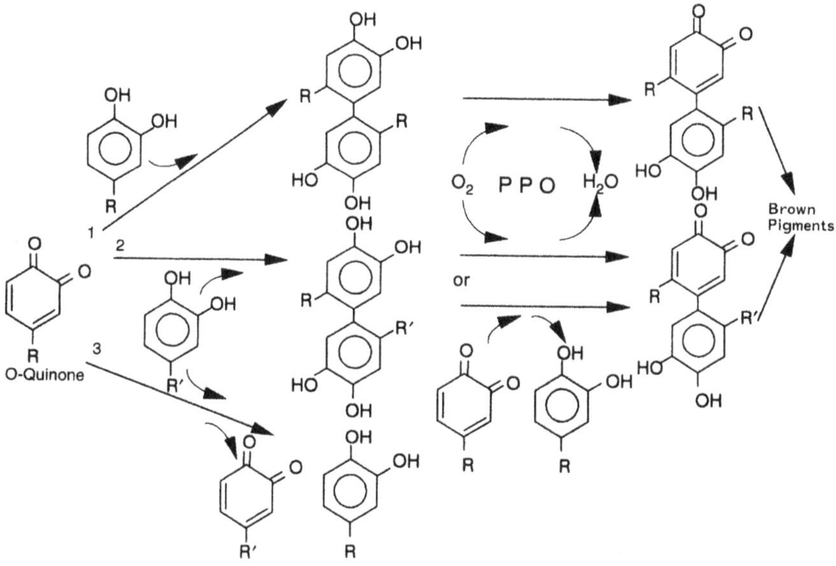

Fig. 4.4. Reactions of *o*-quinones with phenolic compounds; reactions 2 and 3 are able to regenerate the original phenol (from Rouet-Mayer *et al.*, 1993).

When pH is more acidic (< 4), sufiltes act as reducing agent (Cheynier *et al.*, 1989*a*). When pH is > 4, *o*-quinones can give colourless adducts with sulfites. Adducts with cysteine and glutathione has been also identified (Sanada *et al.*, 1972; Cheynier *et al.*, 1986; Dudley and Hotchkiss, 1989; Richard *et al.*, 1991). These compounds have a phenolic structure with a thiol substitution mainly on the 2-position for caffeic acid and its esters (chlorogenic and caffeoyltartaric acids) (Cheynier *et al.*, 1986; Richard *et al.*, 1991), on the 5-position for 4-methylcatechol (Richard *et al.*, 1991), on both position for catechins (Richard *et al.*, 1991), and, on three positions 2,5 and 6 for L-DOPA (Prota, 1988). Following the phenol, cysteine was found to form a single addition compound with 4-methylcatechol and chlorogenic acid, and two products with (+)-catechin and (–)-epicatechin (Richard *et al.*,

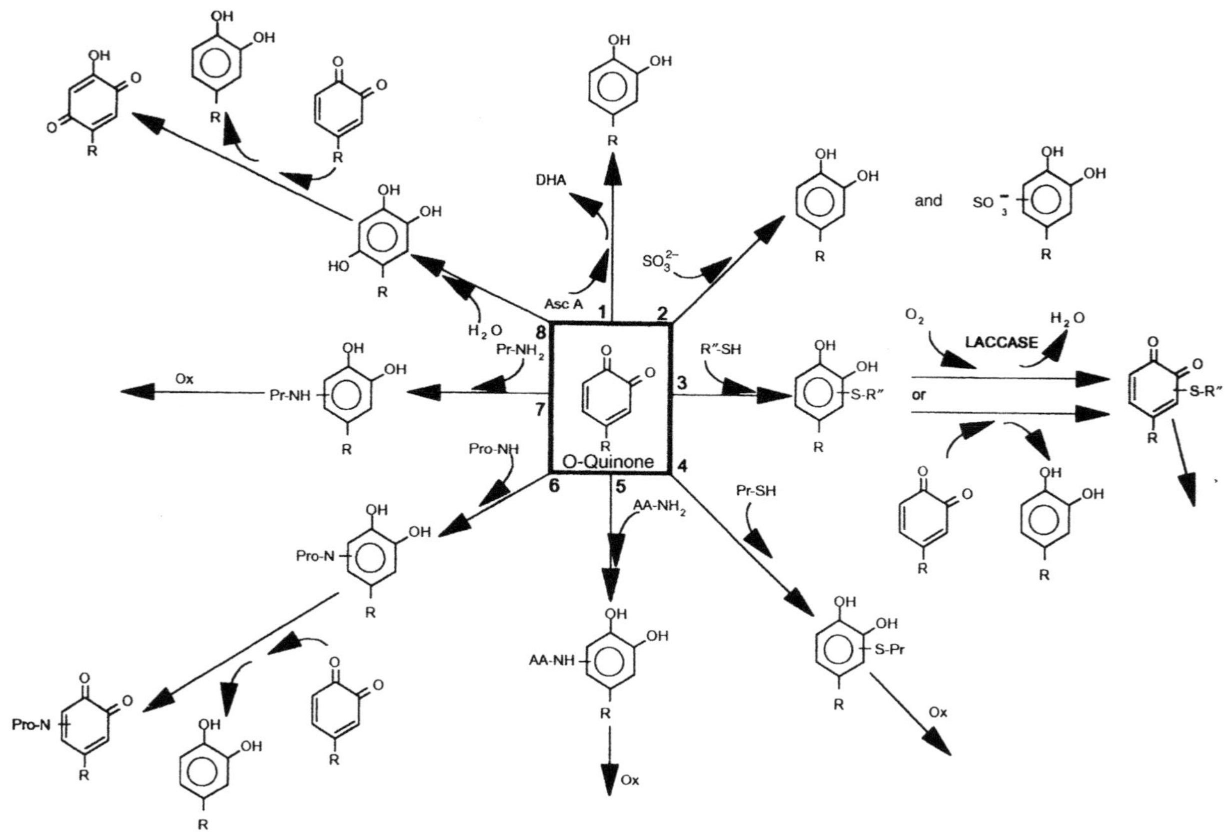

Fig. 4.5. Reactions or *o*-quinones with non phenolic compounds. Ox, further oxidation reactions by oxygen or *o*-quinone; reactions 1, 2, 3, 6, and 8 are able to regenerate the original phenol. Pro-NH, proline; $AA\text{-}NH_2$, amino acids; Asc A, ascorbic acid; DHA, dehydroascorbic acid; R″SH, thiol compounds (cysteine or gluthatione) (from Rouet-Mayer *et al.*, 1993).

1991). Although these adducts have the *ortho*-diphenolic structure, they are not substrates of PPOs, but are competitive inhibitors (Richard-Forget *et al.*, 1992*a*). However, they can be oxidized by laccase (Salgues *et al.*, 1986) and also be co-oxidized with an excess of quinones by coupled reactions leading to highly coloured pigments (Richard-Forget *et al.*, 1992*a*). In more acidic conditions, it has been suggested that the hydroxylation of the quinone of chlorogenic acid and 4-methylcatechol yield a triphenol (Richard-Forget *et al.*, 1992*b*, 1995*a*). This triphenol may participate in further reactions, either as a substrate of PPO or reacting with the *o*-quinone giving the *p*-quinone and regenerating the parent phenol of the *o*-quinone (Fig. 4.5).

o-Quinones can react with the parent phenol giving dimers (Fig. 4.4). These dimers have been suggested to undergo further oxidative reactions leading to large oligomers with different colour intensities. Oszmianki and Lee (1990) have pointed out that catechin produced mainly dimers and polymers with a low polarity while chlorogenic acid produced mainly polymers. Several catechin dimers resulting from PPO-catalysed oxidation were isolated and characterized by NMR and MS (Guyot, 1994). The solutions oxidized at pH below 4 contained mostly colourless products (max 280 nm), whereas yellow compounds (λ_{max} 385–415 nm), eluting later than the colourless ones, predominated at higher pHs (Guyot *et al.*, 1995). The major colourless compounds formed were shown to be dehydrodicatechins of the B-ring-type presenting C–C or C–O interflavanic linkages. One of the two main pigments was dehydrodicatechin A already identified in other oxidation models (Young *et al.*, 1987). The other one presented a new structure of the quinone methide type.

The nature and relative amounts of caffeic acid oxidation products were also shown to be pH dependent. In particular, the major products formed in acidic media, were not observed at pH values above 4.6 (Fulcrand *et al.*, 1994). They were shown to be tetrahydrofuran derivatives resulting from the condensation between caffeic acid and its quinone (or coupling of two semiquinone radicals) followed by the nucleophilic attack of water on the exocyclic double bond of the intermediate dimeric quinone methide (Fulcrand *et al.*, 1994). Oxidation of caffeic acid in the range from 5 to 11 yielded several dimers and trimers. The reactions leading to these products, referred to as caffeicins, also involved the side chain double bond of at least one of caffeic acid units (Cilliers and Singleton, 1991).

The *o*-quinones can also react with another phenolic compound according to one of two pathways: the formation of copolymers, or the regeneration of the original phenol and the oxidation of the other phenol by coupled reaction (Nicolas *et al.*, 1994). In the latter case, it has been shown that among the phenolics present in fruit, some are not substrates of PPO, such as glycosides of flavonols and anthocyanins, but they are degraded (Amiot *et al.*, 1992; 1995). In apple (Richard-Forget, 1992) or in grape musts (Cheynier *et al.*,

1994), the role of caffeoyl derivatives *o*-quinones in the degradation of anthocyanins and flavonols has been clearly demonstrated. In the presence of caffeoyl tartaric acid quinone, cyanidin-3-glucoside, which is an *o*-diphenol, was degraded mostly by coupled oxidation whereas non-*o*-diphenolic malvidin-3-glucoside formed adducts with caffeoyltartaric acid quinone. In solutions containing equimolar amounts of both anthocyanins, both reactions took place but coupled oxidation of cyanidin-3-glucodide occured at a faster rate, partly protecting malvidin-3-glucoside (Sarni *et al.*, 1995).

Enzymatic browning and colour

Many authors have attempted to correlate browning intensity with the main parameters involved, i.e. phenolic content and PPO activity. All these studies have been reviewed by Macheix *et al.* (1990). Studies on apple browning have shown contradictory results (Nicolas *et al.*, 1994). The degree of browning can be predominantly influenced by the phenolic content or the PPO activity, both of which change during growth and maturation (Macheix *et al.*, 1990) and storage (Amiot *et al.*, 1995). The susceptibility of fruit and vegetables to browning illustrates the complex interactions between PPO and phenolic content. The first problem lies in the method used to determine the susceptibilty to browning. Two methods are used:

- absorbance measurements around 400 nm on solutions after crushing the tissues, centrifugation and elimination of the pellet
- tristimulus colorimetry measurements on purées or cut surfaces.

Although these methods are easy to carry out, they do not reflect the real enzymatic browning. In the first method, absorbance measurements estimate only the soluble brown pigments and do not take in account the insoluble ones resulting from polymerization or those remaining bound to the membranes. In addition, the absorbance measurements are usually taken at a single wavelength, although the maximal optical absorption for different pigments may vary from 360 to 500 nm. In the second method, the decrease in luminance δL^* (Burda *et al.*, 1990), the changes in L^*, a^* and b^* (Mastrocolola *et al.*, 1989; Sapers *et al.*, 1990), or the colour vector difference δE^* (Aubert *et al.*, 1992) are all used to measure the degree of browning. The values do not give a precise representation of the oxidized colour. In fact, these values also depend on the optical conditions of measurement and on the physical state of the examined sample. In some cases, such as in apples, the susceptibility of fruit to browning could be adequately determined by simultaneous measurements of soluble (absorbance at 400 nm) and insoluble (lightness) brown pigments (Amiot *et al.*, 1992). Soluble pigments were correlated to the amount of hydroxycinnamic esters degraded (especially chlorogenic acid) whereas insoluble pigments were correlated to the flavan-3-ols degraded ((–)-epicatechin and procyanidin B2).

In an attempt to standardize determinations of browning, model solutions containing chlorogenic acid or (–)-epicatechin were oxidized with purified apple polyphenol oxidase (Amiot, 1990; Goupy *et al.*, 1994*b*, 1995). Oxidized chlorogenic acid gave higher a^* values than oxidized (–)-epicatechin. As reported by Rouet-Mayer *et al.* (1993) and Richard-Forget *et al.* (1992*b*), chlorogenic acid quinone is dull orange-yellow with a maximal absorbance at 420 nm, catechin quinone is bright yellow with a maximum at 380 nm and dihydrophenylalanine turned pink after oxidation with a more bathochromic maximum at 480 nm. Catechin *o*-quinones were much less stable and colored than those formed from chlorogenic acid at the first phase of reaction. However, in the second phase, the secondary products derived from the oxidation of catechin were darker than the *o*-quinones. In model solutions containing both chlorogenic acid and (–)-epicatechin (Table 4.2) (Goupy *et al.*, 1995), the tristimulus colour values and the absorbance values at 400 nm were strongly correlated to initial and degraded amounts of chlorogenic acid and (–)-epicatechin. Multilinear regression analysis indicated that both phenolics contributed to the darkening (decrease in L^*) and yellowness (increase in b^*) (Table 4.2). It has been demonstrated that an increase in chlorogenic acid leads to an increase in a^* (redness), whereas (–)-epicatechin was almost without effect. Amiot *et al.* (1995) also reported a high correlation between δa^* and the initial content of chlorogenic acid in pears. In juices elaborated from immature apples, which are rich in phenolics (chlorogenic acid and (–)-epicatechin), both a^* and b^* increased markedly up to 30 min after processing (Murata *et al.*, 1995). Chlorogenic acid, considered as the best substrate of PPO in pears and in apples, appeared to be the main factor in the development of reddish pigments. Apple phenolic solutions have been also oxidized (Goupy *et al.*, 1995). For a constant PPO activity, oxidized

Table 4.2. Equations relating tristimulus colour parameters L^*, a^*, and b^* to initial (in) and oxidized (ox) chlorogenic acid (CG) and (–)-epicatechin (EP) in binary mixtures obtained by multilinear regression analysis (from Goupy *et al.*, 1995)

Tristimulus colour parameters	Initial amount	Oxidized amount
L^*	$99–21.0\ CG_{in}–25.2\ EP_{in}\ _{r=0.96}$	$98.5–19.6\ CG_{ox}–30.4\ EP_{ox}\ _{r=0.97}$
a^*	$–4.21+27.9\ CG_{in}+0.48\ EP_{in}\ _{r=0.95}$	$–4.57+31.8\ CG_{ox}–2.0\ EP_{ox}\ _{r=0.95}$
b^*	$2.58+51.5\ CG_{in}+98.2EP_{in}\ _{r=0.98}$	$4.62–41.3\ CG_{ox}–118.3\ EP_{ox}\ _{r=0.99}$

amounts could be adequately predicted in relation to the initial content. Due to coupled reactions, oxidized amounts within one phenolic class can be either increased or decreased by increasing the amounts of the other phenolic classes available. By multilinear regression analysis, L^*, a^* and b^* were correlated to both the initial and oxidized amounts in each phenolic class (Table 4.3; Fig. 4.6). Better correlations have been obtained with the amount of oxidized phenolics than the initial content. Amiot *et al.* (1992) have indicated that the susceptibility of apple to browning was strongly correlated to the amounts of oxidized hydroxycinnamic derivatives and flavan-3-ols. Besides these two classes, the flavon-3-ols and the dihydrochalcones appeared to have a significant influence on colour development (Goupy *et al.*, 1995).

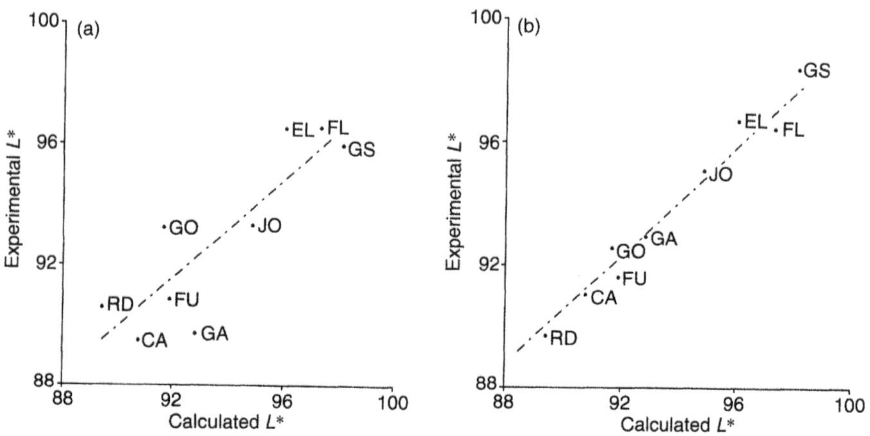

Fig. 4.6. Comparison of experimental and calculated values of L^* using the equations determined with initial (a) or oxidized (b) amounts of phenolic compounds reported in Table 4.3. Apple cultivars CA, Reinette Canada; EL, Elstar; FL, Florina; FU, Fuji; GA, Gala; GO, Golden, GS, Granny Smith; RD, Red Delicious (from Goupy *et al.*, 1995).

Recent advances in the control of browning

As has been reported previously, the biological molecular methods have to be developed to control browning. In addition, different chemical or physical ways of controlling enzymatic browning are still the objects of research. These ways can be divided into three classes depending on whether they mainly affect enzymes, substrates or reaction products, although in some cases two or three targets can be affected at the same time.

Action on enzyme Acidification, alkalinization and thermal treatment are frequently applied to inhibit enzymatic activity. In the case of enzymatic

Table 4.3. Equations relating tristimulus colour parameters L^*, a^*, and b^* to initial (in) and oxidized (ox) amount in each phenolic class in apples obtained by multilinear regression analysis (from Goupy *et al.*, 1995). HD: hydroxycinnamic acid derivatives; FO: flavonols; FA: flavanols and DC: dihydrochalcones

Tristimulus colour parameters	Initial amount	Oxidized amount
L^*	$97.6 – 5.61HD_{in} – 3.23FO_{in} + 2.1\ FA_{in} – 2.39DC_{in}$ $r = 0.87$	$98.1 — 9.29HD_{ox} – 0.04FO_{ox} + 0.53\ FA_{ox} + 0.02DC_{ox}$ $r = 0.99$
a^*	$–0.02 + 2.75HD_{in} + 0.49FO_{in} – 0.74FA_{in} — 0.9DC_{in}$ $r = 0.92$	$–0.47 + 2.99HD_{ox} + 0.72\ FO_{ox} – 0.93\ FA_{ox} – 1.14DC_{ox}$ $r = 0.92$
b^*	$5.74 + 18.3HD_{in} + 16.2FO_{in} – 5.46\ FA_{in} + 14.9DC_{in}$ $r = 0.96$	$3.92 + 38HD_{ox} – 1.42FO_{ox} + 2.77\ FA_{ox} + 4.88DC_{ox}$ $r = 0.99$

browning, alkalinization cannot be applied since phenolic compounds became highly sensitive to oxidation at alkaline pH.

PPOs show optimal activity between 4 and 7 (Vamos-Vigyazo, 1981) and the enzyme seems relatively tolerant to acidic pH. Thus, the control of enzymatic browning only by acidification is very difficult unless a very low pH is used.

Conversely, compared to other enzymes responsible for food-quality degradation, PPO is not a very heat-stable enzyme (Adams, 1991; Vamos-Vigyazo, 1981; Williams *et al.*, 1986). Obviously, since heat treatment cannot be applied to living tissues without causing irreversible damage, its use is limited to processed products such as fruit purées and juices (Zemel *et al.*, 1990). Recently, Lopez *et al.* (1994) have proposed the combined use of ultrasonic waves and moderate pressures to reduce the intensity of the heat treatments needed to inactivate PPO and other damaging enzymes, notably peroxidase and lipoxygenase. Similarly, Chen *et al.* (1992) have shown that PPO from different origins can be inactivated by high pressure carbon dioxide treatment.

According to Mayer and Harel (1979), chemical inhibitors acting on PPO can be divided into two groups those interacting with copper and those affecting the active site for the phenolic substrate.

In the first group, inhibition by azide (Healey and Strothkamp, 1981), cyanide (Duckworth and Coleman, 1970), carbon monoxide (Albisu *et al.*, 1989), tropolone (Kahn and Andrawis, 1985; Valero *et al.*, 1991*a*) and diethyldithiocarbamate (Fujita *et al.*, 1991; Zawistowski *et al.*, 1988), more or less specific to copper, is well known. However, most of these chemical are toxic and therefore cannot be used in the food industry. Inhibition by halide ions has long been recognized for PPO and is widely documented (Ben-Shalom *et al.*, 1977; De Poix *et al.*, 1980; Penafiel *et al.*, 1984; Martinez *et al.*, 1986; Rouet-Mayer and Philippon, 1986; Janovitz-Klapp *et al.*, 1990*a*). Inhibition by halides was strongly pH dependent, increasing as the pH was decreased. It was proposed that halide and copper formed a complex after displacement of a protonated histidine from copper (Ben-Shalom *et al.*, 1977; Penafiel *et al.*, 1984). A mechanism was also postulated whereby halide binds to the enzyme in competition with the phenolic substrate only when the enzyme molecule is in the protonated form (Penafiel *et al.*, 1984). For the apple PPO, sodium chloride was shown to be non competitive with an apparent inhibition constant, K_{iapp}, which decreases with the pH according to the following equation (Janovitz-Klapp *et al.*, 1990*a*) :

$$\mathrm{K}_{\mathrm{iapp}} = K_{\mathrm{i}}\ (K_{\mathrm{a}} / [\mathrm{H}^{+}] + 1)$$

with $pK_a = 3.65$ and $K_i = 2.4$ mM.

Among inhibitors affecting the site for the phenolic substrate, aromatic carboxylic acids of the benzoic and cinnamic acid series have been studied widely during the last three decades. Although most authors found that these compounds were competitive inhibitors of PPO because of their similarities to phenolic substrates, some authors indicated that the type of inhibition was dependent on the substrate used for the assay and was either competitive, non-competitive or mixed (Gunata *et al.*, 1987; Janovitz-Klapp *et al.*, 1990*a*; Kermasha *et al.*, 1993; Menon *et al.*, 1990; Oda *et al.*, 1978; Rivas and Whitaker, 1973; Walker and Wilson, 1975). The degree of inhibition by carboxylic acid was also pH dependent, increasing as the pH decreases. It was proposed that the undissociated form of the acid was responsible for inhibition by forming a complex with copper at the active site (Janovitz-Klapp *et al.*, 1990*a*). Use of cinnamic acids to control browning in fruit juices has been suggested (Walker, 1976), but not for cut surfaces of fruit because of their gradual conversion to PPO substrates by cinnamate hydroxylase and other enzymes involved in the biosynthesis of phenols (Sapers *et al.*, 1989*b*).

The substituted resorcinols, which are also structurally related to phenolic substrates, were recently recognized as PPO inhibitors (McEvily *et al.*, 1991, 1992; Monsalve-Gonzalez *et al.*, 1995; Otwell *et al.*, 1992) and their uses in food have been patented (McEvily *et al.*, 1993). In a structure-activity study on a series of 4-substituted resorcinols, the highest inhibition was obtained for hydrophobic substituents in the 4-position such as 4-hexyl, 4-dodecyl and 4-cyclohexyl resorcinols (McEvily *et al.*, 1992; Monsalve-Gonzalez *et al.*, 1995). In addition, Dawley and Flurkey (1993*a,b*) have shown that 4-hexyl resorcinol at low concentrations was a selective inhibitor of tyrosinase and could differentiate between tyrosinase and laccase activities.

Recently, it was reported that ficin, a sulfhydryl protease extracted from the genus *Ficus*, was able to inhibit shrimp melanosis as well as enzymatic browning of apple and potato slices (Taoukis *et al.*, 1990). It was proposed that the ficin extract inhibited polyphenol oxidase by proteolysis. However, the intervention of other non-enzymatic factors seems probable because the ability of the ficin extract to prevent browning was unaffected by heat denaturation or ultrafiltration (McEvily *et al.*, 1992). Moreover, the ficin extract was shown to contain 2,4-dihydroxydihydrocinnamic acid, 2,4-dihydroxydihydrocinnamoylputrescine, and *bis*-(2,4-dihydroxydihydrocinnamoyl) spermidine (McEvily *et al.*, 1992). All these compounds are analogous 4-substituted resorcinols and inhibitors of mushroom PPO.

Kojic acid (5-hydroxy-2-hydroxymethyl-γ-pyrone) has been shown to competitively inhibit PPO from different origins (Chen *et al.*, 1991*a,b*; Saruno *et al.*, 1979). In addition, this compound can reduce pigments or pigments precursors to colourless compounds (Chen *et al.*, 1991*b*).

Oxalate may inhibit PPO in two ways: upon incubation, it slowly inactivates the enzyme following a first-order mechanism, and it also acts as a competitive inhibitor (Yong *et al.*, 1990; Richard-Forget, 1992). Oxalic acid present in commercial pectins does prevent browning of raw apple juice (Tong *et al.*, 1995). Pure oxalic acid is not, however, an approved food additive, but does occur naturally in some fruit and vegetables, such as rhubarb and spinach (Tong *et al.*, 1995).

Sporix^{TM}, an acidic polyphosphate product, was shown to be a very effective inhibitor of the browning of apple juice when used in combination with ascorbic acid (Sapers *et al.*, 1989*b*). Its effect was pH dependent, decreasing when the pH was increased. It has been attributed to two effects:

- Sporix is a powerful chelating agent as well as an acidulant and therefore would directly inhibit PPO by chelation of its copper
- it would also protect ascorbic acid by inhibiting the other copper-catalysed oxidative reactions, accounting for the apparent synergism obtained with Sporix^{TM}-ascorbic acid combinations (Sapers *et al.*, 1989*b*).

Lastly, a mixed type of inhibition by several natural aliphatic alcohols was reported for grape PPO which seemed mainly related to the hydrophobic chain rather than to the alcohol function (Valero *et al.*, 1990).

Action on substrates Inhibition of enzymatic browning can also be obtained by removal of one of the two substrates (i.e. oxygen or phenol) from the reaction medium. Complete removal of oxygen is the most satisfactory way to control the phenolic oxidation catalysed by PPO. Although this method can be applied to dead tissues, either by creating a physical barrier to oxygen diffusion, for instance by dipping fruits in a sugar syrup or by using an inert atmosphere, it is inapplicable to living tissues because of the risk of metabolic deviations caused by anaerobic conditions. Similarly, vacuum infiltration of solutions into fruit slices resulted in a translucent or waterlogged appearance that would not be acceptable in a fresh product (Ponting and Jackson, 1972).

Concerning the phenolic substrates, two options have been investigated. The first is physical elimination by specific adsorbents. Recently, two studies showed that cyclodextrins were able to inhibit enzymatic browning (Sapers *et al.*, 1989b; Crouzet *et al.*, 1991), and a US patent was obtained on the use of these compounds for raw fruit and vegetable juices (Hicks *et al.*, 1990). β-Cyclodextrin, but not α- or γ-cyclodextrin, was an effective browning inhibitor in Granny Smith apple juice (Sapers *et al.*, 1989*b*). Insoluble β-cyclodextrins and hydroxypropyl-β-cyclodextrin gave results similar to that of β-cyclodextrin. The percentage of inhibition, which increased with the cyclodextrin concentration, appeared to be dependent on the apple variety, probably owing to differences in phenolic composition (Crouzet *et al.*, 1991). The inhibitory effects of cyclodextrins were enhanced when used in combination with ascorbic acid or its derivatives (Sapers *et al.*, 1989*b*). As emphasized by Sapers and Hicks (1989), however, dipping treatments in cyclodextrin solutions were not successful for the control of browning in diced apple, probably as a result of too low a diffusion rate of the cyclodextrins. In a recent study (Billaud *et al.*, 1995), it was shown that inhibition results from the ability of cyclodextrins to form inclusion complexes with polyphenols, preventing their oxidation by PPO. The binding strength of the cyclodextrin–phenol complex is, however, highly dependent on phenolic structure. Therefore, in a multiphenolic solution, addition of cyclodextrin can result in a large or a small decrease in browning or even a slight activation of PPO activity. Moreover, since the balance of available phenolic substrates can be highly modified, cyclodextrin addition can induce great changes in the colour of the oxidized solution (Billaud *et al.*, 1995).

Anionic polysaccharides, such as carrageenan, xylan sulfate and amylose sulfate, also inhibited browning of fresh juices (Tong and Hicks, 1991). Thus, addition of ι-, κ-, or λ-carrageenan alone (0.25%) or in combination (0.05%) with 0.5% citric acid resulted in an almost 100% inhibition of browning of Granny Smith apple juice over a 24 h period at room temperature. Inhibition was also obtained with amylose sulfate (0.025%) and xylan sulfate (0.025%) in combination with 0.5% citric acid. The combination of 0.5% carrageenan and 0.5% citric acid delayed browning of diced apples. The mechanism of inhibition by these carbohydrate polymers is unknown. Besides the possible complexation of phenolics, it has been suggested that the copper cofactor of PPO was chelated by the sulfate groups of the carrageenans, amylose sulfate or xylan sulfate.

Chitosan, a natural poly-ββ-(14)-*N*-acetyl-D-glucosamine derived from shellfish chitin, was able to increase the luminance of apple juice (Imeri and Knorr, 1988; Soto-Peralda *et al.*, 1989; Sapers, 1992). Thus, browning of McIntosh apple juice was inhibited by adding 200 ppm of chitosan, followed by filtration through diatomaceous earth. However, chitosan efficiency appeared to depend on cultivar. The mode of action of chitosan is not clear, but could be attributed to adsorption of polyphenols, pigments, or even enzyme itself by coagulation with PPO-containing particulates. Nevertheless, the potential uses of chitosan as well as of cyclodextrins are restricted to liquid systems and will depend on both their regulatory status and their cost relative to the other antibrowning treatments.

The second way of removing phenolic compounds is by their enzymatic modification. First, the use of *O*-methyltransferase to convert the *o*-dihydroxyphenolics, substrates of PPO, to the corresponding methoxy derivatives (e.g. caffeoyl derivatives converted to feruloyl derivatives) has been suggested (Finkle and Nelson, 1963). An alternative modification involves oxidative ring opening by protocatechuate 3,4-dioxygenase (Kelly and Finkle, 1969). Unfortunately, these elegant processes are impeded by the high cost of the enzymes and do not appear feasible for commercial use.

Action on products Acting on reaction products is the third method of enzymatic browning control. *o*-Quinones can be either reduced back to *o*-diphenols or trapped as colourless addition compounds, as it was described previously. Therefore, many compounds that act on *o*-quinones and especially ascorbic acid (Taüfel and Voigt, 1964), thiol compounds (Voldrich *et al.,* 1995), sulfites (Sayavedra-Soto and Montgomery, 1986), and amino acids (Kahn, 1985) can be used for the control of enzymatic browning. However, their effectiveness in preventing browning is greatly reduced if their use is delayed until after the reaction has started.

Although some authors reported a direct inactivation of PPO by ascorbic acid or its derivatives (Golan-Goldhirsch and Whitaker, 1984; Hsu *et al.,* 1988), the main role of ascorbic acid in browning control corresponds to its ability to reduce the *o*-quinones to diphenols. Ascorbic acid is probably the most frequently used chemical product for the control of browning control in fruit products (El-Shimi, 1993; Santerre *et al.,* 1988; Sapers *et al.,* 1989*a,b,* 1990, 1991; Sapers and Miller, 1993). However, its effect is only temporary, as it is irreversibly oxidized to dehydroascorbic acid during the process and browning can occur after depletion. Therefore, it has been proposed to use an isomer, erythorbic (isoascorbic) acid, or more stable forms of ascorbic acid (Sapers and Hicks, 1989; Sapers and Ziolkowski, 1987).

The mode of action or erythorbic and ascorbic acids is similar in model solutions containing phenolic substrates and enzyme (Hsu *et al.,* 1988; Janovitz-Klapp *et al.,* 1990*a*). Their relative effectiveness as inhibitors of browning is, however, dependent on food products. Ascorbic acid derivatives with increased stability against oxidation have been proposed for preventing the browning of apple products (Sapers *et al.,* 1989*a,b,* 1991). It has been shown that ascorbyl phosphate esters were stable against oxygen and gave stoichiometric release of ascorbic acid when hydrolysed by acid phosphatase (Liao and Seib, 1990). Ascorbic acid-2-phosphate and ascorbic acid-2-triphosphate were better inhibitors when applied as dips to the cut surfaces of Red Delicious apple than ascorbic acid at the same concentrations (Sapers *et al.,* 1989*b*). The effectiveness was again however dependent on the product because the ascorbyl phosphate esters were less effective than ascorbic acid in preventing the browning of apple juice as well as cut surfaces of potato (Sapers and Miller, 1992). The failure of ascorbyl phosphate esters to prevent browning could, in these cases, result from too low an activity of endogenous acid phosphatases due to enzyme inactivation or inadequate conditions (e.g. pH) (Sapers *et al.,* 1991).

In contrast, ascorbyl-6-fatty esters can function directly, because substitution is on C 6, leaving free the ene–diol function of C-2 and C-3. As emphasized by Cort (1982), however ascorbyl palmitate, laurate and stearate did not function as oxygen scavengers in aqueous systems below pH 9 because of their low solubilities. Thus, ascorbyl palmitate, laurate and decanoate added to Granny Smith apple juice at concentrations as high as 1.14 mM (equivalent to 200 ppm of ascorbic acid) were less effective than, or similar to, ascorbic acid initially, but surpassed ascorbic acid after longer periods of storage (Sapers *et al.*, 1989*b*).

Cysteine, among other thiol compounds, is an effective inhibitor of enzymatic browning (Montgomery, 1983; Zawistowski *et al.*, 1987; Molnar-Perl and Friedman, 1990*a,b*; Sapers *et al.*, 1994; Friedman and Bautista, 1995). There is some debate on their mechanism of action. It is generally assumed that their major effect is due to the formation of colourless thiol-conjugates with *o*-quinone. However, the direct inhibition of PPO by cysteine through the formation of stable complexes with copper has also been proposed (Kahn, 1985; Valero *et al.*, 1991*b*). If adducts of cysteine as well as those of glutathione were not substrates of the enzyme (Cheynier and Van Hulst, 1988; Singleton *et al.*, 1985; Richard *et al.*, 1991), Sanada *et al.* (1972, 1976) claimed that the cysteinylcatechol conjugate showed some inhibitory effect on the browning reaction, whereas Singleton *et al.* (1985) indicated that the glutathione derivative of caftaric acid was not an inhibitor as estimated by the oxygen uptake. Conversely, with apple PPO, Richard-Forget *et al.* (1992*a*) showed that cysteinyl addition compounds had competitive inhibitory properties, with an affinity for the enzyme higher than that of their parent phenols. With a sufficient amount of cysteine (i.e. for cysteine : phenolic ratios above 1), the phenol was fully degraded in colourless addition compounds. For cysteine:phenolic ratios below 1, *o*-quinones formed in excess were able to react with the addition compounds, with regeneration of phenolics (i.e. substrates for the enzymatic reaction) and the formation of highly coloured pigments. Therefore, provided this critical concentration level is exceeded, cysteine could, in theory, provide fruit products with permanent protection against enzymatic browning. Regrettably, the required level may sometimes be inconsistent with good organoleptic properties in the final product (McEvily *et al.*, 1992).

Miscellaneous Some studies have also been devoted to 'natural' inhibitors of enzymatic browning. Thus, it was shown that a small peptide obtained form the mycelia of *Dactylium dendroides* inhibited the PPO in apples and also delayed the browning of apple slices (Harel *et al.*, 1967). Similarly, the inhibition properties of honey on different PPOs and on the browning of apple slices, grape juice, and in model systems were attributed to a small peptide with an approximate molecular weight of 600 Da (Oszmianski and Lee, 1990). In these two cases, however the inhibitory compounds have been only partially characterized and their exact mechanism of action is not fully understood, although the honey extract was shown to be non-competitive with epicatechin.

Products of the Maillard reaction caused a strong inhibitory effect on the browning catalysed by apple PPO as well as on the peroxidase activity extracted from the same source (Pitotti *et al.*, 1990 ; Nicoli *et al.*, 1991). It has been postulated that the browning inhibition resulted from both the reducing and the chelating properties of the Maillard reaction products.

Although superoxide dismutase has been patented as an antioxidant in foods (Michelson and Monod, 1975), this enzyme extracted from bovine blood did not inhibit the browning of apple slices or the activity of purified mushroom PPO (Kon, 1980).

Plant phenolics as natural antioxidants

Whereas enzymatic oxidation of phenolic compounds leads to discoloration, oxidation of lipids initiates other changes in food products that affect their nutritional quality, wholesomeness, safety, colour, flavour and texture. All techniques leading to the exclusion of oxygen, as well as refrigeration or freezing, can reduce the rate of autoxidation. However, as very little oxygen is needed to initiate and maintain the oxidation process, it is quite common to combine such methods with the use of antioxidants.

Antioxidants are one of the principal ingredients that protect food quality by preventing oxidative deterioration of lipids. It has been proposed that an ideal food antioxidant should

- be safe
- not impart colour, odour, or flavour
- be effective at low concentrations
- be easy to incorporate
- survive after processing
- be stable in the finished product.

Moreover, epidemiological studies strongly suggest that a high intake of food rich in natural antioxidants, reduces the risk of some but not all cancers, cardiovascular disease and even cataract formation (Huang and Ferraro, 1992; Clydesdale, 1995). For instance, the so-called 'French paradox' has been attributed in part to the routine consumption of red wine, and the low incidence of coronary heart disease has been correlated with alcohol consumption (Renaud and de Lorgeril, 1992). It has recently been shown however that the phenolic antioxidants, which are very abundant in red wine may protect against this disease via the prevention of oxidation mechanisms and of inhibition of platelet aggregation (Kinsella *et al.*, 1993; Kanner *et al.*, 1994; Decker, 1995). Consequently, the future use of plant antioxidant nutrients might well create many challenges in the food industry.

Natural antioxidants in foods may be from endogenous compounds in one or more components of the food; from substances formed from reactions during processing or from food additives isolated from natural sources. Most antioxidants are phenolics compounds present in all fruit and vegetables and widely consumed in the daily diet, either as fresh plant material or in derived food products.

Common plant phenolic antioxidants include flavonoid compounds (flavanes, flavonols, flavanones, flavones), cinnamic acid derivatives, coumarins and tocopherols (vitamin E) (Figs. 4.1 and 4.7) (Larson, 1988). Among these, many phenolic compounds (catechins, caffeic esters, etc.) are both good browning substrates and good antioxidants (Table 4.4). On the contrary, flavonols (quercetin, quercitrin, etc.) are not good browning substrates, but very active antioxidants.

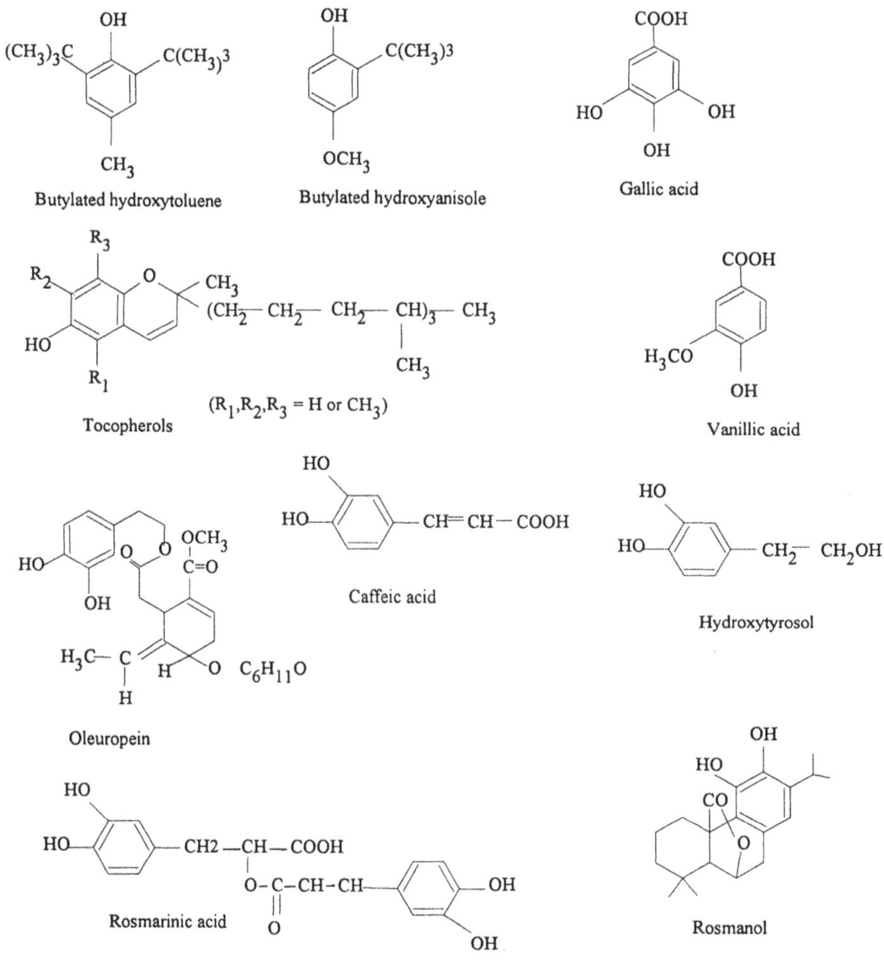

Fig. 4.7. Some examples of phenolic structures having antioxidant activity.

Table 4.4. Antioxidant activity of various phenolic compounds (from Pratt and Hudson, 1990)

Compound[a]	Time (h) to reach peroxide value of 50
Control (stripped corn oil)	110
Quercetin	475
Fisetin	450
Quercitrin	475
Naringenin	198
D-catechin	410
p-Coumaric acid	120
Ferulic acid	145
Caffeic acid	495
Chlorogenic acid	505
Propyl gallate	435

[a] 5 10^{-4} M in stripped corn oil

Natural sources of plant antioxidants have been extensively reviewed (Namiki, 1990; Laughton *et al.*, 1989; Hertog *et al.*, 1992; Ho, 1992; Shahidi *et al.*, 1992). They include different plant organs such as seeds (soybean, peanut, cottonseed, mustard, rapeseed, rice, sesame seeds), fruits (grape, citrus, black pepper, olive), leaves (tea, many spices, rosemary, thyme, oregano) and others (sweet potato, onion, oat seedlings) (Table 4.5).

Table 4.5. Some plant sources rich in phenolic antioxidants

Plants	Compounds
Green tea	Flavan-3-ols, condensed tannins
Soybean and oilseeds	Isoflavones, cinnamic acids
Herbs and spices	Cinnamic acids, flavonoids, phenolic diterpenoids, phenolic amines
Grapes and wines	Condensed tannins, flavonoids, cinnamic acids
Olives and olive oil	Flavonoids, oleuropein, hydroxytyrosol
Potatoes	Phenolic acids
Cereal bran	Ferulic acid sugar esters

Antioxidation of polyunsaturated lipids in food includes three main phases: initiation (production of lipid free radicals), propagation (amplification of production of free radicals) and termination (production of non-radical reactions). Antioxidants are generally classified into two groups

(Gordon, 1990): primary or chain-breaking antioxidants which react with lipid radicals to convert them into more stable products and secondary or preventive antioxidants which reduce the rate of chain initiation or decompose hydroperoxides to non-radical species. Plant phenolics and some of their degradation products are multifunctional and can act as reducing agents, free radical terminators, metal chelators and single oxygen quenchers. Furthermore, phenolic antioxidants not only inhibit the autooxidation of lipids, but sometimes also have the ability to retard their oxidation by inhibiting lipoxygenase activity (Ho, 1992; Okuda, 1993). Oxidative metabolites of arachidonic acid may promote carcinogenesis and phenolic inhibitors of lipoxygenase have been observed to inhibit tumor promotion (Huang and Ferraro, 1992). Four green tea catechins having strong antioxidant activity (Lunder, 1992) also exhibited various degrees of lipoxygenase-inhibitory activities (Ho, 1992; Richard-Forget *et al.*, 1995*b*). Whereas (–)-epigallocatechin gallate, (–)-epicatechin gallate and epigallocatechin are active, (–)-epicatechin is, on the other hand, relatively inactive (Table 4.6). Two of the oxidative dimers of tea catechins, namely theaflavin monogallate B and theaflavin digallate, which are important polyphenols in black tea (Balentine, 1992), have an even stronger lipoxygenase-inhibitor activities than the catechin monomers.

Table 4.6. Inhibition of soybean lipoxygenase by tea polyphenols (from Ho, 1992)

Compound	IC_{50} (μM)
(-)-Epicatechin	140
(–)-Epicatechin gallate	18
(–)-Epigallocatechin	21
(–)-Epigallocatechin gallate	10
Theaflavin	3604
Theaflavin monogallate A	366
Theaflavin monogallate B	0.62
Theaflavin digallate	0.25

Conclusions

Phenolic compounds are present in all fruit and vegetables and in all their derived food products. Phenolic composition is generally modified by oxidative reactions occuring during the processing and storage of food products. Phenolic compounds act both as substrates in enzymatic browning and as antioxidants in lipid oxidation. Concerning enzymatic browning, even though a lot of work has been devoted to polyphenol analyses, to polyphenol oxidases and the control of enzymatic browning, further studies are still neededI:

- to understand non-enzymatic reactions involving *o*-quinones in relation with the oxidized colour
- to define precisely the role of peroxidases
- to find new inhibitory molecules
- to optimize treatments with a combination of existing molecules, verify their safety, and demonstrate their commercial feasibility.

As antioxidants, phenolic compounds are also of a great interest to the food industry. The search and development of natural effective antioxidants has been very active in recent years. Furthermore, although the mutagenicity of phenolic compounds and their eventual prooxidant action (Laughton *et al.*, 1989; Aruoma *et al.*, 1993) have to be checked carefully, human daily intake by consumption of fruit and vegetables is considered to have beneficial effects.

References

Adams, J.B. (1991). *Int. J. Food Sci. Technol.* **26**, 1–20.

Albisu, I., King, R.D. and Kozlov, I.A. (1989). *J. Agric. Food Chem.* **37**, 775–6.

Amiot, M. J. (1990). In *9ème colloque sur les recherches fruitières* (ed. INRA-CTIFL), pp. 279–89.

Amiot, M.J., Tacchini, M., Aubert, S. and Nicolas, J. (1992). *J Food Sci.* **57**, 958–62.

Amiot, M.J., Tacchini, M., Aubert, S. and Oleszek, W. (1995). *J. Agric. Food Chem.* **43**, 1132–37.

Aruoma, O.I. (1993). In *DNA and free radicals* (ed. B. Halliwell and O.I. Aruoma), pp. 315–325. Ellis Horwod, Chichester.

Aubert, S., Amiot, M.J. and Nicolas, J. (1992). *Sci. Alim.* **12**, 625–47.

Bachem, C.W.B., Speckmann, G.J., Van der Linde, P.C.G., Verheggen, F.T.M., Hunt, M.D., Steffens, J.C. and Zabeau, M. (1994). *Biotechnology* **12**, 1101–5.

Balentine, D.A. (1992). In *Phenolic compounds in food and their effects on health I . Analysis, occurrence and chemistry* (ed. C.T. Ho, C.Y. Lee and M.T. Huang), pp. 102–17. ACS Symposium Series 506, Washington.

Ben-Shalom, N., Kahn, V., Harel, E. and Mayer, A.M. (1977). *J. Agric Food Chem.* **28**, 545–50.

Billaud, C., Regaudie, E., Fayad, N., Richard-Forget, F. and Nicolas, J. (1995). In *Enzymatic browning and its prevention in foods* (ed. C.Y. Lee and J.R. Whitaker J.R.), pp. 295–312. ACS Symposium Series 600, Washington.

Boss, P.K., Garoner, R.C., Janssen, B.J. and Ross, G.S. (1995). *Plant Mol. Biol.* **27**, 429–33.

Burda, S., Oleszek, W. and Lee, C.Y. (1990). *J. Agric. Food Chem.* **38**, 945–8.

Cary, J.W., Lax, A.R. and Flurkey, W.H. (1992). *Plant Mol. Biol.* **20**, 245–53.

Chen, J.S., Wei, C.I., Rolle, R.S., Otwell, W.S., Balaban, M.O. and Marshall, M.R. (1991*a*). *J. Agric. Food Chem.* **39**, 1396–401.

Chen, J.S., Wei, C.I. and Marshall, M.R. (1991*b*). *J. Agric. Food Chem.* **39**, 1897–1901.

Chen, J.S., Wei, C.I., Marshall, M.R. and Wei, Y.H. (1992). *J. Agric. Food Chem.* **40**, 2345–9.

Cheynier, V. and Moutounet, M. (1992). *J. Agric. Food Chem.* **40**, 2038–44.

Cheynier, V. and VanHulst, M.W.J. (1988). *J. Agric. Food Chem.* **36**, 10–15.

Cheynier, V.F., Trousdale, E.K., Singleton, V.L., Salgues, M.J. and Wylde, R. (1986). *J. Agric. Food Chem.* **34**, 217–21.

Cheynier, V., Basire M. and Rigaud, J. (1989*a*). *J. Agric. Food Chem,* **37**, 1069–71.

Cheynier, V., Rigaud, J. and Moutounet, M. (1989*b*). *J. Chromatogr.* **472**, 428–32.

Cheynier, V., Rigaud, J. and Moutounet, M. (1990*a*). *Phytochemistry* **29**, 1751–3.

Cheynier, V., Rigaud, J., Souquet, J.M., Duprat, F. and Moutounet, M. (1990*b*). *Am. J. Enol. Vitic.* **41**, 346–9.

Cheynier, V., Souquet, J.M., Kontek, A. and Moutounet, M. (1994). *J. Sci. Food Agric.* **66**, 283–8.

Cilliers, J.J.L. and Singleton, V.L. (1991). *J. Agric. Food Chem.* **39**, 1298–1303.

Clydesdale, F.M. (1995). *Crit. Rev. Food Sci. Nutr.* **35**, 1–173.

Constabel, C.P., Bergey, D.R. and Ryan, C.A (1995). *Proc. Natl. Acad. Sci. USA* **92**, 407–11.

Cort, W.C. (1982). In *Ascorbic acid: Chemistry, metabolism, and uses* (ed. P.A. Seib and B.T.Tolbert), pp. 533–50. ACS, Washington.

Crouzet, J., Bastier, P., Signoret, A. and Mentink, L. (1991). In *Food ingredients Europe*, Conference Proceedings, pp. 293–95. Expoconsult Publishers, Maarssen.

Dawley, R.M. and Flurkey, W.H. (1993*a*). *J. Food Sci.* **58**, 609–10,70.

Dawley, R.M. and Flurkey, W.H. (1993*b*). *Phytochemistry* **33**, 281–4.

Decker, E.A. (1995). *Nutri. Rev.* **53**, 49–58.

Demenuyk, M.N., Nizharadeze, A.N. and Sal'kova, E.G. (1974). *Prikl. Biokhim. Microbiol.* **10**, 659–65.

DePoix, A., Rouet-Mayer, M.A. and Philippon, J. (1980). *Lebensm. Wissen. Technol.* **14**, 105–10.

Duckworth, H.W. and Coleman, J.E. (1970). *J. Biol. Chem.* **245**, 1613–25.

Dudley, E.D. and Hotchkiss, J.H. (1989). *J. Food Biochem.* **13**, 65–75.

El-Shimi, N.M. (1993). *Plant Foods Human Nutrition* **43**, 71–6.

Eskin, N.A.M. (1990). In *Biochemistry of foods*, 2nd edn (ed. N.A.M. Eskin), pp. 401–32. Academic Press, New York,

Finkle, B.J. and Nelson, R.F. (1963). *Nature* **197**, 902–3.

Finley, K.T. (1974). In *The chemistry of quinoid compounds* (ed. S. Patai), pp. 877–1144. Wiley, New York.

Flurkey, W.H. (1986). *Plant Physiol.* **81**, 614–18.

Fraignier, M.P., Marquès, L., Fleuriet, A. and Macheix, J.J. (1995). *J. Agric. Food Chem.* in press.

Friedman, M. and Bautista, F.F. (1995). *J. Agric. Food Chem.* **43**, 69–76.

Fujita, S. and Tono, T. (1988). *J. Sci. Food Agric.* **46**, 115–23.

Fujita, S., Tono, T. and Kawahara, H. (1991). *J. Sci. Food Agric.* **55**, 643–51.

Fulcrand, H., Cheminat, A., Brouillard, R. and Cheynier, V. (1994). *Phytochemistry* **35**, 499–505.

Ganesa, C., Fox, M.T. and Flurkey, W.H. (1992). *Plant Physiol.* **98**, 472–79.

Golan-Goldhirsch, A. and Whitaker, J.R. (1984). *J. Agric. Food Chem.* **32**, 1003–9.
Goodenough, P.W., Kessell, S., Lea, A.G.H. and Loeffler, T. (1983). *Phytochemistry* **22**, 359–63.
Gordon, M.H. (1990). In *Food antioxidants* (ed. B.J.F. Hudson), pp. 1–18. Elsevier Applied Science, London.
Goupy, P., Macheix, J.J., Nicolas, J. and Varoquaux, P. (1994*a*). *Sci. Alim.* **14**, 751–62.
Goupy, P., Amiot, M.J., Aubert, S. and Nicolas, J. (1994*b*). In *17th International Conference on Polyphenols* (ed. R. Brouillard, M. Jay and A. Scalbert), pp. 183–4. INRA Editions, Paris.
Goupy, P., Amiot, M.J., Richard-Forget, F., Duprat, F., Aubert, S. and Nicolas, J. (1995). *J. Food Sci.* **60**, 497–501, 505.
Gunata, Y.Z., Sapis, J.C. and Moutounet, M. (1987). *Phytochemistry* **26**, 1573–5.
Guyot, S. (1994). Obtention et caractérisation de produits d'oxydation enzymatique de la (+)-catéchine et de l'acide t-cafeoyltartrique. Inhibition d'enzymes impliquées dans la dégradation des polysaccharides pariétaux. Thèse Doctorat, University of Rennes.
Guyot, S., Cheynier, V., Souquet, J.M. and Moutounet, M. (1995). *J. Agric. Food Chem* **43**, 2458–62.
Harel, E., Mayer, A.M., Marcus, Z. and Avigad, G. (1967). *Israel J. Bot.* **16**, 38–41.
Healey, D.F. and Strothkamp, K.G. (1981). *Arch. Biochem. Biophys.* **211**, 86–91.
Hertog, M.G.L., Hollman, P.C.H. and Katan, M.B. (1992). *J Agric. Food Chem.* **40**, 2379–83.
Hicks, K.B., Sapers, G.M. and Seib, P.A. (1990). Process for Preserving Raw Fruit and Vegetable Juices using Cyclodextrins and Compositions Thereof, US Patent. 4,975,293.
Ho, C.T. (1992). In *Phenolic compounds in food and their effects on health II. Antioxidants and cancer prevention* (ed. M.T. Huang, C.T. Ho and C.Y. Lee), pp. 2–7. ACS Symposium Series 507, Washington.
Hsu, A.F., Shieh, J.J., Bills, D.D. and White, K. (1988). *J. Food Sci.* **53**, 765–71.
Huang, M.T. and Ferraro, T. (1992). In *phenolic compounds in food and their effects on health II. Antioxidants and cancer prevention* (ed. M.T. Huang, C.T. Ho and C.Y. Lee), pp. 8–34. ACS Symposium Series 507, Washington.
Hunt, M.D., Eanneatta, N.T., Yu, H.F., Newman, S.M. and Steffens, J.C. (1993). *Plant Mol. Biol.* **21**, 59–68.
Hutcheson, S.W., Buchanan, B.B. and Montalbini, P. (1980). *Plant Physiol.* **66**, 1150–4.
Imeri, A.G. and Knorr, D. (1988). *J. Food Sci.* **53**, 1707–9.
Janovitz-Klapp, A., Richard, F. and Nicolas, J. (1989). *Phytochemistry* **28**, 2903–7.
Janovitz-Klapp, A., Richard, F.C., Goupy, P.M. and Nicolas, J. (1990*a*). *J. Agric. Food Chem.* **38**, 926–31.
Janovitz-Klapp, A., Richard, F.C., Goupy, P.M. and Nicolas, J. (1990*b*). *J. Agric. Food Chem.* **38**, 1437–41.
Kahn, V. (1977). *J. Sci. Food Agric.* **28**, 233–39.
Kahn, V. (1985). *J. Food Sci.* **50**, 111–15.
Kahn, V. and Andrawis, A. (1985). *Phytochemistry* **24**, 905–8.
Kanner, J., Frankel, E., Granit, R., German, B. and Kinsella, J.E. (1994). *J. Agric. Food Chem.* **42**, 64–69.
Kelly, S.H. and Finkle, B.J. (1969). *J. Sci. Food Agric.* 20, 629–32.

Kermasha, S, Goetghebeur, M., Monfette, A., Metche, M. and Rovel, B. (1993). *J. Agric. Food Chem.* **41**, 526–31.

King, R. and Flurkey, W.H. (1987). *J. Sci. Food Agric.* **41**, 231–40.

Kinsella, J.E., Frankel, E., German, B. and Kanner, J. (1993). *Food Technol.* April, 85–89.

Kon, S. (1980). *J. Food Sci.* **45**, 1066–7.

Larson, R.A. (1988). *Phytochemistry* **27**, 969–78.

Larson, R.A. (1995). *Arch. Insect Biochem. Physiol.* **29**, 175–86.

Laughton, M.J., Halliwell, B., Evans, P.J. and Hoult, J.R.S. (1989). *Biochem. Pharmacol.* **38**, 2859–65.

Liao, M.L. and Seib, P.A. (1990). *J. Agric. Food Chem.* **38**, 355–66.

Lopez, P., Sala, F.J., De la Fuente, J.L., Condon, S., Raso, J. and Burgos J. (1994). *J. Agric. Food Chem.* **42**, 252–56.

Lunder, T.L. (1992). In *Phenolic compounds in food and their effects on health II. Antioxidants and cancer prevention* (ed. M.T. Huang, C.T. Ho and C.Y. Lee), pp. 114–20. ACS Symposium Series 507, Washington.

Macheix, J.J., Fleuriet, A. and Billot, J. (1990). *Fruit phenolics*. CRC Press, Boca Raton.

Macheix, J.J., Sapis, J.C. and Fleuriet, A. (1991). *Crit. Rev. Food Sci. and Nutr.* **30**, 441–86.

Marquès, L., Fleuriet, A., Cleyet-Marel, J.C. and Macheix, J.J. (1994). *Phytochemistry* **36**, 1117–21.

Marquès, L., Fleuriet, A. and Macheix, J.J. (1995*a*). *Plant Physiol. Biochem.* **33**, 193–200.

Marquès, L., Fleuriet, A. and Macheix, J.J. (1995*b*). In *Enzymatic browning and its prevention in foods* (ed. C.Y. Lee and J.R. Whitaker), pp. 99–101. ACS Symposium Series 600, American Chemical Society, Washington.

Martinez, J.H., Solano, F., Penafiel, R., Galindo, J.D., Iborra, J.L. and Lozano, J.A. (1986). *Comp. Biochem. Physiol.* **83B**, 633–6.

Mason, H.S. and Peterson, E.W. (1965). *Biochem. Biophys. Acta* **111**, 134–46.

Mastrocolola, D., Severini, C. and Lerici, C.R. (1989). In *Engineering and food: physical properties and process control* (ed. W.E.L. Spiess and H. Schubert), Vol 1, pp. 662–70. Elsevier Applied Science, London.

Mayer, A.M. (1987). *Phytochemistry* **26**, 11–20.

Mayer, A.M. and Harel, E. (1979). *Phytochemistry* **18**, 193–215.

Mayer, A.M. and Harel, E. (1981). *In Recent advances in the biochemistry of fruits and vegetables* (ed. J. Friend and M.J.C. Rhodes), pp. 161–80. Academic Press, London.

Mayer, A.M. and Harel, E. (1991). In *Food enzymology* (ed. P.F. Fox), pp. 373–98. Elsevier Applied Sciences, London.

McEvily, A.J., Iyengar, R. and Otwell, W.S. (1991). *Food Technol.* **45**, 80–6.

McEvily, A.J., Iyengar, R. and Otwell, W.S. (1992). *CRC Crit. Rev. Food Sci. Nutr.* **32**, 253–73.

McEvily, A.J., Iyengar, R. and Gross, A. (1993). Compositions and methods for inhibiting browning in foods and beverages. U.S. Patent 5,202,141.

Menon, S., Fleck, R.W., Yong, G. and Strothkamp, K.G. (1990). *Arch. Biochem. Biophys.* **280**, 27–32.

Metche, M. (1986). *Bull. Liais. Groupe Polyphénols* **13**, 292–8.

Michelson, A.M. and Monod, J. (1975). Superoxide dismutases and their applications as oxidation inhibitors. U.S. Patent 3,920,521.

Molnar-Perl, I. and Friedman, M. (1990a). *J. Agric. Food Chem.* **38**, 1648–51.
Molnar-Perl, I. and Friedman, M. (1990b). *J. Agric. Food Chem.* **38**, 1652–6.
Monsalve-Gonzales, A., Barbosa-Canovas, G.V., McEvily, A.J. and Iyengar, R. (1995). *Food Technol.* April, 110–18.
Montgomery, M.W. (1983). *J. Food Sci.* **48**, 951–2.
Moore, B.M. and Flurkey, W.H. (1990). *J. Biol. Chem.* **265**, 4982–8.
Murata, M., Kurokami, C., Homma, S. and Matsuhashi, C. (1993). *J. Agric. Food Chem.* **41**, 1385–90.
Murata, M., Tsurutani, M., Tomita, M., Homma, S. and Kaneko, K. (1995). *J. Agric. Food Chem.* **43**, 1115–21.
Namiki, M. (1990). *Crit. Rev. Food Sci. and Nutr.* **29**, 273–300.
Newman, S.M., Eannetta, N.T. Yu, H., Prince, J.P., de Vicente, M.C. Tanksley, S.D. and Steffens, J.C. (1993). *Plant Mol. Biol.* **21**, 1035–51.
Nicolas, J.J., Cheynier, V., Fleuriet, A. and Rouet-Mayer, M.A. (1993). In *Polyphenolic phenomena* (ed. A. Scalbert), pp. 165–75. INRA Editions, Paris.
Nicolas, J.J., Richard-Forget, F., Goupy, P.M., Amiot, M.J., Aubert, S.Y. (1994). *Crit. Rev. Food Sci. and Nutr.* **34**, 109–57.
Nicoli, M.C., Elizalde, B.E., Pitotti, A. and Lerici, C.R (1991). *J. Food Biochem.* **15**, 169–84.
Oda, Y., Oh-E, S., Saito, G. and Matsuoka, H. (1978). *Agric. Biol. Chem.* **42**, 253–8.
Okuda, T. (1993). In *Polyphenolic phenomena* (ed. A. Scalbert), pp. 221–35. INRA Editions, Paris.
Oleszek, W., Lee, C.Y. and Price, K.R. (1989). *Acta Soc. Bot. Poloniae* **58**, 273–83.
O'Malley, D.M., Whetten, R., Bao, W., Chen, C.L. and Sederoff, R.R. (1993). *Plant J.* **4**, 751–7.
Oszmianski, J. and Lee, C.Y. (1990). *J. Agric. Food Chem.* **38**, 1892–95.
Oszmianski, J. and Lee, C.Y. (1991). *J. Agric. Food Chem.* **39**, 1050–2.
Otwell, W.S., Iyengar, R. and McEvily, A.J. (1992). *J. Aquatic Food Product Technol.* **1**, 53–65.
Penafiel, R., Galindo, J.D., Solano, F., Pedreno, E., Iborra, J.L. and Lozano, J.A. (1984). *Biochim. Biophys. Acta* **788**, 327–32.
Pierpoint, W.S. (1969). *Biochem. J.* **112**, 609–17.
Piretti, M.V., Serrazanetti, G.P. and Paglone, G. (1977). *Ann. Chim.* **67**, 395–413.
Pitotti, A., Nicoli, M.C., Sensidoni, A. and Lerici, C.R. (1990). In *Engineering and food: physical properties and process control* (ed. W.E.L. Spiess and H. Schubert), Vol 1, pp. 671–81. Elsevier Applied Science, London.
Ponting, J.D. and Jackson, R. (1972). *J. Food Sci.* **37**, 812–14.
Pratt, D.E. and Hudson, B.J.F. (1990). In *Food antioxidants* (ed. B.J.F. Hudson), pp. 171–91. Elsevier Applied Science, London.
Prota, G. (1988). *Medicinal Res. Rev.* **8**, 525–6.
Rathjen, A.H. and Robinson, S.P. (1992). *Plant Physiol.* **99**, 1619–25.
Renaud, S. and de Lorgeril, M. (1992). *Lancet* **339**, 1523–6.
Richard, F., Goupy, P.M., Nicolas J.J., Lacombe, J.M. and Pavia, A.A. (1991). *J. Agric. Food Chem.* **39**, 841–7.
Richard-Forget, F. (1992). Recherches sur le brunissement enzymatique. Etudes sur l'oxydation de phénols et sur l'inhibition de la polyphénoloxydase isolée de la Pomme (*Malus sylvestris,* var. Red Delicious). Thèse Doctorat, University of Paris 7.
Richard-Forget, F., Goupy, P.M. and Nicolas, J.J. (1992*a*). *J. Agric. Food Chem.* **40**, 2108–13.

Richard-Forget, F., Rouet-Mayer, M.A., Goupy, P.M., Philippon, J. and Nicolas, J.J. (1992*b*). *J. Agric. Food Chem.* **40**, 2114–22.

Richard-Forget, F., Amiot, M.J., Goupy, P. and Nicolas, J. (1995*a*). In *Enzymatic browning and its prevention in foods*, (ed. C.Y. Lee and J.R.Whitaker), pp. 144–158. ACS Symposium Series 600, Washington.

Richard-Forget, F., Gauillard, F., Hughes, M., Thiry, J.M., Boivin, P. and Nicolas, J. (1995*b*). *J. Food Sci.*, **60**, 1325–29.

Rivas, N.J. and Whitaker, J.R. (1973). *Plant Physiol.* **52**, 501–7.

Roberts, A.C. and McWeeny, D.J. (1972). *J. Food Technol.* **7**, 221–38.

Robinson, D.S. (1991). In *Oxidative enzymes in foods* (ed. D.S. Robinson and N.A.M. Eskin), pp. 1–47. Elsevier, London.

Robinson, S.P. and Dry, I.B. (1992). *Plant Physiol.* **99**, 317–23.

Robinson, S.P., Loveys, B.R. and Chacko, E.K. (1993). *Aust. J. Plant Physiol.* **20**, 99–107.

Rouet-Mayer, M.A. and Philippon, J. (1986). *Phytochemistry* **25**, 2717–19.

Rouet-Mayer, M.A., Philippon, J. and Nicolas, J. (1993). In *Encyclopedia of food science, food technology and nutrition* (ed. R. McRae, R.K. Robinson and M.J. Sadler), Vol 1, pp. 499–506. Academic Press, London.

Salgues, M., Cheynier, V., Gunata, Z. and Wylde, R. (1986). *J. Food Sci.* **51**, 1191–4.

Sanada, H., Suzue, R., Nakashima, Y. and Kawada, S. (1972). *Biochim. Biophys. Acta* **261**, 258–66.

Sanada, H., Nakashima, Y., Suzue, R. and Kawada, S. (1976). *J. Nutri. Sci. Vitamin.* **22**, 389–96.

Sanchez-Ferrer, A. Bru, R. and Garcia-Carmona, F. (1989). *Plant Physiol.* **91**, 1481–7.

Sanchez-Ferrer, A. Bru, R. and Garcia-Carmona, F. (1990). *Anal. Biochem.* **184**, 279–82.

Sanchez-Ferrer, A. Laveda, F.. and Garcia-Carmona, F. (1993*a*). *J. Agric. Food Chem.* **41**, 1583–6.

Sanchez-Ferrer, A. Laveda, F. and Garcia-Carmona, F. (1993*b*). *J. Agric. Food Chem.* **41**, 1219–24.

Santerre, C.R., Cash, J.N. and VanNorman, D.J. (1988). *J. Food Sci.* **53**, 1713–16, 1736.

Sapers, G.M. (1992). *J. Food Sci.* **57**, 1192–93.

Sapers, G.M. and Hicks, K.B. (1989). In *Quality factors of fruits and vegetables: Chemistry and technology* (ed. J.J. Jen), pp. 29–43. ACS Symposium Series 405, Washington.

Sapers, G.M. and Miller, R.L. (1992). *J. Food Sci.* **57**, 1132–5.

Sapers, G.M. and Miller, R.L. (1993). *J. Food Sci.* **58**, 1076–8.

Sapers, G.M. and Ziolkowski, M.A. (1987). *J. Food Sci.* **52**, 1732–3, 1747.

Sapers, G.M., Garzarella, L. and Pilizota, V. (1990). *J. Food Sci.* **55**, 1049–53.

Sapers, G.M., El-Atawy, Y.S., Hicks, K.B. and Garzarella, L. (1989*a*). *J. Food Sci.* **54**, 1096–7.

Sapers, G.M., Hicks, K.B., Phillips, J.G., Garzarella, L., Pondish, D.L., Matulaitis, R.M., McCormack, T.J., Sondey, S.M., Seib, P.A. and El-Atawy, Y.S. (1989*b*). *J. Food Sci.* **54**, 997–1002, 1012.

Sapers, G.M., Miller, R.L., Douglas Jr., F.W. and Hicks, K.B. (1991). *J. Food Sci.* **56**, 419–22, 30.

Sapers, G.M., Miller, R.L., Miller , F.C., Cooke, P.H. and Choi, S.W. (1994). *J. Food Sci.* **59**, 1042–7.

Sarni, P., Fulcrand, H., Souillol, V., Souquet, J.M. and Cheynier, V. (1995). *J. Sci. Food Agric.* **69**, 385–91.

Saruno, R., Kato, F. and Ikeno, T. (1979). *Agric. Biol. Chem.* **43**, 1337–39.
Sayavedra-Soto, L.A. and Montgomery, M.W. (1986). *J. Food Sci.* **51**, 1531–6.
Shahar, T. Hennig, N. Gutfinger, T., Hareven, D. and Liftshitz, E. (1992). *Plant Cell* **4**, 135–47.
Shahidi, F., Janitha, P.K. and Wanasundara, P.D. (1992). *Crit. Rev. Food Sci. and Nutr.* **32**, 67–103.
Sherma, R.C. and Ali, R. (1980). *Phytochemistry* **19**, 1597–1600
Shomer, I. Ben-Shalom, N., Harel, E., Mayer, A.M. and Harel., E. (1979). *Ann. Bot.* **44**, 261–3.
Singleton, V.L. (1987). *Amer. J. Enol. Vitic.* **38**, 69–77.
Singleton, V.L., Salgues, M., Zaya, J. and Trousdale, E. (1985). *Am. J. Enol. Vitic.* **36**, 50–6.
Smith, D.M. and Montgomery, M.W. (1985). *Phytochemistry* **24**, 901–4.
Sommer, A., Ne'eman, E., Steffens, J.C., Mayer, A.M. and Harel, E. (1994). *Plant Physiol.* **105**, 1301–11.
Soto-Peralda, N.V., Müller, H. and Knorr, D. (1989). *J. Food Sci.* **54**, 495–6.
Steffens, J.C., Kowalski, S.P. and Yu, H. (1990). In *The molecular and cellular biology of the potato* (ed. M.V.W.O. Park), pp. 103–12. CAB, Wallingford, Oxon.
Sterjades, R., Dean, J.F.D. and Erikson, K.E.L. (1993). *Plant Physiol.* **99**, 1162–68.
Taoukis, S.L., Labuza, T.P., Lillemo, J.H. and Lin, S.W. (1990). *Lebensm. Wissen. Technol.* **23**, 52–4.
Taüfel, K. and Voigt, J. (1964). *Z. Lebensm. Unters. Forsch.* **126**, 19–24.
Taylor, A.J. and Clydesdale, F.M. (1987). *Food Chem.* **24**, 301–13.
Taylor, S.L., Highley, N.A. and Bush, R.K. (1986). *Adv. Food Res.* **30**, 1–76.
Tong, C.B.S. and Hicks, K.B. (1991). *J.Agric. Food Chem.* **39**, 1719–22.
Tong, C.B.S., Hicks, K.B., Osman, S.F., Hotchkiss, A. and Haines, R.M. (1995). *J.Agric. Food Chem.* **43**, 592–7.
Valero, E. and Garcia-Carmona, F. (1992). *Plant Physiol.* **98**, 774–6.
Valero, E., Varon, R. and Garcia-Carmona, F. (1988). *J. Food Sci.* **53**, 1482–85.
Valero, E., Varon, R. and Garcia-Carmona, F. (1990). *J. Agric. Food Chem.* **38**, 1097–100.
Valero, E., Varon, R. and Garcia-Carmona, F. (1991*a*). *Biochem. J.* **277**, 869–74.
Valero, E., Garcia-Moreno, M., Varon, R. and Garcia-Carmona, F. (1991*b*). *J. Agric. Food Chem.* **39**, 1043–6.
Vamos-Vigyazo, L. (1981). *Crit. Rev. Food Sci. Nutr.* **15**, 49–127.
Varoquaux, P. and Wiley, R.C. (1994). In *Minimally processed refrigerated fruits and vegetables* (ed. R.C. Wiley), pp. 226–8. Chapman & Hall, London.
Vaughn , K.C. and Duke, S.O. (1984). *Physiol. Plant.* **60**, 106–12.
Volk, R., Harel, E., Mayer, A.M. and Gan-Zvi, E. (1977). *J. Exp. Bot.* **28**, 820–30.
Voldrich, M., Dupont, D., Dobias, J. and Philippon, J. (1995). *Lebensm. Wissen. Technol.* **28**, 213–17.
Walker, J.R.L. (1976). *J. Food Technol.* **11**, 341–45.
Walker, J.R.L. and Wilson, E.L. (1975). *J. Sci. Food Agric.* **26**, 1825–1.
Wedzicha, B.L. (1984). In *Chemistry of sulphur dioxide in foods* (ed. B.L. Wedzicha), Vol. 1, pp. 1–85. Elsevier Applied Sciences, London.
Williams, D.C., Lim, M.H., Chen, A.O., Pangborn, R.M. and Whitaker, J.R. (1986). *Food Technol.* **40**(6), 130–40. Vol. 1, pp. 1–85, Elsevier Applied Sciences, London.
Yong, G., Leone, C. and Strothkamp, K.G. (1990). *Biochemistry* **29**, 9684–90.

Young, D.A., Youg, E., Roux, D.G., Brandt, E.V. and Ferreira, D. (1987). *J. Chem. Soc. Perkin Trans* **I**, 2345–51.
Yu, H.F., Kowalski, S.P. and Steffens, J.C. (1992). *Plant Physiol.* **100**, 1885–90.
Zawistowski, J., Blank, G. and Murray, E.D. (1987). *Can. Inst. Food Sci. Technol. J.* **20**, 162–5.
Zawistowski, J., Biliaderis, C.G. and Murray, E.D. (1988). *J. Food Biochem.* **12**, 1–22.
Zawistowski, J., Biliaderis, C.G. and Eskin, N.A.M. (1991). In *Oxidative enzymes in foods* (ed. D.S.R. Robinson and N.A.M. Eskin), pp. 217–73. Elsevier Applied Sciences, London.
Zemel, G.P., Sims, C.A., Marshall, M.R. and Balaban, M. (1990). *J. Food Sci.* **55**, 562–63.

5. Astringency

MICHAEL N. CLIFFORD

Food Safety Research Group, School of Biological Sciences, University of Surrey, Guildford, Surrey GU2 5XH, UK

Introduction

The ASTM Committee E-18 on Sensory Evaluation has defined astringency as 'the complex of sensations due to shrinking, drawing or puckering of the epithelium as a result of exposure to substances such as alums or tannins' (ASTM, 1989). The *Shorter Oxford English Dictionary* defines 'astringent' as 'having the power to draw together the organic tissues; binding, constrictive, styptic' (SOED, 1944). Two important, possibly crucial, features of astringency which are not immediately apparent from the foregoing definitions, but which are well-known to all who have encountered it in the mouth, are its lingering nature and tendency to become more pronounced on repeated exposure (on a relatively short time scale such as the repeated sipping of wine).

There has been debate as to whether astringency is a primary taste akin to sweet, bitter, salt and sour, with the astringent signals transduced via the chorda tympani nerve (Schiffman *et al.*, 1991) or a tactile sensation arising from increased friction/reduced lubrication of the tongue and soft palate and transduced by the trigeminal nerve (Green, 1993).

Breslin *et al.* (1993) have demonstrated that volunteers report a feeling of astringency when tannic acid solution is applied under the upper lip, an area devoid of receptors for the primary tastes, suggesting that the classic taste receptors located on the tongue and soft palate are not essential to the phenomenon. The failure to adapt to the stimulus on repeated exposure, i.e. the tendency under these circumstances for astringency to become stronger rather than weaker as does occur with the primary tastes (and the sense of smell), also suggests that these classic taste receptors are not an absolute prerequisite and that astringency is a tactile sensation. Although as scientists we may make this distinction, to the general public astringency remains a taste.

Proponents of the 'tactile hypothesis' explain the sensation as arising when an astringent substance destroys the lubricating power of the saliva by precipitating the secreted proteins. Binding direct to epithelial proteins may also be important especially when the ratio of astringent substances to salivary proteins is particularly high. However, the suggestion that astringents close the lumen of the salivary ducts by cross-linking seems not to be correct since salivary flow rate has been observed to increase (Fischer *et al.*, 1994) rather than decrease as this hypothesis would require.

The mouth, tongue and salivary glands

The inner surface of the lips and cheeks are covered with a mucous membrane in which the epithelium is of the stratified squamous non-keratinizing type. The tongue consists of interlacing bundles of striated muscle with a thin mucous membrane on the lower surface and a thicker one on the upper surface. The anterior two-thirds of the tongue is studded with papillae, mainly of three types: filiform, fungiform and vallate. The latter two contain taste buds. The senses of touch, pain and temperature are monitored by the trigeminal nerve whereas the facial, glossopharyngeal and chorda tympani innervate the taste buds.

In humans, some 95% of total saliva volume is secreted by three major sets of glands which are situated at some distance from the oral cavity and connected by extraglandular ducts. The largest, the parotids, are situated below and in front of the ears between the skin and the masseter muscle. A serous secretion is released through Stensen's ducts which emerge opposite the upper second molars. The submandibular glands are located on the floor of the mouth and have a mixed mucous and serous secretion released beneath the tongue via Wharton's ducts. The smallest major glands, the subman-sublingual, lie superior to the submandibular and produce a mainly mucous secretion. The sublingual glands connect with the oral mucosa via ducts which join those of the submandibular glands and numerous small ducts which open onto the floor of the mouth.

Many minor glands open into the oral cavity either directly or via short ducts. These include the labial, lingual (including Von Ebner's), palatal, buccal and pharyngeal. The majority of the taste buds are sheltered in the epithelial folds of the circumvallate and foliate papillae where saliva is supplied predominantly by Von Ebner's gland. It has therefore been suggested that Von Ebner's gland and its secretions might be more important in modulating the perireceptor events and hence transduction of primary tastes than has previously been appreciated (Schmale *et al.*, 1993).

Salivary proteins and glycoproteins

The total solids content of human saliva is said to be only 0.5% w/v of which 0.2% w/v is inorganic constituents and the balance organic, mainly protein.

Bennick (1982) reports protein concentrations of 1.0–3.5 mg/ml. In humans approximately 70% of these are constitutive proline-rich proteins and glycoproteins (PRP and PRG) but there is a general belief, apparently derived primarily from animal (rodent) studies, that the concentration reflects the approximate level of polyphenols in the diet. In rodents such as rats and mice it has been demonstrated that tannins mimic β-agonists and induce parotid hypertrophy and increase synthesis of salivary PRP. This effect of tannins can be abolished by β-antagonists, but the exact mechanism whereby the tannins achieve the stimulation is not known and the author has never seen data that unequivocally demonstrate this effect in humans. These PRP and PRG have a repetitive primary structure consisting usually of 21 amino acid residues in which there are substantial amounts of proline, glycine and glutamine/glutamic acid, few aromatic, basic or sulphur-containing residues and no hydroxyproline. These proteins are usually subdivided into acidic, basic, glycosylated and low molecular mass (< 3000 Da) subgroups.

Acidic PRP

According to Bennick (1982), in humans the acidic PRP account for about 30% of total salivary protein, i.e. about 40% of total PRP. Originally four proteins were reported, A, B, C and D. A and C are now known to be the two major acidic proteins accounting for about 28% of parotid PRP (Bennick *et al.*, 1979). Protein A (11 945 Da) contains 106 residues (24 proline). There are 13 proline residues among the 32 residues in the C-terminal region but with a maximum of 4 in succession (Wong *et al.*, 1979). There is a negatively charged N-terminal region (11 dicarboxylic acids, 2 phosphoserines, 1 phenylalanine) which can be released by tryptic digestion as the TX peptide.

Protein C contains 150 residues. The N-terminal region embraces the 106 residues of Protein A. The additional 44 C-terminal residues are rich in proline, glycine and glutamine. It includes a hexaproline sequence and five repeats of -Pro-Gln-Gly-Pro-Pro- (Wong *et al.*, 1979). Thrombin and salivary kallikrein slowly convert Protein C to Protein A plus a CTZ peptide. Mouse submaxillary gland protease achieves the same conversion much more rapidly and it is thought that this protease may be important in postribosomal transformation of acidic salivary PRP (Wong and Bennick, 1980).

Isoforms of A and C are known in which the glutamine residue at position 5 is replaced with glutamic acid. The existence of other similar isoforms is suspected. Protein B is now known to be identical with statherin, an acidic phosphoprotein (5380 Da; pI=4.2) which is relatively rich in tyrosine, proline and glutamic acid and Protein D is one of a family of five histidine-rich proteins (HRP; histatins) produced by the submandibular and parotid glands (Sabatini *et al.*, 1993). The biological activities of the submandibular PRP are thought to be associated with the N-terminal 30 residues (common to both A and C). These are:

(a) calcium binding;
(b) hydroxyapatite binding; and
(c) inhibiting formation of hydroxyapatite from more basic calcium phosphate.

This suggests a major role in maintaining the integrity of the teeth. Sequence analysis suggests a distant relationship with the troponin superfamily of proteins.

The submandibular PRP bind calcium strongly ($K_d = 2 \times 10^{-4}$). There are 6–7 sites in Protein A but only 3 in Protein C, probably because the longer polypeptide chain of C allows the positively charged N-terminal residues to interact with the negatively charged calcium-binding sites. The phosphopeptides referred to above (Madapallimattam and Bennick, 1990) are more potent inhibitors *in vitro* of hydroxyapatite formation than their acidic PRP precursors, and bind to tooth enamel *in vivo*.

Mehansho *et al.* (1983) have suggested that an additional function of submandibular PRP is to bind dietary tannin, consistent with a random rod-like structure (axial ratio >25) as judged by CD spectroscopy (Bennick, 1982).

Basic PRP

According to Bennick (1982), submandibular PRP form about 23% of total human salivary protein, i.e. some 30% of total PRP, whereas after induction in rats by exposure to a tannin-rich diet (or β-agonists such as isoprenaline) they represent some 90% of the total. Proline, glutamine and glycine account for some 80% of all amino acid residues but proline alone accounts for some 40% compared with some 25% in the submandibular PRP. Some amino acid sequences in the submandibular PRP are common to the parotid PRP but their induction by tannin suggests that the submandibular PRP might be particularly important in binding tannins, at least in rats. All submandibular PRP have C-terminal arginine, suggesting kallikrein processing of common precursor(s). According to some workers, carbohydrate may be present as glucosamine and galactosamine. At least nine submandibular PRP have been isolated from human parotid saliva.

The six submandibular PRP secreted by isoprenaline-treated rats have a p*I* >10 and masses in the range 15–18 kDa. These contained no phosphate and very little carbohydrate and had a random rod-like structure (axial ratio >25) as judged by CD spectroscopy (Muenzer *et al.*, 1979). So far as the author is aware, tannin binding is the only function proposed for the basic PRP.

Mucins and proline-rich glycoproteins (PRG)

Mucins are high-mass glycoproteins. The PRG found in human submandibular and sublingual saliva are characterized by a protein core (apomucin) enriched with serine, threonine, proline, glycine and glutamic acid and *O*-glycosidically (serine or threonine) linked galactose, fucose, *N*-acetyl-glucosamine and/or sialic acid and *N*-glycosidically linked mannose. The 'naked' regions are enriched with hydrophobic amino acids and cysteine. There are similarities between salivary mucins and intestinal mucins.

According to Bennick (1982), PRG account for some 17% of total salivary protein in humans, i.e. some 25% of total PRP. A PRG having substantial glycosylation (36 400 Da; 40% carbohydrate) has been isolated from human parotid saliva (Levine *et al.*, 1969). IEF separated this into several bands having p*I* > 9. Spectroscopic examination (CD; NMR) of eight fragments from human PRG indicate poly-L-proline II helix as the predominant conformation (Loomis *et al.*, 1991).

Oho *et al.* (1992) have isolated from human parotid saliva a PRG with an apparent mass by SDS–PAGE of 78 kDa, 34% carbohydrate (rich in *N*-linked mannose and *N*-acetyl-glucosamine) and a proline-rich (proline, glycine and glutamine = 77%) core protein of 51 kDa. This PRG binds strongly to hydroxyapatite and to a protein from submandibular saliva having an apparent mass of 95 kDa. A low-mass mucin, which is less glycosylated and contains predominantly mannose residues has been isolated from human submandibular and sublingual saliva in two isoforms (MG2a and MG2b) differing primarily in their sialic acid and fucose contents (Reddy *et al.*, 1992; 1993).

Rathman *et al.* (1990) isolated from submandibular, sublingual and minor salivary glands a PRG which is barely detectable in the parotid secretions and which accordingly was named extraparotid glycoprotein (EPGP). EPGP behaves as if it were 40 kDa on gel filtration but separates into 19 kDa and 20 kDa fractions on SDS–PAGE and at least four protein bands using IEF (p*I* = 4.8; 5.0; 5.2; and 5.4). EPGP contains approx 13% carbohydrate wherein most of its heterogeneity resides. EPGP is found also in seminal plasma, tears, nasal mucus and sweat (Schenkels *et al.*, 1991) and resembles the secretory actin-binding protein known also as prolactin-inducible protein or gross cystic disease fluid protein-15. Its function in mucosal secretions has not been determined, but binding to hydroxyapatite and micro-organisms have been observed. EPGP is quite distinct from the PRG secreted by Von Ebner's gland (Schenkels *et al.*, 1994).

According to Blaker *et al.* (1993) the major secretion of human Von Ebner's gland is an 18 kDa protein with considerable homology to two proteins secreted by the rat (Kock *et al.*, 1994*a*). These Von Ebner's gland proteins (VEGP) belong to the lipocalin superfamily of hydrophobic ligand-binding proteins (which includes rat odorant binding protein II, retinol binding protein and β-lactoglobulin). Kock *et al.* (1994*b*), by gene transfer, have introduced the rat VEGP to the mouse which does not naturally secrete them. Transgenic mice possessing the proteins have much greater tolerance of the bitter substance denatonium benzoate than mice lacking these proteins, and this has led to the suggestion that VEGP transport the bitter compound away from the bitter receptor although earlier it had been suggested that VEGP might transport bitter substances to the receptor. However, according to Spielman (1990) attempts to demonstrate binding of PRP to bitter substances such as quinine using equilibrium dialysis have failed. The possibility that salivary PRG in general and VEGP in particular have a role in astringency will be explored below.

The rat parotid secretes constitutively a 200 kDa glycoprotein containing 23 repeats of [-Asp-Gln-Gly-(Asn)-Gln-Thr-Gln-Pro-Arg-Pro-Pro-His-Pro]. This is identical to the core of the isoprenaline-induced rat submandibular 158 kDa glycoprotein which differs only in its glycosylation (Miao *et al.*, 1995).

Low molecular mass PRP

Human saliva contains a large number of phosphopeptides derived from acidic PRP after secretion from the gland. These represent 0.5% of the acidic PRP in parotid saliva but 11% of the acidic PRP in whole saliva (Madapallimattam and Bennick, 1990), implying that proteolysis occurs after secretion. Perinpanayagam *et al.* (1995) characterized the low molecular mass peptides (< 3 kDa) of human saliva by amino acid sequencing and concluded that probably they were derived by proteolysis of the previously characterized PRP.

Inter- and intraperson variations in saliva composition

The salivary PRP are coded by six genes that span approximately 700 kb of DNA. Four of the PRP genes (PRB1–4) code for basic and glycosylated PRP and two (PRH1 and PRH2) for submandibular PRP. The PRP genes have been localized on chromosome 12p13.2 (O'Hanlon *et al.*, 1988; Azen, 1993). There is growing evidence for many PRP phenotypes (Hay *et al.*, 1994) and polymorphisms (Schlesinger *et al.*, 1994) including ethnic variations (Shintani *et al.*, 1990; Azen, 1993) between Caucasians, Chinese, Indians, Japanese and Malays. Among the submandibular PRP these may involve an unusual but well-controlled post-translational proteolysis affecting only a single amino acid residue and specific proteolysis may also produce many submandibular PRP from precursor proteins, but either of these events may be prevented by specific mutations. A mutation leading to single cysteine substitutions in PRP may lead to disulphide homo- or heterodimers through interaction with salivary peroxidase.

There would appear also to be other biochemical and physiological factors able to superimpose further variations. For example, our own studies have recorded whole saliva protein contents generally in the range 1.0–3.5 mg/ml but with isolated values as high as 5.6 mg/ml in one male subject and 8.2 mg/ml in one female subject. In an attempt to assess whether dietary tannin level does modulate the secretion of salivary protein samples of whole saliva provided by Mormon adults and children were compared with equivalent samples from non-Mormon adults and children. For free-living non-Mormon adults the values ranged from 1.7 to 3.1 (8.2) mg/ml. The mean value for non-Mormon children was 0.45 mg/ml. The equivalent values for Mormon adults and children were mean values of 0.5–0.8 mg/ml. In view of the small number of volunteers participating in this study one must be cautious in extrapolating to the wider population. Nevertheless, it is interesting to note that for Mormon adults and children who eschew tannin-containing foods and beverages, the saliva protein contents were similar and, although similar to the values for non-Mormon children (also on low-tannin diets), were much lower than those for non-Mormon adults who were known to consume tea, beer, wine, etc. If it were to be established that chronic exposure to tannins did indeed induce PRP production in humans it might follow that professional tea and wine assessors are somewhat less sensitive to astringency than the customers for whom they are grading. A study of three female volunteers over two menstrual cycles each has demonstrated considerable variations in saliva protein content, protein profile by RP-HPLC and pH value but has failed to establish any cyclical pattern (see Figs 5.1–5.3). It is not possible to exclude a cyclical variation, but if it exists, then it seems as though other factors dominate.

The pH value of saliva is greatly influenced by the bicarbonate content and bicarbonate secretion is dependent on stimulation by food. Acids in particular increase bicarbonate secretion. Studies with pilocarpine-stimulated rats have given saliva pH values in the range 8.1–9.2 which after dialysis to remove the bicarbonate had values of 6.1–6.5 (Matsuo and Yamamoto, 1992). Clifford and Sproule (unpublished data) working with free-living volunteers have measured values in the range 4.12–8.61. The significance of such variations is not clear, but might influence the extent to which phenols and tannins are oxidized in the mouth, and the extent to which PRP are precipitated.

In view of the clearly demonstrated effects of β-agonists and β-antagonists on the secretion of salivary PRP in rodents, one must assume that the clinical use of such drugs will have similar effects in humans. This point does not, however, seem to have been investigated, but could be of some significance, since β-antagonists are extensively prescribed.

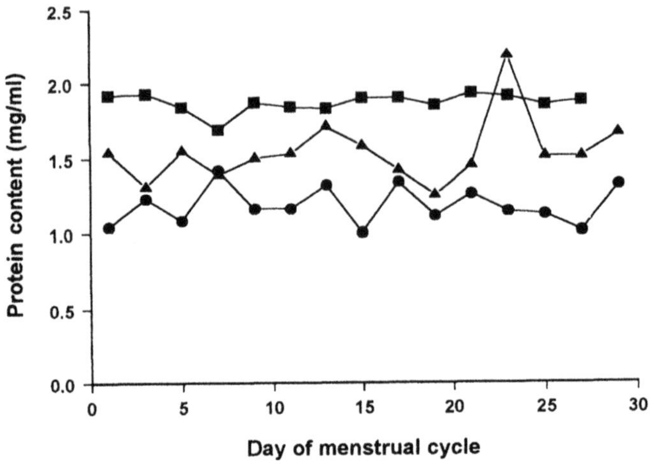

Fig. 5.1. Variation in saliva protein content (mg/ml as measured with Biuret using BSA as standard) over one menstrual cycle for each of three volunteers.

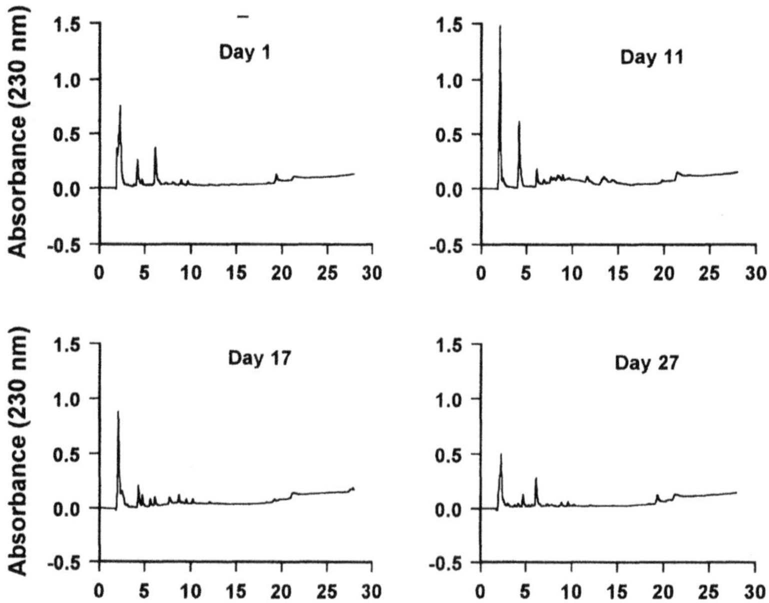

Fig. 5.2. Variation in the reversed phase chromatographic profile of saliva proteins over one menstrual cycle for one volunteer.

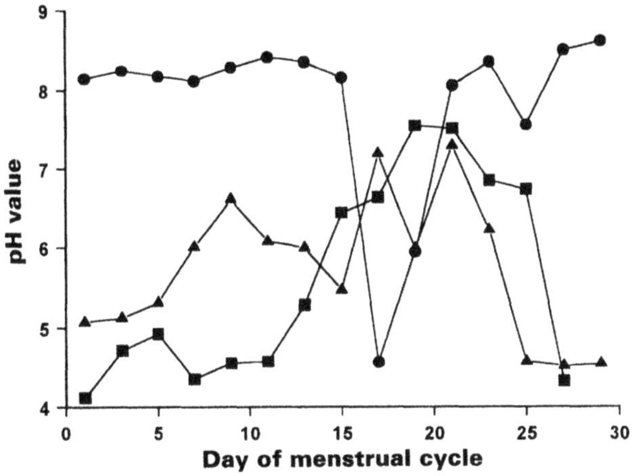

Fig. 5.3. Variation in saliva pH value over one menstrual cycle for each of three volunteers.

Astringent substances

Although alums may be useful investigative probes and a convenient reference standard they do not normally occur in food or beverages and there can be no doubt that the classic dietary astringents are the vegetable tannins. Tannins have been defined as naturally occurring, structurally complex, water-soluble secondary metabolites having molecular masses in the range 500 to at least 3–4000 and which possess 12–16 phenolic hydroxyls and 5–7 aromatic rings per 1000 relative molecular mass (Haslam, 1993), but there is now strong evidence that some condensed tannins may be larger (Prieur *et al.*, 1994; Helfer and Mattice, 1995). The distinctive structural feature of a tannin responsible for its astringency is said to be 'the accumulation, within a molecule of moderate size, of a substantial number of phenolic groups, many of which are associated with a 1,2-dihydroxy or 1,2,3-trihydroxy orientation within a phenyl ring' (Haslam, 1981). Commercial tannic acid is the tannin most often adopted as a standard because of its ready availability, but unfortunately, the commercial product varies in composition with biological origin, commercial source and batch (Makkar *et al.*, 1993). Purified β-1,2,3,4,6-penta-*O*-galloyl-D-glucose is to be preferred, but is structurally distinct from the tannins most commonly found in foods and beverages and there is no entirely satisfactory standard readily available.

The structure of tannins found in the human diet is based on two broad themes (Haslam, 1993). One of these, the ellagic acid esters of glucose which are known as ellagitannins, are found in, e.g. pomegranate, persimmon, chestnuts and fruits of Rosaceae. The second group is based structurally on the flavanol skeleton and consists of two broad subclasses:

- the condensed tannin oligomers and associated monomers which are found in most fruits, legumes and pulses, beer, cider, wine and some spices (Clifford, 1986);
- flavanol-derived pigments produced by enzymic (polyphenol oxidases, peroxidases), coupled oxidations and condensation reactions; these are known also as complex phenols or derived tannins and are found particularly in red wine, black tea, cider, beer, etc. (Robertson, 1992; Opie *et al.*, 1993).

Significant advances have been achieved recently in both the isolation and characterization of these complex phenols (Bailey *et al.*, 1992; Powell *et al.*, 1994; Powell, 1995) and recent studies at the University of Surrey suggest that the thearubigins of black tea have masses in the range 1500–2500 Da (Powell, 1995) but so far little is known about their relative astringency.

Many workers suggest that cross-linking of salivary proteins is essential for their efficient precipitation, and thus for a molecule to be astringent it must contain at least two di- or trihydroxyphenyl moieties. Increasingly, taste panel studies are producing evidence that molecules possessing only one dihydroxyphenyl moiety, e.g. 5-caffeoylquinic acid (Naish *et al.*, 1993), (+)-catechin and (–)-epicatechin (Kallithraka *et al.*, 1995; Thorngate and Noble, 1995) are also astringent. Moreover, there are reports of astringency elicited by non-phenolic acids, e.g. tartaric or citric acids (Guinard *et al.*, 1986*a*; Lawless *et al.*, 1994) and by certain dipeptides (Kirimura *et al.*, 1969) and it is now clear that sensory astringency is not as close to the leather tanning phenomenon as once thought. This point will be discussed further below.

Frequently tasters when presented with a classic astringent also use terms such as ‘bitter’, ‘sour’ and ‘metallic’, or even ‘like sucking a brass door knob’ to describe the resultant sensation. More recently, using trained taste panels, the astringent sensation has been subdivided into rough, dry and puckery notes (Lawless *et al.*, 1994) and it has been suggested that these three notes are generated independently (Kallithraka *et al.*, 1995).

PRP–tannin complexation

Extensive studies by Haslam and colleagues have done much to enhance our understanding of the factors controlling protein–tannin interactions and the significance of these interactions for astringency. Such studies have been reviewed by Luck *et al.* (1994) from which the following key points are extracted. Proline is a unique amino acid in which the side chain has undergone ring closure. Accordingly, proline residues in proteins place greater restrictions than any other amino acid on the conformations which can be adopted immediately adjacent to a proline residue(s) with the result that proline-rich proteins have much more open and flexible structures (axial ratios $>25:1$) than typical globular proteins. This makes the peptide bonds more readily accessible to di/trihydroxyphenyl residues in tannins and thus increases the potential for complexation by some two orders of magnitude. Elegant NMR studies (Murray and Williamson, 1994; Murray *et al.*, 1994) using synthetic proline-rich peptides corresponding to the repeat sequence of the salivary PRP have demonstrated that isolated proline residues or the N-terminal residue in a sequence of prolines have the greatest affinities and may be considered as 'sticky patches' which capture the tannin.

The association of PRP and tannin is entropic, being driven by hydrophobic forces, i.e. the hydrophobic regions of the receptor and ligand associate in such a manner as to exclude from their surfaces the layers of partially ordered water. Subsequently hydrogen bond networks are formed cooperatively and directed from the tannin towards the carbonyls of prolyl peptide groups. If the complex is sufficiently hydrophobic, or becomes so following aggregation, then it precipitates. The tannin–protein precipitates formed by many proteins, including gelatin which is itself proline-rich, can be resolubilized in the presence of excess protein. In contrast, those formed by salivary PRP are not readily resolubilized in the presence of excess salivary PRP (Haslam *et al.*, 1992). This observation could be important with reference to the tendency for astringency to increase on repeated exposure to the stimulus despite the fact that there is an increased salivary flow (Guinard *et al.*, 1986*b*; Lawless *et al.*, 1994), although it should not be assumed that the increased flow necessarily indicates an increased supply of PRP therein. The lack of this information is a rather serious deficiency in our knowledge.

Asquith *et al.* (1987) using 158 kDa rat submandibular PRG have demonstrated that the carbohydrate residues not only increase the affinity of the protein core for tannins but also the solubility of the complex. Since the carbohydrate residues *per se* after hydrolysis and purification had no affinity for the tannins it was assumed that the increased affinity was a function of effects on the conformation of the protein core. The increased solubility of the complexes was attributed to increased hydrophilicity/reduced hydrophobicity and possibly electrostatic repulsion associated with the sialic acid residues. The relative PRG content of the 'typical proline-rich protein preparation from human saliva' used by Haslam *et al.* (1992) is not known. In retrospect it seems important to study the interaction of tannins and PRG with and without the admixture of PRP.

Factors modulating tannin–PRP precipitation

A number of studies have demonstrated that individual tannins differ in their ability to precipitate one or other proteins or in the sensation of astringency that they produce. For example it has been observed that tannins that are relatively soluble in water, e.g. the ellagitannins vescalagin or castalagin are much poorer precipitators than the comparatively water-insoluble β-1,2,3,4,6-penta-*O*-galloyl-D-glucose. It is presumed that poorer solubility in water corresponds to greater hydrophobicity and a greater potential for hydrophobic interaction with the protein. However, the ellagitannins generally and the examples quoted above in particular, lack conformational flexibility relative to the gallotannins and this fact also impairs the ability of the ellagitannins to precipitate proteins. The condensed tannins also have structures of low flexibility due to the restrictions imposed by the 4→6 and/or 4→8 interflavan bonds and generally are considered to be less astringent than the hydrolysable tannins for a given number of di/trihydroxyphenyl binding sites (Beart *et al.*, 1985; Spencer *et al.*, 1988).

As a generalization, it has been said that, subject to being soluble in water, tannins of higher mass and/or having more di/trihydroxyphenyl groups per molecule, are better able to cross-link, are more efficient precipitators of protein and are more astringent. However, in the gallotannins, protein precipitation ability increases progressively from two to five galloyl residues and declines again thereafter due possibly to steric hindrance. For (−)-epicatechin-derived condensed tannins, Lea and Arnold (1978) reported that the astringency increased progressively from the dimer up to at least the heptamer. This series of oligomers would adopt a rigid helical structure in solution (Helfer and Mattice, 1995). It is interesting to note that the progressively increased astringency does not parellel the relative hydrophobicity as judged by the reversed phase HPLC capacity factors (EC = 0.875; EC-4β→8-EC = 0.856; EC-4β→8-EC-4β→8-EC = 0.877; EC-4β→8-EC-4β→8-EC-4β→8-EC = 0.879) (Santos-Buelga *et al.*, personal communication) suggesting that there are complexities here which at present are not fully appreciated. Haslam *et al.* have shown that flavanol gallates have a much greater ability to precipitate caffeine and/or proteins than the corresponding non-gallated flavanols. Since the gallated flavanols have two di/trihydroxyphenyl residues, thus making them similar to dimeric condensed tannins, this is not surprising, but it is interesting to note that the majority of the interaction occurs via the gallate ester rather than via the B-ring (Spencer *et al.*, 1988). A similar pattern of behaviour has been observed for theaflavin and theaflavin mono- and digallates from black tea (Powell *et al.*, 1992; Luck *et al.*, 1994) and a portion of the thearubigins (Powell *et al.*, 1992). Some smaller phenols (e.g. methyl gallate, 5-CQA) also form complexes with caffeine and some at least precipitate readily. The tertiary amide structure of caffeine resembles that of proline and the mechanisms of interaction are identical (Haslam *et al.*, 1992).

Other proteins, and especially other proteins with hydrophobic regions and/or a greater than usual proline content, e.g. K-casein, have significant affinitiy for tannins. If these proteins are hydrophilic, e.g. K-casein or extensively ionized, e.g. α_s-caseins or β-casein, they may sequester tannins and form water-soluble complexes of low astringency, and this is the case when milk is added to black tea beverage (Luck *et al.*, 1994). The behaviour of the phosphoproteins is influenced significantly by cations, especially Ca^{2+}, even at physiological concentrations (0.012 M). The phosphoproteins of milk and the casein complex itself, are much more extensively precipitated by β-1,2,3,4,6-penta-*O*-galloyl-D-glucose in the presence of 0.012 M Ca^{2+} than in distilled water (Luck *et al.*, 1994) thus partially removing the caseins' protective effects. Since the submandibular PRP in saliva bind to teeth it is reasonable to assume that the concentration of Ca^{2+} in the mouth, whether from saliva or food/beverage, might have some influence on the strength of the astringent sensation. This possibility does not seem to have been investigated in taste panel studies.

Some polysaccharides, including xanthan, gellan, gum arabic, and soluble pectin, which may occur naturally in foods and beverages or be incorporated as additives, are able individually or in mixtures to inhibit the precipitation of proline-rich gelatin by β-1,2,3,4,6-penta-*O*-galloyl-D-glucose (Luck *et al.*, 1994). The extent to which, for example, pectin modulates astringency in wine, cider, etc. seems not to have been investigated, but some effect could be anticipated and investigation seems desirable.

Although for leather tanning multiple di/trihydroxyphenyl residues are essential, it is now clear that this is not the case for the astringent sensation as judged by trained taste panellists. Phenols possessing only one dihydroxyphenyl residue, e.g. (+)-catechin and (–)-epicatechin (Kallithraka *et al.*, 1995; Thorngate and Noble, 1995), 5-*O*-caffeoylquinic acid (Naish *et al.*, 1993), and, according to Peleg and Noble (1995) even some phenols containing two isolated hydroxyls (gentisic acid) or a single hydroxyl (salicylic acid) have distinct astringency. While simple phenols at high concentration such as pyrogallol (1 molal; 126 g/kg) are capable of precipitating BSA (McManus *et al.*, 1981), the levels at which the astringency of simple phenols has been detected (5-CQA = 0.7 g/l; (+)-catechin and (–)-epicatechin 0.5 g/l; benzoic acids 2–3 g/l (17 mM) are probably much too low to precipitate PRP from the dilute solutions normally secreted, but observation of red shifts in λ_{max} suggest that they do bind (Naish *et al.*, 1993). Dicaffeoylquinic acids (diCQA; 516 Da) have two 1,2-dihydroxyphenyl residues and are thus able to cross-link. They are more astringent than 5-CQA as judged by triangle tests and by their ability to precipitate haemoglobin. DiCQA–CQA mixtures are less astringent to panellists and less able to precipitate haemoglobin although there is obvious denaturation of the protein and the supernatant is muddy brown rather than clear red. The author rationalized these observations by suggesting that both CQA and diCQA bind to the protein, but, as the proportion of cross-links decline relative to the number of distal carboxylate anions, so the propensity to precipitate declines.

As discussed in a previous publication dealing specifically with 5-CQA but more or less relevant to other simple phenols (Naish *et al.*, 1993), the prospect of cross-linking is much reduced. In seeking an alternative explanation, the possibility of 5-CQA being oxidized in the mouth, and of the oxidation products having some role in the sensations generated, was considered. Since then, it has been noticed (Clifford *et al.*, unpublished data) that mixtures of 5-CQA and saliva go yellow and the 5-CQA is partially isomerized (yielding 3-CQA and 4-CQA) and hydrolysed (yielding caffeic acid and presumably quinic acid although this was not detected on the chromatograms). The mechanisms are not certain, although high salivary pH values as sometimes encountered (see Fig. 5.3) would be expected to facilitate isomerization, hydrolysis and oxidation. A role for the little-studied salivary peroxidase, found at a few micrograms per millilitre in human saliva (Mansson-Rahemtulla *et al.*, 1988), also deserves consideration since peroxidases generally are of low specificity and at least some isoforms have the ability to oxidize phenols.

At present no definite explanation can be offered to account for these astringent sensations or for those elicited by certain γ-L-glutamyl dipeptides (Kirimura *et al.*, 1969) or by tartaric and citric acids (Guinard *et al.*, 1986*a*; Lawless *et al.*, 1994). Since the panellists were all trained to recognize astringency as generated by accepted, traditional astringents it seems unlikely that simple confusion was responsible.

Despite the diversity of structure among these small astringent compounds, a feature common to all is their ability to elicit one or more basic tastes and other notes in addition to the astringent sensation. For example the astringent dipeptides were also sour or sour and bitter (Kirimura *et al.*, 1969); the benzoic acids were also sweet, sour and bitter (Peleg and Noble, 1995); (+)-catechin and (–)-epicatechin were bitter (Kallithraka *et al.*, 1995; Thorngate and Noble, 1995); 5-CQA was bitter and sour (Kellard *et al.*, 1987); and tartaric and citric acid were sour (Guinard *et al.*, 1986*a*; Lawless *et al.*, 1994). Even some traditional astringents, e.g. oligomeric condensed tannins, are noticeably bitter with bitterness peaking at the tetramer whereas astringency increased at least up to the heptamer (Lea and Arnold, 1978) and it seems reasonable to assume that the bitterness arises through binding to the bitter receptor and some interaction with Von Ebner's gland protein. A recent and completely unexpected observation is that a trimeric condensed tannin Seligueain A ((–)-epiafzelchin-4β8, 2β*O*7-(–)-epiafzelchin-4β8-(+)-

afzelchin) is sweeter than sucrose (Baek *et al.*, 1993). The ability to bind, even poorly, to the bitter or sweet receptor could explain the observations reported by Schiffman *et al.* (1991) of stimulation of the chorda tympani nerve, i.e. it is recording the bitter element or other basic taste as appropriate, rather than the astringency *per se*. So far as the author is aware, this is the first report of the taste associated with a derivative of (epi)afzelchin and it should be noted that such compounds lack the di/trihydroxyphenyl B-ring characteristic of the typical condensed tannins. Although much less studied, screening by HPLC for pelargonidin after using Porter's reagents suggests that these monomers might occur more extensively than once thought (Powell *et al.*, 1995).

Sourness is almost certainly associated with the titratable acidity of these substances and the amount of bicarbonate required to restore the resting pH value. Incidentally, an overshoot, elevating the pH value, would predispose to the oxidative changes referred to above.

The other notes observed include metallic associated with 5-CQA (Kellard *et al.*, 1987) and more recently dryness, roughness and puckeriness associated with tannic acid, tartaric acid and alum (Lee and Lawless, 1991) and it was considered that they might be independent of astringency. Our studies (Kallithraka *et al.*, 1995) with (+)-catechin and (–)-epicatechin in model solutions and added to wine support the suggestions made by Lee and Lawless (1991) with taste panellists detecting dryness and roughness (but not puckeriness). The intensity of dryness did not parallel the intensity of astringency, suggesting that dryness contributes to astringency but is not subsumed within it.

As an aside, it is interesting to note that foods and beverages rarely contain gallotannins. Ellagitannins, although more common, are still relatively restricted, being found in, for example, pomegranate, persimmon, chestnuts and fruits of Rosaceae. The commonest dietary tannins by far are those based on the flavanol skeleton, i.e. condensed tannins and particularly the poorly characterized derived tannins, which, so far as one can judge, are the least astringent on a mass basis. Is this a coincidence, or have the gallotannin-rich commodities been eschewed as human food, leaving us with the more palatable materials?

Sensory studies of astringency in beverages or complex model systems

There can be no doubt that studies using model systems, and/or in which taste panels report on the sensation associated with ingesting isolated pure tannins, have been extremely helpful in leading to our present level of understanding and doubtless will continue to be important in furthering it. However, it must not be forgotten that there is a need to understand the response of consumers to chemically complex foods and beverages in real-life situations. This immediately complicates the issue. Substances other than tannins will be ingested simultaneously, and obviously any effects due to these other substances must be considered.

Caffeine present in green tea complexes with, and possibly precipitates, the astringent flavanol gallates. In black tea it complexes with theaflavin gallates and some thearubigins, precipitating a significant proportion as 'tea cream' (Powell *et al.*, 1992). Not only does the precipitation remove from the beverage a fraction of the astringent substances, but the soluble complexes are thought to be less astringent than the free phenol and less bitter than the free caffeine, though there is scope for further systematic work in this area.

Recent studies, particularly those by Pangborn, Noble and colleagues in California have begun to shed some light on these complexities with reference to wine. Increasing the content of tartaric acid in model wine solutions (containing varying levels of tannic acid, sugar and ethanol) or increasing the citric acid content of white wines (to avoid complications associated with the precipitation of potassium hydrogen tartrate) produced (Guinard *et al.*, 1986*a*) significant increases in astringency. It was suggested that this increase arose because at lower pH values there were more unionized phenolic groups (less phenolate anions) favouring increased binding to salivary proteins. In view of the pK_a of phenolic hydroxyls (>pH 8.0) and the pH range considered (2.6–3.8) the increase in the magnitude of the common ion effect seems too small to account for the increase in astringency observed. An alternative explanation might be an increased precipitation of the submandibular PRP as their pI is approached. Any reduction in precipitation of the submandibular PRP ($pI > 10$) would be marginal in this pH range. Lowering the pH value of red wines which had a greater tannin content than the white wines had no detectable effects on perceived astringency possibly because any increase was less than the minimum detectable (just noticeable difference) by the panellists.

Similar studies in which the sucrose concentration was varied demonstrated significant reductions in the maximum intensity and total duration of astringency as sucrose concentration was increased from 0 to 200 g/l at two levels of wine tannins (Ishikawa and Noble, 1995) confirming observations made for the astringency of alum (Breslin *et al.*, 1993) and for the mouth drying effect and bitterness of tannic acid (Lyman and Green, 1990). Ishikawa and Noble (1995) also established clear differences in the perception of astringency by volunteers differing in saliva flow rate ($\approx$1–4 ml/min). Those with low flow rate reported greater astringency and greater duration of the astringent aftertaste than those with higher flow rates. It was presumed, but not demonstrated unequivocally, that a greater flow of saliva corresponded to an increased supply of salivary PRP and that this was sufficient to maintain a greater level of lubrication. Spielman (1990) has reported that 'In a recent study concerning stimulation (species not stated) with sweet, sour, salty and bitter compounds the relative proportion of salivary proline-rich proteins was unaffected by the nature of the taste stimuli' but gives no data for the absolute content.

In an earlier study, the total duration of astringency (Y) was observed to increase exponentially ($Y = ae^{bx}$) with repeated ingestion, the magnitude of the coefficient (b) increasing as a function of the tannin content of the wine (Guinard *et al.*, 1986*b*). In this study the increased flow of saliva failed to counteract the additional astringent ingested on each repeat but once again no data are available for the PRP content of the saliva. Mice in which PRP synthesis has been induced by chronic treatment with the β-agonist isoproterenol are much more tolerant of tannic acid in their drinking water (Glendinning, 1992). Hagermann and Robbins (1993) have observed that mammals accustomed to a diet rich in, for example, linear condensed tannins produce salivary proteins better able to bind linear condensed tannins than, for instance, ellagitannins in *in vitro* assays and are less willing to accept ellagitannin-containing browse. It is possible therefore that increased secretion in humans might also reduce sensitivity to astringency, at least where the astringents are of a type to which exposure has occurred previously. This is somewhat at odds with the established belief that PRP–tannin interactions are relatively non-specific. In future studies of astringency, salivation, etc. in humans it is important that protein content, degree of glycosylation, pH value, peroxidase activity and PRP-binding specificity be more closely examined.

Perspectives

Over the last twenty years, a lot has been learned about the peculiar salivary proteins, and about the interactions of tannins and other macromolecules. In what might be described as the first phase of these studies much use has been made of model systems. The knowledge so gained improved our understanding of the interaction of salivary proteins and tannins, and hence of astringency in fruits and vegetables. The second phase has seen a move to sensory studies, sometimes with animals but mostly with trained volunteers and it is now clear that although astringency has certain features in common with the tanning of hides to produce leather, there are also significant differences. These studies also have highlighted some surprising gaps in our knowledge. The next phase must address these. It would, for example, be helpful to know:

- how tannins/tannin-rich preparations activate the β-receptor
- how small mass phenols and non-phenols elicit astringency
- whether prescribed medication influences the secretion of PRP in humans

- whether increased saliva flow is accompanied by increased secretion of salivary PRP, and if so, under what conditions
- whether exposure to tannins influences the extent to which PRP are secreted in humans
- whether the extent of variation in the PRP : PRG and PRP : VEGP ratios influence the intensity or temporal aspects of astringency or bitterness
- whether there is a role for salivary peroxidase and phenol oxidation in eliciting astringency
- whether in complex foods and beverages interactions between Ca^{2+} and caseins, Ca^{2+} and pectins, pectins and tannins or caffeine and tannins influence the perception of astringency.

Acknowledgements

The author is grateful for the assistance he has received from his many co-workers. Dr Ann Noble at the University of California is thanked for kindly providing access to her data base of references and for supplying copies of her own papers in advance of publication. Dr C. Santos-Buelga and colleagues at the University of Salamanca are thanked for determining RP-HPLC capacity factors of certain wine tannins.

References

Asquith, T. N., Uhlig, J., Mehansho, H., Putman, L., Carlson, D. M. and Butler, L. G. (1987). *J. Agric. Food Chem.* **35**, 331–4.

ASTM (1989). Standard definitions of terms relating to sensory evaluation of materials and products. In *Annual Book of ASTM Standards, Vol. 15.07*, p. 2. American Society for Testing and Materials, Philadelphia.

Azen, E. A. (1993). *CRC Crit. Rev. Oral Biol. Med.* **4**, 479–85.

Baek, N. I., Chung, M. S., Shamon, L., Kardono, L. B., Tsauri, S., Padmavinata, K., Pezzuto, J. M., Soejarto, D. D. and Kinghorn, A. D. (1993). *J. Nat. Prod.* **56**, 1532–8.

Bailey, R. G., Nursten, H. E. and McDowell, I. (1992). *J. Sci. Food Agric.* **59**, 365–75.

Beart, J. E., Lilley, T. H. and Haslam, E. (1985). *Phytochemistry* **24**, 33–9.

Bennick, A. (1982). *Mol. Cell. Biochem.* **45**, 83–99.

Bennick, A., Cannon, M. and Madapallimattam, G. (1979). *J. Biol. Chem.* **256**, 4741–6.

Blaker, M., Kock, K., Ahlers, C., Buck, F. and Schmale, H. (1993). *Biochim. Biophys. Acta* **1172**, 131–7.

Breslin, P. A. S., Gilmore, M. M., Beauchamp, G. K. and Green, B. G. (1993). *Chem. Senses* **18**, 405–7.

Clifford, M. N. (1986). In *Interactions of Food Components* (ed. G.G. Birch and M.J. Lindley), pp. 143–64. Elsevier Applied Science, London.

Fischer, U., Boulton, R. B. and Noble, A. C. (1994). *Food Quality and Preference* **5**, 55–64.

Glendinning, J. I. (1992). *Chem. Senses* **17**, 1–12.

Green, B. G. (1993). *Acta Physiologica* **84**, 119–25.

Guinard, J., Pangborn, R. M. and Lewis, M. J. (1986*a*). *J. Sci. Food Agric.* **37**, 811–17.

Guinard, J., Pangborn, R. M. and Lewis, M. J. (1986*b*). *Am. J. Enol. Vitic.* **37**, 184–9.

Hagerman, A. E. and Robbins, C. T. (1993). *Can. J. Zool.* **71**, 628–33.

Haslam, E. (1981). In *The Biochemistry of plants* (ed. E.E. Conn), pp. 327–61. Academic Press, London.

Haslam, E. (1993). In *Polyphenolic phenomena* (ed. A. Scalbert), pp. 23–32. INRA Editions, Paris.

Haslam, E., Lilley, T. H., Warminski, E., Liao, H., Cai, Y., Martin, R., Gaffney, S. H., Goulding, P. N. and Luck, G. (1992). In *Phenolic compounds in foods and their effects on health. 1 Analysis, occurrence and chemistry* (ed. C.-T. Ho, C.-Y. Lee and M.-T. Huang), pp. 8–48. American Chemical Society, Washingron, DC.

Hay, D. I., Ahern, J. M., Schluckebier, S. K. and Schlesinger, D. H. (1994). *J. Dent. Res.* **73**, 1717–26.

Helfer, C. A. and Mattice, W. L. (1995). *Trends in Polymer Science* **3**, 117–23.

Ishikawa, T. and Noble, A. C. (1995). *Food Quality and Preference* **6**, 27–33.

Kallithraka, S., Bakker, J. and Clifford, M.N. (1995). In *Proceedings of the Second Davis Pangborn Symposium, Davis, California, July 1995*. In press.

Kellard, B., Clifford, M. N. and Birch, G. G. (1987). In *Douzième Colloque Scientifique International sur le Café*, pp. 254–9. Association Scientifique Internationale du Café, Paris.

Kirimura, J., Shimizu, A., Kimizuka, A., Ninomiya, T. and Katsuya, N. (1969). *J. Agric. Food Chem.* **17**, 689–95.

Kock, K., Ahlers, C. and Schmale, H. (1994*a*). *Eur. J. Biochem.* **221**, 905–16.

Kock, K., Morley, S. D., Mullins, J. J. and Schmale, H. (1994*b*). *Physiol. Behav.* **56**, 1173–7.

Lawless, H. T., Corrigan, C. J. and Lee, C. B. (1994). *Chem. Senses* **19**, 141–54.

Lea, A. G. H. and Arnold, G. M. (1978). *J. Sci. Food Agric.* **29**, 478–83.

Lee, C. B. and Lawless, H. T. (1991). *Chem. Senses* **16**, 225–38.

Levine, M.J., Weill, J.C. and Ellison, S. (1969). *Biochem. Biophys. Acta* **188**, 165–7.

Loomis, R. E., Gonzalez, M. and Loomis, P. M. (1991). *Int. J. Pept. Protein Res.* **38**, 428–39.

Luck, G., Liao, H., Murray, N. J., Grimmer, H. R., Warminski, E. E., Williamson, M. P., Lilley, T. H. and Haslam, E. (1994). *Phytochemistry* **37**, 357–71.

Lyman, B. J. and Green, B. G. (1990). *Chem. Senses* **15**, 151–64.

Madapallimattam, G. and Bennick, A. (1990). *Biochem. J.* **270**, 297–304.

Makkar, H. P. S., Blümmel, M., Borowy, N. K. and Becker, K. (1993). *J. Sci. Food Agric.* **61**, 161–6.

Mansson-Rahemtulla, B., Rahemtulla, F., Baldone, B. C., Pruitt, K. M. and Hjerpe, A. (1988). *Biochemistry* **27**, 233–9.

Matsuo, R. and Yamamoto, T. (1992). *Brain Res.* **583**, 71–80.

McManus, J. P., Davis, P. K., Lilley, T. H. and Haslam, E. (1981). *J. Chem. Soc. Chem. Commun.* 309–11.

Mehansho, H., Hagerman, A. E., Clements, S., Butler, L. G., Rogler, J. C. and Carlson, D. M. (1983). *Proc. Natl. Acad. Sci. USA* **80**, 3948–52.

Miao, Y. J., Subramaniam, N. and Carlson, D. M. (1995). *Eur. J. Biochem.* **228**, 343–50.
Muenzer, J., Bidstein, C., Gleason, M. and Carlson, D. M. (1979). *J. Biol. Chem.* **254**, 5623–8.
Murray, N. J. and Williamson, M. P. (1994). *Eur. J. Biochem.* **219**, 915–21.
Murray, N. J., Williamson, M. P., Lilley, T. H. and Haslam, E. (1994). *Eur. J. Biochem.* **219**, 923–35.
Naish, M., Clifford, M. N. and Birch, G. G. (1993). *J. Sci. Food Agric.* **61**, 57–64.
O'Hanlon, K., Weissbecker, K., Cortessis, V., Spence, M. A. and Azen, E. A. (1988). *Cytogenet. Cell Genet.* **49**, 315–17.
Oho, T., Rahemtulla, F., Mansson-Rahemtulla, B. and Hjerpe, A. (1992). *Int. J. Biochem.* **24**, 1159–68.
Opie, S. C., Clifford, M. N. and Robertson, A. (1993). *J. Sci. Food Agric.* **63**, 435–8.
Peleg, H. and Noble, A. C. (1995). *Chem. Senses* 1995 **20**, 393–400.
Perinpanayagam, H. E., Van Wuyckhuyse, B. C., Ji, Z. S. and Tabak, L. A. (1995). *J. Dent. Res.* **74**, 345–50.
Powell, C. (1995). The polyphenolic pigments of black tea. PhD Thesis. University of Surrey, Guildford, UK.
Powell, C., Clifford, M. N., Opie, S. C., Ford, M. A., Robertson, A. and Gibson, C. L. (1992). *J. Sci. Food Agric.* **63**, 77–86.
Powell, C., Clifford, M. N., Opie, S. C. and Gibson, C. L. (1994). In *XVIIe Journeés Internationales Groupe Polyphénols.* INRA Editions, Paris.
Powell, C., Clifford, M. N., Opie, S. C. and Gibson, C. L. (1995). *J. Sci. Food Agric.* **68**, 33–8.
Prieur, C., Rigaud, J., Cheynier, V. and Moutounet, M. (1994). *Phytochemistry* 36, 781–4.
Rathman, W. M., van Zeyl, M. J., van den Keijbus, P. A., Veerman, E. C. and Amerongen, A. V. N. (1990). *J. Biol. Bucc.* **18**, 9–18.
Reddy, M. S., Bobek, L. A., Haraszthy, G. G., Biesbrock, A. R. and Levine, M. J. (1992). *Biochem. J.* **287**, 639–43.
Reddy, M. S., Levine, M. J. and Paranchych, W. (1993). *CRC Crit. Rev. Oral Biol. Med.* **4**, 315–23.
Robertson, A. (1992). In *Tea: cultivation to consumption.* (ed. K. C. Wilson and M. N. Clifford), pp. 555–602. Chapman & Hall, London.
Sabatini, L. M., Ota, T. and Azen, E. A. (1993). *Mol. Biol. Evol.* **10**, 497–511.
Schenkels, L. C. P. M., Rathman, W. M., Veerman, E. C. and Amerongen, A. V. N. (1991). *Biol. Chem. Hoppe Seyler* **372**, 325–9.
Schenkels, L. C. P. M., Schaller, J., Walgreen Weterings, E., Schadee Eestermans, I. L., Veerman, E. C. and Amerongen, A. V. N. (1994). *Biol. Chem. Hoppe Seyler* **375**, 609–15.
Schiffman, S. S., Suggs, M. S., Sostam, A. L. and Simon, S. A. (1991). *Physiol. Behav.* **51**, 55–63.
Schlesinger, D. H., Hay, D. I., Schluckebier, S. K. and Ahern, J. M. (1994). *Pept. Res.* **7**, 242–7.
Schmale, H., Ahlers, C., Blaker, M., Kock, K. and Spielman, A. I. (1993). *Ciba. Found. Symp.* **179**, 167–80.
Shintani, M., Minaguchi, K., Lim, K. A., Hashimoto, M. and Suzuki, K. (1990). *Hum. Hered.* **40**, 89–98.
SOED (1944). *The Shorter Oxford English Dictionary* 3rd edn, Vol. 1. Oxford University Press, Oxford.

Spencer, C. M., Cai, Y., Martin, R., Gaffney, S. H., Goulding, P. N., Olato, D., Lilley, T. H. and Haslam, E. (1988). *Phytochemistry* **27**, 2397–409.
Spielman, A. I. (1990). *J. Dent. Res.* **69**, 838–43.
Thorngate, J. H. and Noble, A. C. (1995). *J. Agric. Food Chem.* **67**, 521–5.
Wong, R.S.C. and Bennick, A. (1980). *J. Biol. Chem.* **255**, 5943–8.
Wong, R.S.C., Hoffman, T. and Bennick, A. (1979). *J. Biol. Chem.* **254**, 4800–7.

6. Fruit aroma precursors with special reference to phenolics

J. CROUZET, M. SAKHO and D. CHASSAGNE

Laboratoire de Genie Biologique et Sciences des Aliments, Equipe de Microbiologie et Biochimie Industrielles associée à l' INRA, Université de Montpellier II, 34095 Montpellier Cedex 05, France

Introduction

Odorous compounds present in food reach the olfactory mucosa receptors in two different ways. The first is by directly sniffing volatile molecules into the nose, where they interact with the receptors producing the sensation known as odour. The second is where volatile substances released from food in the mouth reach the olfactory epithelium by a retronasal route, producing the sensation known as aroma. The combined sensations of odour, aroma, and taste give the overall 'flavour' of a food.

The odour and aroma fractions are composed of molecules belonging to different classes of organic compounds. These include phenolics, which have a low molecular weight (less than 400) and a relative hydrophobicity. The molecules are present in low concentrations (1 mg/T to 1 mg/kg), are able to escape from foods and are present in the vapour phase.

A natural product aroma consists of several hundred volatile compounds. There are more than 650 in coffee aroma, 170 in passion fruit, 150 in mango and 120 in apricot. The compound that gives a food product its distinctive aroma is known as a character-impact compound (Nursten, 1979). In certain cases this compound determines the character of the food product on its own, examples being isopentyl acetate in banana, benzaldehyde in almond, vanillin in vanilla and eugenol in clove. Most food products, however, depend on a mixture of compounds for their aromatic character. The character-impact compound for raspberry, for instance, is considered to be 4-(4′-hydroxyphenyl)-butan-2-one, known as raspberry ketone, but cis-3-hexenol, damascenone and α- and β-ionone are also important contributors to the overall raspberry aroma. Aroma generally results from the contribution of a more or less important number of volatile compounds, each of them being responsible for a typical contribution or note. For example, the combination in the banana of elimicine with eugenol, methoxyeugenol and eugenyl methyl ether contributes to a full-bodied, mellow, ripe-fruit aroma (Nursten, 1970). 4-Vinyl phenol, 4-methoxy-5-vinyl phenol, in association with ethyl hexanoate and 2-methyl butanoic acid, are the major contributors to cloudberry aroma (Pyysallo *et al.,* 1977).

Volatile compounds in fruit and vegetables are also present as non-volatile precursors, and are released from these precursors during maturation, pretreatment, or processing by enzymatic or chemical reactions. As an example, vanillin, the characteristic aroma compound of vanilla, is developed during the fermentation process by hydrolysis of vanillin glucoside catalysed by vanilla β-glucosidase (Arana, 1943) .

This paper investigates glycosidically bound compounds, particularly phenolic compounds, present in apricot, mango and purple passion fruit. The role played by *p*-hydroxycinnamoyl derivatives as volatile phenol precursors is also discussed.

Glycosidically bound volatile compounds

Among aroma precursors, glycosidically bound volatile compounds, including phenolics, have been identified in several fruits (Pyysallo *et al.,* 1977; Fleuriet and Macheix, 1980; Pabst *et al.,* 1990; Schwab *et al.,* 1990; Krammer *et al.,* 1991; Wu *et al.,* 1991; Marlatt *et al.,* 1992; Adedeji *et al.,* 1992; Koulibaly *et al.,* 1992; Suarez *et al.,* 1991; Crouzet *et al.,* 1995). These bound compounds can be more important than free compounds, as for instance in aromatic grape cultivars or purple passion fruits; they are non-volatile, and generally bitter. They are present in plant matter, especially in fruits, as glucosides or diglycosides and in these compounds a glucose unit is generally bound to the aglycone moiety. α-L-Arabinofuranosyl-β-D-glucopyranosides, α-L-arabinopyranosyl-β-D-glucopyranosides (vicianosides), α-L-rhamnopyranosyl- β-D-glucopyranosides (rutinosides), β-D-apiofuranosyl- β-D-glucopyranosides, β-D-glucopyranosyl-β-D-glucopyranosides (gentiobiosides) and β-D-xylopyranosyl-β-D-glucopyranosides have been identified in fruits to date. Aglycones are aliphatic, aromatic, terpenic and sesquiterpenic alcohols or polyhydroxylated compounds, norisoprenoids, phenols, hydroxyesters and acids. According to the nature of the aglycone moiety, free volatile or polyhydroxylated compounds can be released by enzymatic or acid hydrolysis; free volatiles are also generated by acid-catalysed rearrangement of terpenic alcohols and polyhydroxylated compounds.

Heterosidic fraction isolation

The glycosidic fraction is best obtained from fruit juice by adsorption on a C_{18} reversed phase (Williams *et al.,* 1982) or on Amberlite XAD-2 (Gunata *et al.,* 1985). After elimination of water-soluble compounds by washing with distilled water and volatile compounds by washing with pentane, the glycosidically bound compounds are recovered using methanol as elution solvent.

Chemical analysis

Volatile compounds are released by chemical hydrolysis of the glycosidic fraction. It should be noted that rearrangement reactions can occur under the acidic conditions used, especially in the case of terpenic compounds, and this should be taken into account when considering the results. Information concerning the composition glycosidic moiety is obtained by TLC and GC of trimethylsilyl derivatives of saccharides released by chemical hydrolysis.

Complementary data concerning the nature, cyclic structure, and linkage between the saccharidic units is obtained by GC and GC–MS of partially methylated alditol acetates. The results obtained for purple passion fruit heterosidic extract are given in Table 6.1 (Chassagne *et al.*, 1995*a*).The identification of 2,3,4 tri-*O*-methyl rhamnose and of 2,3,4 tri-*O*-methyl arabinose indicates the presence of terminal rhamnofuranose and arabinopyranose units linked to a glucopyranoside unit, and the identification of 2,3,4,6 tetra-*O*-methyl glucose besides 2,3,4 tri-*O*-methyl glucose indicates the presence of gentiobiose.

Table 6.1. Structural analysis of the glycosidic moiety of passion fruit glycosidically bound compounds (partially methylated alditol acetates)

Methyl ethers	Relative molar per cent
2, 3, 4- Tri-*O*-methyl rhamnose	22.2
2, 3, 4- Tri-*O*-methyl arabinose	6.1
2, 3, 4, 6- Tetra-*O*-methyl glucose	27.6
2, 3, 4- Tri-*O*-methyl glucose	44.1

Enzymatic hydrolysis

According to Gunata *et al.* (1988) glycoside enzymatic hydrolysis proceeds from the terminal unit of the saccharidic sequence. Glucosides released by the action of specific glycosidases—rhamnofuranosidase, arabinofuranosidase, arabinopyranosidase, apiofuranosidase—are then substrates of β-glucosidase, and free volatile compounds appear.

These enzymes are generally present as secondary activities in commercial fungal preparations, particularly in pectolytic enzymes. Sweet almond glucosidase (emulsin) also contains an arabinopyranosidase activity as well as the main β-glucosidase activity. Unfortunately these enzymes are inhibited by glucose, in particular β-glucosidase, and their efficiency is variable (Chassagne *et al.*, 1995*a*).

A non-glucose-inhibited endo-β-glucosidase has, however, been isolated from an *Aspergillus niger* mutant strain grown on a medium containing the flavonol rutin (quercetin 3-rhamnosyl-glucoside) as the sole carbon source, and used for flavour enhancement of muscat wine and passion fruit juice (Shoseyov *et al.*, 1988).

Volatile compounds isolated from apricot, mango and passion fruit hydrolysed heterosidic extracts are given in Table 6.2. Eugenol was identified in the three fruits studied, vanillin in mango and apricot and isoeugenol, methyl salicylate, 4-vinyl phenol, 4-ethyl phenol, 4-allyl phenol, 4-propenyl phenol and 4-vinyl guaiacol in passion fruit. The presence of cinnamic and ferulic acid on one hand, and 4-alkyl or alkenyl phenols on the other, is worthy of note and will be discussed later.

Table 6.2. Phenolic compounds identified in purple passion fruit, mango and apricot after enzymatic hydrolysis of heterosidic extracts

Phenolic compounds	Passion fruit	Mango	Apricot[a]
Methyl salicylate	+		
Phenol			+
Thymol		+	
Carvacrol		+	
Eugenol	+	+	+
Isoeugenol	+		
4-Vinyl guaiacol			+
4-Vinyl phenol			
4-Ethyl phenol	+		+
4-Allyl phenol	+		
4-Propenyl phenol	+		
Vanillin		+	+
Cinnamic acid	+		
Ferulic acid[b]		+	

[a]Krammer *et al.* (1991).
[b]Adedeji *et al.* (1992).

The sensory value of volatile phenols is already well known. Vanillin contributes to the flavour of several natural extracts, and is widely used at low concentrations in many aroma and perfume preparations. Another good example is eugenol: this commonly found compound, largely responsible for the aroma of cloves, was used in synthetic banana aroma preparations before its identification in banana fruit.

4-Vinyl phenol and 4-methoxy-5-vinyl phenol have been described in cloudberry aroma as being responsible for fruity and fresh notes, and it is possible that their increase during cooking contributes to the aroma of cloudberry jam (Pyysallo *et al.*, 1977).

4-Ethyl phenol, 4-vinyl phenol, 4-ethyl guaiacol and 4-vinyl guaiacol are present in wine at low concentrations (Etievant, 1991) and are considered to contribute to the aroma, particularly in the case of red wines. Their presence at a high level could be responsible for unpleasant phenolic off-flavours (Dugelay *et al.*, 1993).

The quantities of bound phenolic compounds vary according to the species of passion fruit and are more abundant in *Passiflora edulis* f *flavicarpa* (yellow passion fruit) juice than in *Passiflora edulis* (purple passion fruit) juice (Fig. 6.1). Variations in individual bound phenol content were found to depend on the origin of the fruit (Fig. 6.2), these variations being due to the nature of the variety, as well as climatic and culture conditions (Chassagne, 1995).

Study of derivatized heterosidic compounds by gas chromatography–mass spectrometry

Direct analysis of glycosidic extracts by gas chromatography–mass spectrometry after derivatization—per-*O*-methylation, trifluoracetylation (TFA) or trimethylsilylation (TMS)—has been used (Schwab *et al.*, 1990; Voirin *et al.*, 1992*a*,*b*; Chassagne, 1995). According to Voirin *et al.* (1992*a*), TFA derivatives of monoterpenyl glycosides are more suitable for qualitative and quantitative analysis using electron impact ionization than TMS derivatives.This method was used for the study of mango and passion fruit glycosidically bound compounds.

Gas chromatography–electron impact mass spectrometry of TFA derivatives

The characteristic fragment ions defined by Voirin *et al.* (1992*a*) were used to determine the sacharidic moiety, while characteristic fragment ions of the aglycone were used to tentatively identify the compound.

As shown in Fig. 6.3, fragment ions at *m*/*z* 205, 193, 179 and 177 are characteristic of a glucose unit, and fragment ions at *m*/*z* 164, 149, 103, 81 and 71 can be attributed to eugenol. The presence of eugenyl glycoside in mango and passion fruit was confirmed using an authentic sample obtained by synthesis (Chassagne, 1995).

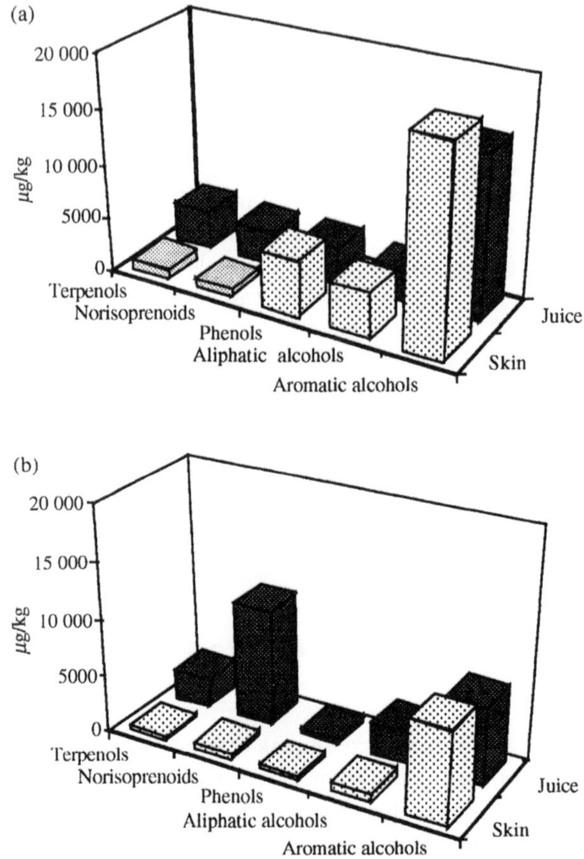

Fig. 6.1. Glycosidically-bound compounds in *P. edulis* (a) and *P. edulis* f *flavicarpa* (b).

Similarly (Fig. 6.4) fragment ions at *m/z* at 193 and at *m/z* 435, 207 and 179 indicate the presence of a glucose and a rhamnose unit respectively, and fragment ions at *m/z* 152, 120, 109, 81 and 69 characterize methyl salicylate. The presence of methyl salicylate rutinoside in the TFA heterosidic fraction obtained from passion fruit was confirmed by comparing the natural product retention index (RI = 2234) to that of the synthetic compound (RI = 2224).

Gas chromatography–chemical ionization mass spectrometry of TFA derivatives

Unsatisfactory results, except for some terpenyl and aryl derivatives, were obtained when chemical ionization was used in positive mode (Voirin *et al.*, 1992*a,b*). In contrast, the high electron affinity of polyhalogenated organic compounds, such as TFA derivatives, made CI–MS in negative mode very attractive (Budzikiewicz, 1986), and this method was used to qualitatively analyse passion fruit heterosidic extracts (Chassagne *et al.*, 1995*b*).

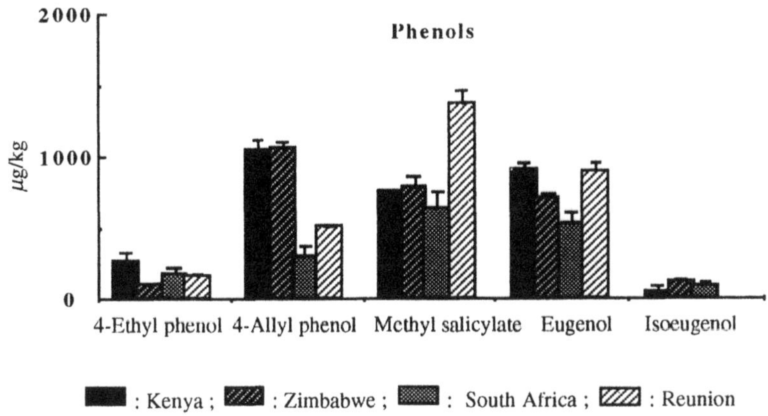

Fig. 6.2. Bound phenolic compounds in purple passion fruit of several origins.

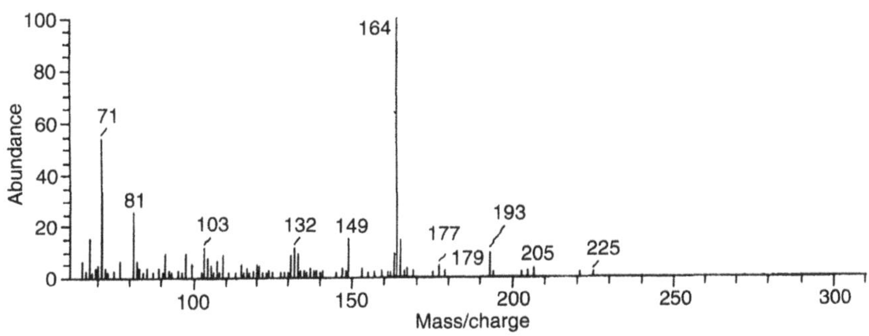

Fig. 6.3. TFA Eugenyl glucoside EI mass spectrum.

The presence of specific fragment ions: M^-, $[M\text{-}TFAO]^- = [M\text{-}113]^-$, $[M + TFAO]^- = [M + 113]^-$ in the experimental conditions used, unequivocally determine the mass of each glycoside, for example 1036 for the TFA methyl salicylate glycoside (Fig. 6.5). Certain data concerning the saccharidic moiety are also obtained. Ions at *m/z* 563 [GlcTFA$\sim$O]$^-$, 667, 544 and 451 are specific to a glucose unit, ions at *m/z* 901[RutTFA$\sim$O]$^-$, 882, 789, 674 and 451 characterize rutinosides, whereas ions at *m/z* 887 [AraTFA$\sim$GlcTFA$\sim$]$^-$, 868,

675, 660 and 437 are present in the NCI spectra of arabinosyl (or apiosyl) glucosides. Structural information on the aglycone moiety are limited, since the molecular mass can only be obtained by difference between the molecular mass of the glycoside and of the saccharidic moiety, respectively.

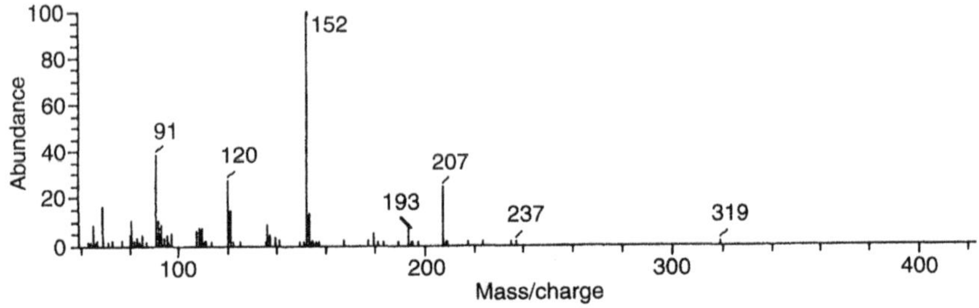

Fig. 6.4. TFA Methyl salicylate rutinoside EI mass spectrum.

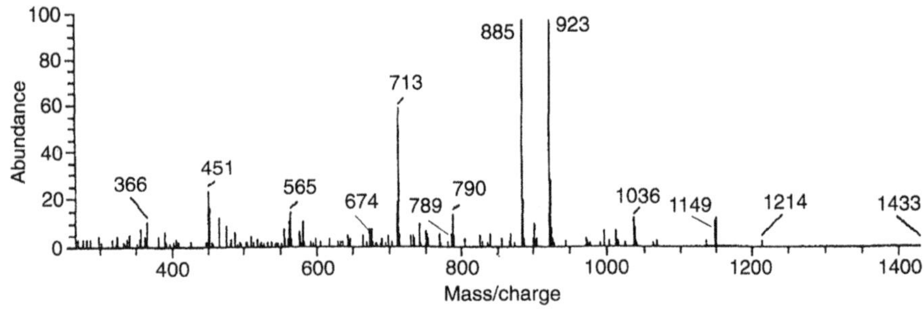

Fig. 6.5. TFA Methyl salicylated rutinoside NCI mass spectrum.

As in the case of EI–MS analysis of TFA derivatives, it was only possible to perform a tentative identification of natural glycosides as synthetic compounds are needed for formal identification. The glycosidically bound phenolic compounds identified in mango and passion fruit are listed in Table 6.3. Eugenyl glucoside has been previously reported in *Melissa officinialis* (Mulkens and Kapetanidis, 1988) and in sour cherry (Krammmer *et al.*, 1991) whereas the other compounds, with the exception of vanilla glucoside, have not to the best of our knowledge been previously reported.

Apart from the need to have reference compounds, another limitation of derivatizated heterosidic compound direct analysis is that poor results are obtained for higher molecular weight compounds such as polyhydroxylated norisoprenoids (Baumes *et al.*, 1994). Owing to these problems, fractionation of heterosidic extracts was performed.

Table 6.3. Glycosidically-bound phenolic compounds identified or tentatively identified in purple passion fruit and mango.

Bound compound	Mango	Passion fruit
Methyl salicylate glucoside		+
Eugenyl glucoside	+	+
Vanillyl glucoside	+	
Methyl salicylate rutinoside		+
Eugenyl rutinoside		+[a]

[a]tentatively identified

Study of heterosidic extract after fractionation

Several chromatographic methods have been used to fractionate the heterosidic extract: gel chromatography (Pabst *et al.*, 1990; Salles *et al.*, 1990), preparative over pressure layer chromatography (OPLC) (Salles *et al.*, 1990), HPLC (Schwab *et al.*, 1990; Pabst *et al.*, 1990; Chassagne *et al.*, 1995*c*) and countercurrent chromatography (Strauss *et al.*, 1987; Winterhalter, 1990).

Gel chromatography on Fractogel TSK HW-40 S and preparative OPLC was used for apricot glycosidically-bound volatile compound fractionation. The small amounts of pure fractions obtained were analysed by mass spectrometry using chemical ionization (NI–CI) or desorption chemical ionization in negative mode (NI–DCI). As indicated in Fig. 6.6 (Reyné *et al.*, 1992), information concerning the molecular weight $(M\text{-}H)^-$ at *m/z* 461, the nature of the sugars constituting the saccharidic moiety and their sequence: *m/z* 325 $[Rut{\sim}O]^-$, *m/z* 163 $[Rha{\sim}0]^-$ and *m/z* 315 $[O{\sim}Glu{\sim}O]^-$ were obtained just as for geranyl (or neryl) rutinoside, although the isomers cannot be distinguished by this technique. The relative abundance of the ionic species detected in low energy collisionally activated decomposition (CAD) spectra, and more particularly fragment ions at *m/z* 179 and 161 for glucosides and fragment ions at *m/z* 315, 205, 161, 163 and 101 for rutinosides, however, varies according to the nature of the aglycone moiety (Cole *et al.*, 1989). From these data, obtained using synthetic compounds, empirical rules have been defined and have enabled aromatic and terpenyl glycosides (Salles *et al.*, 1991) to be tentatively identified. Isomers such as linalyl oxide and dienediol glycosides can be differentiated using ND_3 as reagent gas according to the presence of acidic protons and resulting shift of the parent ion in the NI-DCI mass spectrum, from 331 to 334 (Md_4-D) for linalyl oxide derivatives (four acidic protons) and from 331 to 335 (Md_5-D) for dienediol derivatives (five acidic protons) (Salles *et al.*, 1991).

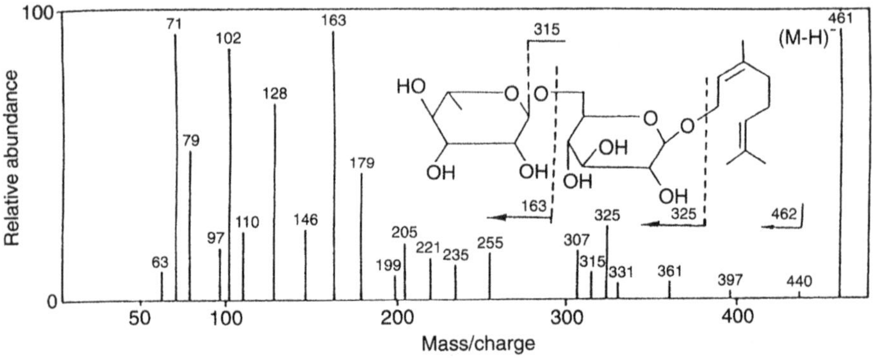

Fig. 6.6. Negative ion desorption chemical ionization spectrum of geranyl β-D-glucoside (reactant gas: ammonia).

The identity of the different compounds tentatively identified by MS or MS–MS was established by comparing their retention in reversed phase HPLC with that of authentic samples. In the absence of reference standards some compounds, such as linalyl arabinoglucoside, were identified after treatment with an *A. niger* pectinase preparation in the presence of δ-gluconolactone (a β-D-glucosidase inhibitor), to sequentially release the arabinose units. The decrease of the peak initially present and the increase of a peak with the same retention time as linalyl glucoside, indicated the presence of arabinoglucoside in apricot (Salles *et al.*, 1991). Purified arabinofuranosidase and rhamnofuranosidase were subsequently used (Reyné *et al.*, 1992).

Although phenolic compounds have been detected among the volatile products released by enzymatic hydrolysis of the apricot glycosidic pool (Krammer *et al.*, 1991), bound phenolics have not been found in the studied fractions. Identification using ^{1}H and ^{13}C NMR was not possible, owing to the small quantities of product recovered for each OPLC fraction.

NMR techniques associated with EI–MS and NCI–MS were used for the structural analysis of linalool, benzyl alcohol and 3-methyl-but-2-en-1-ol arabinopyranosyl glucosides (Fig. 6.7). These compounds were isolated after partial enzymatic hydrolysis by the action of a glycosidase preparation devoid of arabinopyranosidase activity (Hemicellulase REG 2) on passion fruit heterosidic extract. A second separation on Amberlite XAD-2 was performed to remove the aglycones and sugars released, and the fraction recovered by methanol elution fractionated using semipreparative HPLC on RP 18 phase (Chassagne *et al.*, 1995*c*). The ^{1}H NMR spectrum of **2a** (Fig. 6.8) revealed the existence of aromatic [δ 7.43 (5H)] and aliphatic protons [δ 4.91 and 4.73] corresponding to a benzyl aglycone moiety, and two

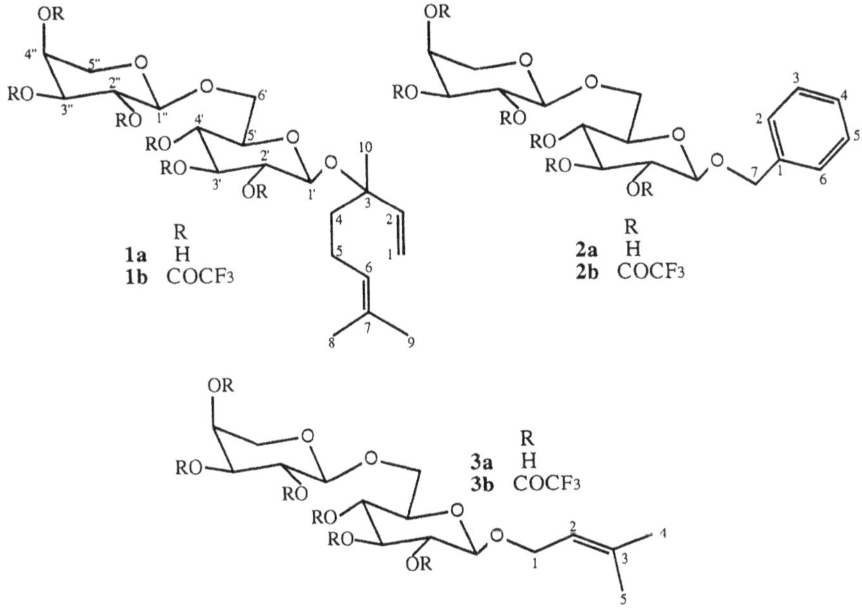

Fig. 6.7. Linalyl, benzyl alcohol and 3-methyl-but-2-en-1-ol arabinopyranosyl glucosides identified in purple passion fruit.

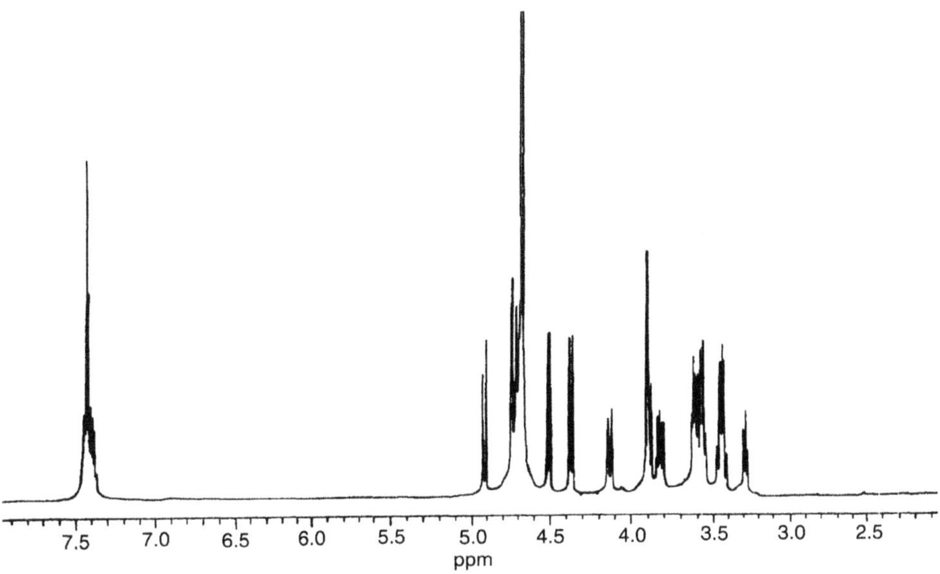

Fig. 6.8. ^{1}H NMR spectrum of benzyl α-l-arabinopyranosyl-β-D-glucopyranoside.

anomeric protons [δ 4.51 and 4.38]. The ^{1}H and ^{13}C NMR spectra indicate the presence of a β-D-glucopyranosyl unit and an α-L-arabinopyranosyl unit in agreement with a previous report (Mizutani *et al.*, 1980).

Using similar methods Pabst *et al.* (1990) identified 4-(4′-hydroxyphenyl) butan-2-one 4′-*O*-β-D-glucopyranoside and 4-(3′,4′-dihydroxyphenyl)-butan-2-one 3′-*O*-β-D-glucopyranoside in raspberry fruit.

Cinnamoyl derivatives as volatile phenol precursors

The formation of volatile phenols from *p*-hydroxy cinnamoyl compounds is well known. For example 4-vinyl phenol and 4-vinyl guaiacol are produced during the alcoholic fermentation, *p*-hydroxycinnamic and ferulic acids being present in grapes, particularly in the skin. 4-Ethyl phenol and 4-ethyl guaiacol are mainly produced during malolactic fermentation by the action of bacteria (Dubois, 1993, 1994). More recently it has been shown that high levels of volatile phenols can be obtained by the combined action of a cinnamoyl esterase, detected in several enzymatic preparations used in wine making, and yeast cinnamic acid decarboxylase. The substrates of these enzymes have been identified as *p*-coumaroyl and feruloyl esters of tartaric acid (Dugelay *et al.*, 1993).

According to Pyysallo *et al.* (1977), cinnamic acids are also involved in the formation of 4-vinyl phenol, 4-ethyl phenol and 4-allyl phenol in cloudberry fruit.

In our study we identified these phenols, as well as cinnamic acid, after enzymatic hydrolysis of passion fruit heterosidic extract (see Table 6.2). Furthermore, it has been previously reported (Pyysallo *et al.*, 1977) that the concentration of 4-vinyl phenol and 2-methoxy-5-vinyl phenol increases 4 to 5 times when cloudberries are heated for 1h at 100 °C. These results indicate that volatile phenol can be produced according to pathways similar to those described by Williams (1993) for terpenic compounds (Fig. 6.9).

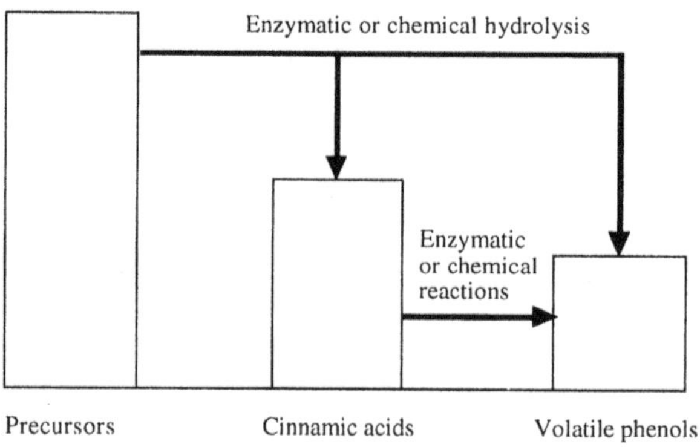

Fig. 6.9. Formation of cinnamic acids and volatile phenols from cinnamoyl heterosidic compounds.

This hypothesis is supported by the identification of some phenolics and several cinnamic acid derivatives: *p*-coumaric, caffeic, ferulic and sinapic acid as tomato heterosidic compound aglycones (Fleuriet and Macheix, 1980). More recently *p*-coumaric acid and 3-(4-hydroxyphenyl) propionic acid-β-D-glucopyranosides have been isolated and identified in the tomato using ^{1}H and ^{13}C NMR spectroscopy and CI-MS (Krammer *et al.*, 1995).

Perspectives

For many years, the phenolic volatile compounds present in fruits, fermented products and smoke flavours have been considered to result from oxidative lignin decomposition through various chemical or biochemical pathways or from the decarboxylation of coumaric acid derivatives.

During the last fifteen years, important research concerning glycosidically bound compounds has been carried out. The interest of scientists working in this field, however, has essentially focused on aroma precursors having a monoterpene or a C_{13} norisoprenoid aglycone moiety.The role played by phenolic and coumaric acid glycosides as aroma precusors was pointed out only recently and research concerning these compounds should be increased.

More particularly, analytical methods adapted to the isolation and identification of phenolic glycosides need to be developed as in some cases, these compounds are lost during the prepurification steps of the crude heterosidic extract.

Topics for further research include:

- The physiological role of glycosides in plants: are they transport derivatives or, more probably, detoxified forms? According to the toxicity of phenolics towards membranes, these compounds are, along with terpene alcohols, good candidates for further study.
- The glycosylation reaction. Only hypotheses concerning the very efficient glycosylating activity are available at present, and studies on the mechanism of the reaction are needed.
- The interactions betwen phenolic compounds in both free and bound forms and other volatile compounds, which are are of major interest.

Acknowledgements

The authors thank Drs G. Martin (URA CNRS 472, Nantes) and G. Rémaud (Eurofins, Nantes) for expert ^{1}H NMR spectroscopy, Drs C. Bayonove and R. Baumes (INRA, Montpellier) for mass spectrometry facilities and the Institut des Sciences de la Nutrition et de l'Alimentation, Montpellier for financial support.

References

Adedeji, J., Hartman, T.G., Lech, J. and Ho, C.T. (1992). *J. Agric. Food Chem.* **40**, 659–61.

Arana, F. (1943). *Food Res.* **8**, 343–51.

Baumes, R., Aubert, C., Gunata, Z., de Moor, W., Tapiero, C. and Bayonove, C. (1994). *J. Essent. Oil Res.* **6**, 587–99.

Budzikiewicz, H. (1986). *Mass Spectrom. Rev.* **5**, 345–80.

Chassagne, D., Crouzet, J., Bayonove, C. and Baumes, R. (1995*a*). In *Bioflavour' 95. Analysis, precursor studies, biotechnology* (ed. P. Etievant and P. Schreier), pp. 217–22. INRA Editions, Paris.

Chassagne, D., Crouzet, J., Baumes, R., Lepoutre, J.P. and Bayonove, C. (1995*b*). *J. Chromatogr.* **964**, 441–51.

Chassagne, D., Crouzet, J., Bayonove, C. Brillouet, J.M. and Baumes, R. (1995*c*). *Pytochemistry* **41**, 1497–500.

Chassagne, D. (1995). Doctoral Thesis, Université de Montpellier II.

Cole, R., Tabet, J.C., Salles, C., Jallageas, J.C. and Crouzet, J. (1989). *Rapid Commun. Mass Spectrom.* **3**, 59–63.

Crouzet, J. Chassagne, D. and Sakho, M. (1995). 210th ACS National Meeting. Chicago.

Dubois, P. (1993, 1994). *Rev. Fr. Oenol.* **144**, 63–72, **145**, 27–40, **146**, 39–50.

Dugelay, I., Gunata, Z., Bitteur, S., Sapis, J.C., Baumes, R. and Bayonove , C. (1993). In *Progress in flavour precursor studies* (ed. P. Schreier and P. Winterhalter), pp. 189–193. Allmed Pub. Corp.

Etievant, P. (1991). In *Volatile compounds in foods and beverages* (ed. H. Maarse), pp. 483–546. M. Dekker, New York.

Fleuriet, A. and Macheix, J.J. (1980). *Phytochemistry* **19**, 1955–8.

Gunata, Z., Bayonove, C., Baumes, R. and Cordonnier, R. (1985). *J. Chromatogr.* **331**, 83–90.

Gunata, Z., Bitteur, S., Brillouet, J.M., Bayonove, C. and Cordonnier, R. (1988). *Carbohyhydr. Res.* **184**, 139–49.

Koulibaly, A., Sakho, M. and Crouzet, J. (1992). *Lebensm. Wiss. u-Technol.* **25**, 374–9.

Krammer, G., Winterhalter, P., Schwab, M. and Schreier, P. (1991). *J. Agric. Food Chem.* **39**, 778–81.

Krammer, G., Buttery, R.G. and Takeoka, G. (1995). In *Fruit flavor* (ed. R. Roussef and M.Leahy), pp.164–81. American Chemical Society, Washington DC.

Marlatt, C., Ho, C.T. and Chien, M. (1992). *J. Agric. Food Chem.* **40**, 249–52.

Mizutani, K., Kasai, R. and Tanaka, O. (1980). *Carbohydr. Res.* **87**, 19–26.

Mulkens, A. and Kapetanidis, I. (1988). *J. Nat. Prod.* **51**, 496–8.

Nursten, H.E. (1970). In *The Biochemistry of fruits and their products*, Vol. 2. (ed. A.C. Hulme), pp. 239–68. Academic Press, New York.

Nursten, H.E. (1979) In *Progress in flavour research* (ed. D.G. Land and H.E. Nursten), pp. 337–65. Applied Science Publishers, London.

Pabst, A., Barron, D., Adda, J. and Schreier, P. (1990). *Phytochemistry* **29**, 3853–8.

Pyyssalo,T., Torkelli, H. and Honkanen, E. (1977). *Lebesm. Wiss. u-Technol.* **10**, 145–7.

Reyné, V., Salles, C. and Crouzet, J. (1992). In *Food science and human nutrition.* (ed. G. Charalambous), pp. 99–114. Elsevier, Amsterdam.

Salles, C., Jallageas, J.C., Fournier, F. Tabet, J.C. and Crouzet, J. (1991). *J. Agric. Food Chem.* **39**, 1979–83.

Salles, C., Jallageas, J. C. and Crouzet, J. (1990). *J. Chromatogr.* **552**, 255–65.

Schwab, W., Scheller, G. and Schreier, P. (1990). *Phytochemistry* **29**, 607–12.

Shoseyov, O., Bravdo, B., Ikan, R. and Chet, I. (1988). *Phytochemistry* **27**, 1973–88.

Strauss, R., Gooley, P., Wilson, B. and Williams, P. (1987). *J. Agric. Food Chem.* **35**, 519–24.

Suarez, M., Duque, C., Wintoch, H. and Schreier, P. (1991). *J.Agric. Food Chem.* **39**, 1643–5.

Voirin, S., Baumes, R., Gunata, Z., Bitteur, S. and Bayonove, C. (1992*a*). *J. Chromatogr.* **590**, 313–28.

Voirin, S., Baumes, R., Sapis, J.C. and Bayonove, C. (1992*b*). *J. Chromatogr.* **595**, 269–81.

Williams, P.J. (1993). In *Flavor science. Sensible principles and techniques* (ed. T. Acree and R. Teranishi). American Chemical Sociey, Washington DC.

Williams, P.J., Strauss, C.R., Wilson, B. and Massy-Westropp, R.A. (1982). *J. Chromatogr.* **135**, 471–80.

Winterhalter, P. (1990). *J. Agric. Food Chem.* **38**, 452–5.

Wu, P., Kuo, M.C., Hartman, T.G., Rosen, R.T. and Ho, C.T. (1991). *J. Agric. Food Chem.* **39**, 170–2.

7. Aroma biochemistry of fruits and vegetables

CARLOS SANZ, JOSE M. OLIAS and ANA G. PEREZ

Department of Physiology and Technology of Plant Products, Instituto de la Grasa, CSIC, Padre García Tejero 4, 41012 Seville, Spain

Introduction

Aroma compounds contribute heavily to the overall sensory quality of fruit and vegetables. Aroma is a complex mixture of a large number of volatile compounds whose composition is specific to species and often to variety. In some cases the typical fruit and vegetable aromas can be ascribed to a specific compound, although in general the overall aroma quality is the sum of a multitude of compounds. At present, it is difficult to ascertain the relative contribution of each volatile component to a characteristic or a desirable aroma. During the last three decades, physicochemical methods of analysis, such as spectrophotometry, nuclear magnetic resonance, and specially gas–liquid chromatography coupled to mass spectrometry allowed the isolation and identification of a large number of volatile compounds from plant aromas. However, rather less work has been done on elucidating the aroma formation mechanisms. These valuable studies are only a beginning to the process of understanding and regulating the aroma quality of fruits and vegetables. In this sense, biotechnological applications for aroma-determining substances in the food industry may support studies leading to a deeper knowledge of aroma biochemistry, since chemical synthesis cannot completely fulfil the requirements of specific chirality found among some aroma impact compounds, or the increasing demand for natural compounds by consumers.

Generally speaking, the characteristic aroma of fruits is mainly based on their ester composition, while sulfur compounds determine vegetable aromas. However, this does not imply that metabolic pathways are total different: some of them are common to both fruits and vegetables.

In this review, our aim is to present a general vision of the currently known or hypothesized biochemistry of fruit and vegetable aromas, through a description of the biochemical pathways within the metabolism of the three main classes of compounds in relation to aroma biogeneration: fatty acids, amino acids, and carbohydrates.

Fatty acid metabolism

Fatty acids seem to be the major precursors of volatile compounds responsible for the aroma of most plant products. In general, they are catabolized through two main oxidative pathways, β-oxidation and the lipoxygenase (LOX) pathway. A distinction should be made between primary aroma, which is generated by intact fruits and vegetables through intracellular pathways, and secondary aroma, which is due to disruption of cell integrity, either by crushing, slicing, or even during mastication (Drawert, 1974; Schreier, 1984). It has been suggested that β-oxidation is the main metabolic pathway producing primary aroma in fruits, whereas the LOX pathway may account for the widest variety of aroma compounds from fatty acids in disrupted plant tissues (Schreier, 1984).

β-*Oxidation*

It is assumed that during fruit ripening there is a predominance of catabolic processes (Tressl and Albrecht, 1986) over anabolic ones, establishing that although *de novo* synthesis of fatty acids could occur, aroma compounds are generated via β-oxidation. This has been inferred from results of early incubation experiments, with either 'cold' or labelled fatty acids, with different fruit tissues such as pear (Jennings, 1967; Jennings *et al.*, 1964; Jennings and Tressl, 1974), banana (Tressl and Drawert, 1971, 1973), and apple (Paillard, 1979). In certain cases, α-oxidation could also be involved (Tressl and Drawert, 1973).

Figure 7.1 shows a general scheme of the β-oxidation process (Conn and Stumpf, 1963) as it has been demonstrated to act in animals and microorganisms, and as it is suppossed to work in fruits in relation to aroma development. Unfortunately, this hypothesis is based on results obtained from the analysis of substrates and products in incubation experiments, identifying the presumed intermediates of the pathway, and not from a deep study of enzymes involved in the β-oxidation cycle. Briefly, fatty acids, in fact acyl-CoA derivatives, are metabolized to shorter chain acyl-CoAs by loosing two carbons at every round of the cycle involving, in the simplest case of saturated fatty acids, the following enzymes in this order:

- acyl-CoA dehydrogenase
- enoyl-CoA hydratase
- 3-hydroxyacyl-CoA dehydrogenase
- acetyl-CoA acetyltransferase, or thiolase (the enzyme responsible for the scission that gives rise to acetyl-CoA and an acyl-CoA shorter by two carbons).

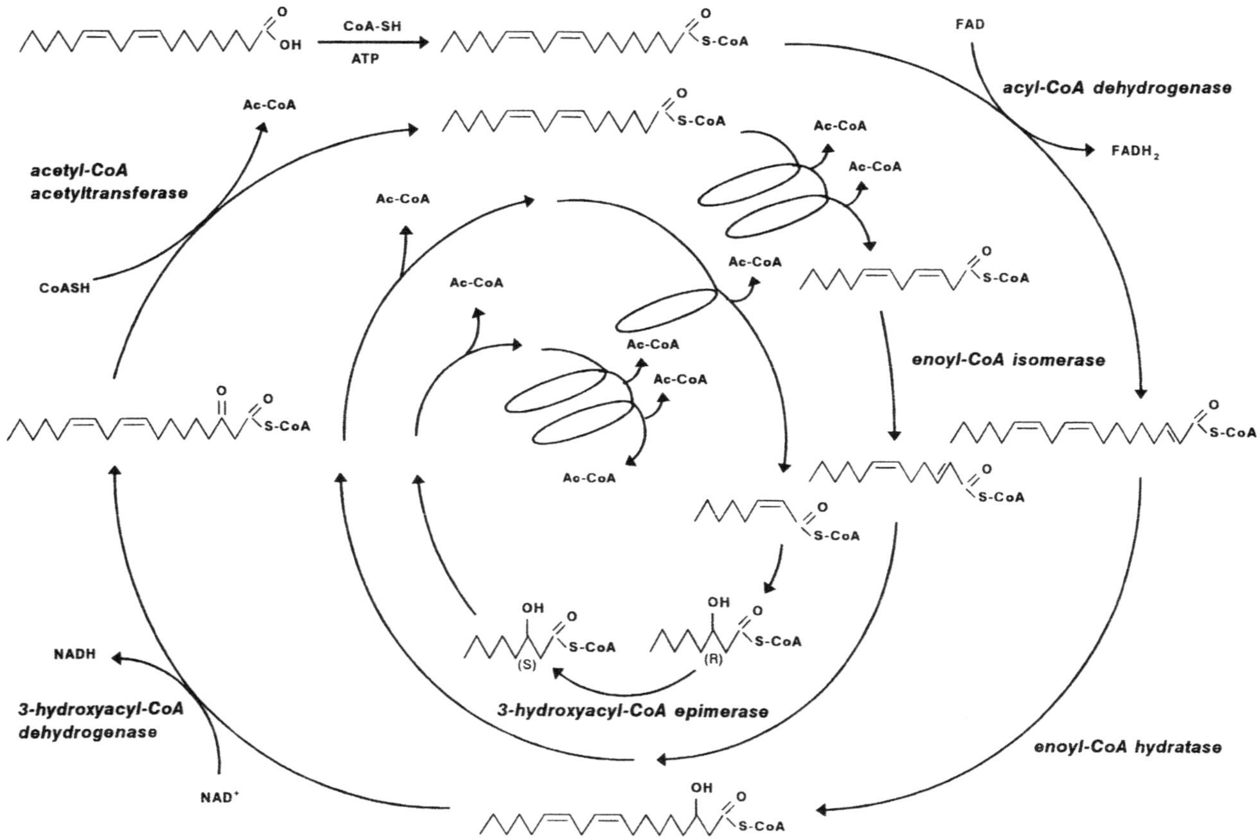

Fig. 7.1. Scheme of β-oxidation for linoleic acid.

The two dehydrogenation steps require electron acceptors such as FAD in the first case, and NAD by 3-hydroxyacyl-CoA-dehydrogenase, while free coenzyme A is needed for the scission step. The resulting acyl-CoAs produced by β-oxidation are found, in most cases, as esters in plant aromas, after the action of the ester-forming enzyme alcohol acyltransferase. Paillard (1979) demonstrated that varietal differences in aroma composition (acetic and butyric acid esters) in apples depend on the last stage in the β-oxidation of fatty acids, that is, the transformation of butyrate into acetate, particularly in Golden Delicious.

The presence of double bonds in the fatty acid molecule involves the actuation of auxiliary enzymes. After three cycles of β-oxidation on oleoyl-CoA, the resultant product is a *(Z)*-3-enoyl-CoA that is isomerized by enoyl-CoA isomerase to *(E)*-2-enoyl-CoA. This is the natural substrate for enoyl-CoA hydratase. For polyunsaturated fatty acids, such as linoleic acid, a second auxiliary enzyme is necessary to complete the oxidation. The process is similar to that for oleoyl-CoA producing five acetyl-CoA molecules and a C_8 *(Z)*-2-enoyl-CoA. This molecule can be hydrated by enoyl-CoA hydratase to form (*R*)-(–)-3-hydroxyacyl-CoA, not the (*S*)-(+)-enantiomer produced during the β-oxidation of saturated fatty acids. The second auxiliary enzyme is the 3-hydroxyacyl-CoA epimerase that gives rise to the (*S*)-(+)-isomer, specific substrate of 3-hydroxyacyl-CoA dehydrogenase of the normal β-oxidation cycle. These auxiliary enzymes could be implicated in the formation of the different enantiomeric compositions found among 3-hydroxyacid esters in tropical fruits (Tressl *et al.*, 1985).

LOX pathway

As pointed out above, it is generally accepted that by plant cell disruption LOX pathway functions actively, producing a variety of C_6 and C_9 volatile compounds. This metabolic pathway is not the only one working in plant homogenates, as shown by Galliard and Matthew (1976). Disruption of cucumber fruit tissue initiates the enzymatic deacylation of endogenous lipids whose reaction products by α-oxidation were consistent with the composition in long-chain aldehydes found in the volatile fraction from cucumber homogenates (Kemp *et al.*, 1974). On the other hand, LOX pathway actuation should not be absolutely restricted to disrupted tissues. During fruit maturation, LOX pathway enzymes and substrates would have different subcellular locations, so that no emission of volatiles by this pathway would be possible. In the course of ripening, cell walls and membranes become more permeable to different substrates, increasing the chance of reaction and so the possibility of a role of LOX pathway in fruit aroma. This has been suggested by Dettweiler *et al.* (1990) to explain the occurrence of C_8-diols in apples, and by Brackmann *et al.* (1993) as a more likely alternative for fatty acid metabolism than β-oxidation in the same fruit. That would explain why deprivation of oxygen by controlled atmosphere storage reduces the capacity of aroma production in different apple varieties (Knee and Hatfield, 1981; Lidster *et al.*, 1981; Streif and Bangerth, 1988). In this sense, the LOX pathway would be regulated by the intracellular oxygen content. This possible involvement of the LOX pathway in fruit aroma biosynthesis could also account for the high amount of hexanol produced by apple discs when incubated with linoleic acid (C-18:2) compared to that produced with stearic acid (C-18) (Paillard, 1979). In fact, this hypothesis was put forward by Heinz and Jennings (1966) in their classical work where pear aroma formation exemplifies the production of aroma via β-oxidation. They proposed LOX as a plausible alternative for the production of the observed C_8 and C_{10} ester moieties.

Many of the alcohols, aldehydes, acids and esters found in fruit and vegetable aromas are generated from the oxidative degradation of linoleic and linolenic acids. Numerous authors have pointed out the importance of enzymes from the LOX pathway to aroma biogeneration in fruits (Kazeniac and Hall, 1970; Jadhav *et al.*, 1972; Tressl and Drawert, 1973; Stone *et al.*, 1975; Galliard and Phillips, 1976; Galliard *et al.*, 1976*a*, 1977; Galliard and Matthew, 1977; Phillips *et al.*, 1979; Schreier and Lorenz, 1981; Crouzet *et al.*, 1984; Hatanaka *et al.*, 1986; Olías *et al.*, 1993).

The LOX pathway has long been recognized as the responsible for the generation of 'green' odour notes in plant products. Hatanaka (1993) reviews catalytic and mechanistic aspects related to the enzyme system involved in the formation of green odour. Although he focused mainly on green leaves, his work serves as an excellent guide to understanding the LOX pathway in any plant material. Apart from green leaves, this biosynthetic system has been demonstrated in numerous fruits such as cucumbers (Galliard and Phillips, 1976; Galliard *et al.*, 1976*a*,*b*; Phillips and Galliard, 1978; Phillips *et al.*, 1979), tomatoes (Galliard and Matthew, 1977; Galliard *et al.*, 1977), grapes (Crouzet *et al.*, 1984) or olives (Olías *et al.*, 1993).

An overall view of aroma formation through the LOX pathway is shown in Fig. 7.2. This enzymatic oxidative degradation is preceded by the action of acylhydrolases which liberates, among other things, polyunsaturated fatty acids from triacylglycerols, phospholipids or glycolipids. LOX action is the key in this pathway. This enzyme has been purified and characterized in many plant systems (Vick and Zimmerman, 1987), and gives rise to the specific hydroperoxidation of polyunsaturated fatty acids with a (1*Z*,4*Z*)-pentadiene structure. The *Z* double bond attacked by oxygen moves into conjugation with the neighbouring *Z* double bond and assumes an *E* configuration. Depending upon the plant source and the isozymes present, oxygen incorporation can occur preferentially at C-9, e.g. tomato (Galliard and Matthew, 1977), at C-13, e.g. grape (Cayrel *et al.*, 1983) and apple (Feys *et al.*, 1982), or at either C-9 or C-13 in a non-specific manner, e.g. olive (Olías *et al.*, 1993). In either case, the hydroperoxide group has an *S* enantioconfiguration. Typical LOX contains a non-haem iron atom in the active site that alternates between Fe^{2+} and Fe^{3+} states during catalysis, having a molecular weight around 100 kDa.

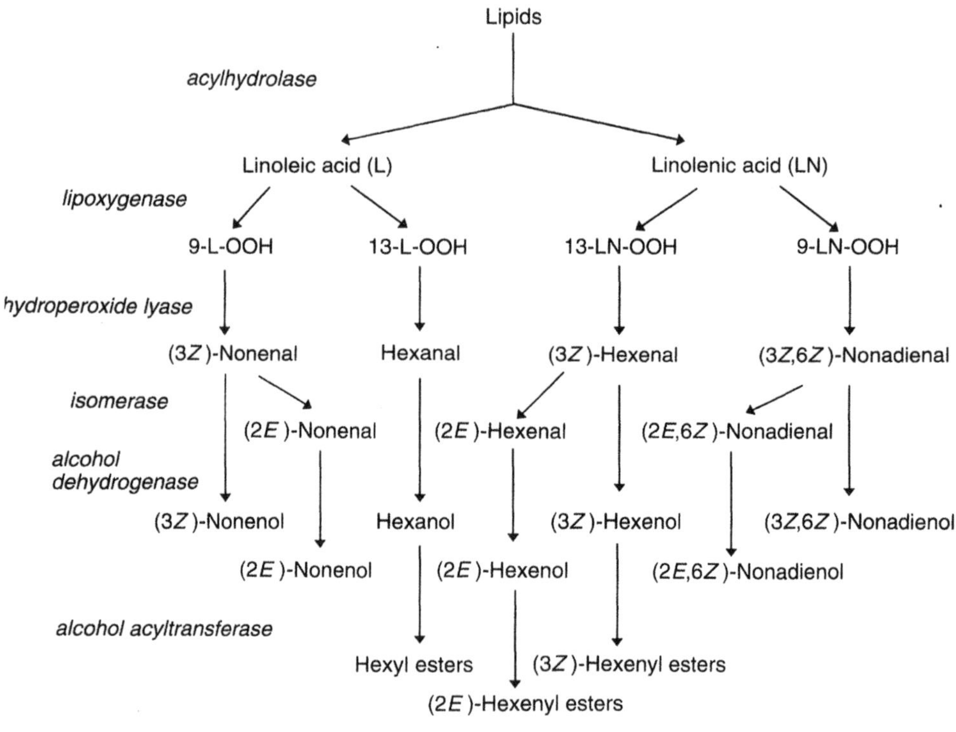

Fig. 7.2. Enzymatic activities and products involved in the LOX pathway.

Also, highly important in the LOX pathway is the enzyme that cleaves these fatty acid hydroperoxides to aldehydes and oxoacids—hydroperoxide lyase. Tressl and Drawert (1973) demonstrated the existence of an enzymatic system responsible for the production of aldehydes and oxoacids that contributes to the aroma of ripening banana extracts. Later, this system was identified in cucumber (Galliard and Phillips, 1976; Galliard *et al.*, 1976*a*), tomato (Galliard and Matthew, 1977; Schreier and Lorenz, 1982), pear (Kim and Grosch, 1981), apple (Schreier and Lorenz, 1982), and in our laboratory in olive (Olías *et al.*, 1993). Based on susbstrate specificity, hydroperoxide lyases are divided into three classes: 9-, 13- and non-specific hydroperoxide lyases. This substrate specificity determines aroma composition in many plant products, despite the specific action of LOX. Thus, it has been found that pear hydroperoxide lyase belongs to the 9-type (Kim and Grosch, 1981), although pear LOX forms mainly 13-hydroperoxides (Kim and Grosch, 1978). In this sense, we have found olive hydroperoxide lyase to have a preference for 13-hydroperoxides from linoleic and linolenic acids, despite olive LOX forming mainly 9-hydroperoxide. This specificity of olive hydroperoxide lyase is reflected in the almost total absence of C_9 carbonyls and the high content of C_6 alcohols, aldehydes and esters in the volatile composition of virgin olive oil aroma (Olías *et al.*, 1993). A quite similar situation was shown by Galliard *et al.* (1976*a*) in cucumber, where LOX produces hydroperoxides in a ratio of 13 : 9/85 : 15 (Wardale and Lambert, 1980), whereas cucumber hydroperoxide lyase is a non-specific type, explaining the presence of the highly important C_9 carbonyls in cucumber aroma. Varietal differences among hydroperoxide lyase specifities could also occur. Thus, Engel and Tressl (1983) found large amounts of C_6 aldehydes in Alphonso mango aroma, while only C_9 aldehydes were detected in Baladi mango. Mushrooms have an unusual hydroperoxide lyase that uses the 10-hydroperoxide of linoleic acid as substrate to produce 1-octen-3-ol, the compound that characterizes mushroom aroma (Wurzenbergen and Grosch, 1984).

Hydroperoxide lyase is a membrane-bound enzyme, partially purified from different fruits (Kim and Grosch, 1981; Schreier and Lorenz, 1981, 1982). Gardner and Plattner (1984) have studied model chemical reactions that mimic hydroperoxide lyase activity and proposed an amino acid acting as a Lewis acid at the active site. Oxygen isotope labelling experiments with this enzyme reaction (Hatanaka *et al.*, 1986) supported the proposed mechanism as the oxo group of 12-oxododecenoic acid arose from a hydroperoxide oxygen atom, whereas the aldehyde oxo group originated from water.

Products from 13-hydroperoxylinoleic acid are hexanal and 12-oxo-(9*Z*)-dodecenoic acid. When 13-hydroperoxylinolenic acid is the substrate, the same oxoacid is obtained and the aldehyde is (3*Z*)-hexenal. The products of hydroperoxide lyase with 9-hydroperoxides of linoleic and linolenic acids are 9-oxononanoic acid and (3*Z*)-nonenal or (3*Z*,6*Z*)-nonadienal, respectively. In most plants, compounds with a (3*Z*)-enal structure are isomerized by a (3*Z*,2*E*)-enal isomerase to the (2*E*)-enal form. This enzyme has been partially purified in cucumber fruit (Phillips *et al.*, 1979), and identified its activity in tomato (Galliard and Matthew, 1977) and olive (Olías *et al.*, 1993).

Before or after the isomerization of unsaturated carbonyls, alcohol dehydrogenase (ADH) acts to produce the corresponding alcohols on most plants (Potty and Bruemmer, 1970*b*, Rhodes, 1973; Yamashita *et al.*, 1978*a*; Bartley and Hindley, 1980; Bicsak *et al.*, 1982; Crouzet *et al.*, 1984). ADH is a metalloprotein (probably containing Zn^{2+}) possessing sulfydryl groups in the catalytic site. It is present in a dimeric form with two probably identical subunits of molecular weight 45 kDa, and requiring NAD(P)H for the activity. Alcohols resulting from ADH activity are natural substrates for the ester-forming enzyme alcohol acyltransferase. The occurrence and characteristics of this enzyme will be outlined below.

Lactone biosynthesis

A third major metabolic pathway leading to aroma formation from fatty acids is the one resulting in lactones. Lactones are important aroma components in fruits such as peach, apricot, coconut, mango or strawberry. Studies on milk products have elucidated precursors and intermediary compounds for lactone formation. In the plant kingdom, especially in fruits, only hypotheses can be drawn for lactone biosynthesis. Most authors agree that an oxidative degradation of fatty acids is involved through β-oxidation (Tressl *et al.*, 1988), and that oxo- and hydroxyacids are the precursors. However, the existence of at least three metabolic ways (reduction of oxoacids, unsaturated fatty acid hydration, and 3-hydroxyacid elongation) in the biosynthesis of δ-octalactone in pineapple (Engel *et al.*, 1989) gives an idea of the complexity of these biosynthesis hypotheses. Albrecht *et al.* (1992) carried out a study in order to elucidate the biosynthetic pathways leading to the production of γ- and δ-lactones. To overcome disadvantages in the use of fruit slices, the yeast *Sporobolomyces odorus*, which produces and excretes several of these compounds, was used as a model. They demonstrated that the biosynthesis of γ- and δ-lactones combines hydroperoxidation of linoleic acid and β-oxidation, as well as some reduction steps of the hydroperoxide group. They even proposed a stereospecific LOX mediated reaction leading to (10*S*)-hydroperoxy-(8*E*,12*Z*)-octadecadienoic acid in the biosynthesis of (*R*)-(6*Z*)-dodecen-4-olide. This LOX type could be involved in mushroom aroma biosynthesis, as mentioned above. On the other hand, participation of a 13-LOX in the production of δ-decalactone is inferred from the results of this study, producing either 13-hydroperoxy- or 13-oxo-(9*Z*,11*E*)-octadecadienoic acid. Thus, lactone biosynthesis could put in contact the two major metabolic pathways producing aroma compounds from fatty acids, LOX pathway and β-oxidation.

Amino acid metabolism

As relevant as fatty acids for aroma biogenesis, amino acids also represent an important source of volatile compounds contributing to the aroma of fruits and vegetables. They should be considered from two different points of view:

as direct precursors, metabolized to produce aroma compounds; or as indirect precursors, transformed through different metabolic pathways into new amino acid derived compounds that, although not contributing to aroma *per se*, give rise to volatile aroma compounds after enzymatic transformation upon cell disruption. As will be mentioned below, the former mechanism is typically found in fruits, whereas the latter is mainly present in vegetables.

Amino acids as direct precursors

In general, the metabolism of amino acids acting as direct precursors generates alcohols, carbonyls, acids and esters, either aliphatic, branched, or aromatic. These compounds contribute to, and in some cases determine, the primary aroma of many fruits. Free amino acid content variations have been demonstrated to occur during ripening, when characteristic aroma is produced in most fruits. Thus, a free amino acid pool could account for the different aroma profile pattern found in fruits (Tressl and Drawert, 1973; Pérez *et al.*, 1992). Amino acids such as alanine, valine, leucine, phenylalanine or aspartic acid have been shown to be aroma precursors in banana (Myers *et al.*, 1969, 1970; Tressl and Drawert, 1973) and tomato (Yu *et al.*, 1968*a*,*b*,*c*).

Radioactive labelling studies have proved that these amino acids are transformed following the scheme shown in Fig. 7.3. This aroma compound biosynthesis is quite similar to that first postulated by Ehrlich in 1907 in yeast (Ehrlich pathway), whose enzymatic basis was established by Sentheshanmuganathan (1960), comprising three enzymatic activities: aminotransferase, decarboxylase, and ADH.

Tressl and Drawert (1973) studies showed that ^{14}C-leucine is converted into labelled 3-methylbutanol, 3-methylbutanoic acid and 3-methylbutyl esters. A first step of transamination is inferred from the experimental data as glutamic acid is produced from 2-oxoglutarate. This hypothesis was validated after identification of the corresponding intermediary 2-oxoacid from the amino acid (Tressl and Drawert, 1973; Yu and Spencer, 1969, 1970), and the isolation of alanine 2-oxoglutarate aminotransferase in tomato (Yu and Spencer, 1969, 1970; Rech and Crouzet, 1974; Gazeu-Reyjal and Crouzet, 1976). This enzyme seems to be partially bound to membrane in mitochondria, exhibits a molecular weight of 100 kDa, and requires pyridoxal-5-phosphate (PALP) as cofactor, although a tight PALP–apoenzyme linkage is demonstrated as no added PALP is needed for *in vitro* maximum activity. The mechanism of action seems to be in good agreement with those elucidated for other plant and animal aminotransferases (Givan, 1980).

The next step is the decarboxylation of the 2-oxoacid formed after amino acid transamination. Drawert (1975) postulated the 2-oxoacid oxidation to

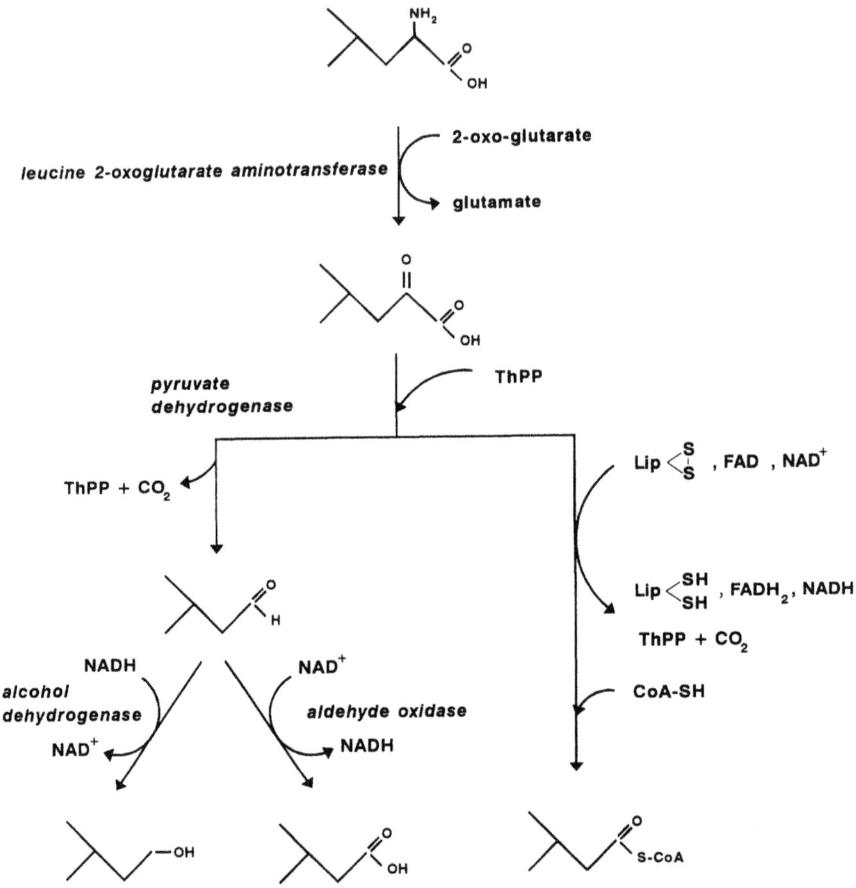

Fig. 7.3. Proposed pathway for the formation of volatiles from amino acids (adapted from Drawert, 1975).

occur by an enzymatic complex similar to pyruvate dehydrogenase (decarboxylating) or 2-oxoglutarate dehydrogenase from the Krebs cycle, involving as cofactors thiamin pyrophosphate (ThPP), lipoic acid, FAD, NAD and coenzyme A. This complex would produce 3-methylbutanoyl-CoA from leucine, substrate for alcohol acyltransferase to 3-methylbutanoate esters. The involvement of ThPP as the solo cofactor would give rise to 3-methylbutanal, which is reduced to alcohol by ADH using NADH as cofactor. This alcohol was the labelled moiety found in banana esters. However, during labelled leucine experiments in banana, no labelled aldehyde was found (Tressl *et al.*, 1970). This suggests that 3-methylbutanal would be quickly reduced to alcohol or oxidized to 3-methylbutanoic acid by an oxidoreductase requiring NAD as cofactor.

Yoshioka *et al.* (1982) proposed 2-oxoisocaproate decarboxylase, as well as alcohol acyltransferase, as the most important enzymes in the development of isoamyl acetate biosynthetic pathway in banana. However, the final products of decarboxylation could depend on the species. Thus, 2-oxovaleric acid is known to act as a powerful alkylating agent in strawberry tissue after decarboxylation (Yamashita *et al.*, 1978*b*; Drawert and Berger, 1982), but apple disc incubations with 2-oxovaleric acid produced more butanoate esters (67%) than butyl esters (21%) through decarboxylation (Berger and Drawert, 1984).

These enzymatic transformations on valine have been demonstrated to participate in the biosynthetic route to asparagusic acid methyl ester in asparagus tissues (Tressl *et al.*, 1977). This compound possesses an aroma character like raw asparagus. On the other hand, phenylalanine can also be metabolized through this pathway to 2-phenylacetyl-CoA by transamination and subsequent decarboxylation. This compound is converted into esters of a variety of alcohols, or reduced to 2-phenylethanol and transformed into 2-phenylethyl esters (Tressl and Albrecht, 1986).

Aromatic amino acids, such as tyrosine or phenylalanine, may also serve as important precursors leading to a family of compounds with phenolic and spicy odour notes, through a different metabolic pathway (Fig. 7.4). Schreier (1984) suggested that cinnamic acid from this pathway is an intermediate for many aromatic compounds identified in food aroma. This compound and its *p*-hydroxy derivative (*p*-coumaric acid) are the starting intermediates for this pathway which has been postulated to operate in banana (Tressl and Albrecht, 1986), involving different enzymatic activities which have been isolated in higher plants (Pridham, 1965; Vaughan and Butt, 1967, 1969).

Recently, Borejsza-Wysocki and Hrazdina (1994) established the biosynthetic path for *p*-hydroxyphenylbutan-2-one (pHPB) formation in raspberry fruits (see Fig. 7.4). pHPB, also known as raspberry ketone, is responsible for the characteristic aroma of this fruit. The biosynthesis of pHPB deviates from the pathway leading to benzoic acid and phenolic derivatives at the *p*-coumaric acid stage. It consists of two enzymatic steps identified in raspberry crude extracts. The first one is a condesation reaction of *p*-coumaryl-CoA with malonyl-CoA catalysed by *p*-hydroxyphenylbut-3-ene-2-one synthase, releasing coenzyme A and carbon dioxide. Secondly, a NADPH-reductase gives rise to pHPB.

Pyrazines are heterocyclic, nitrogen-containing compounds which have been detected in many processed foodstuffs, often associated with roasty and nutty aroma notes (Maga, 1982). The 3-alkyl-2-methoxy derivatives are widely distributed in the plant kingdom (Bramwell *et al.*, 1969; Buttery *et al.*, 1969; Murray *et al.*, 1970; Murray and Whitfield, 1975). Most 3-alkyl-2-methoxypyrazines identified in plants have isopropyl, *s*-butyl, and isobutyl as alkyl moieties. Thus, 3-isobutyl-2-methoxypyrazine seems to be the

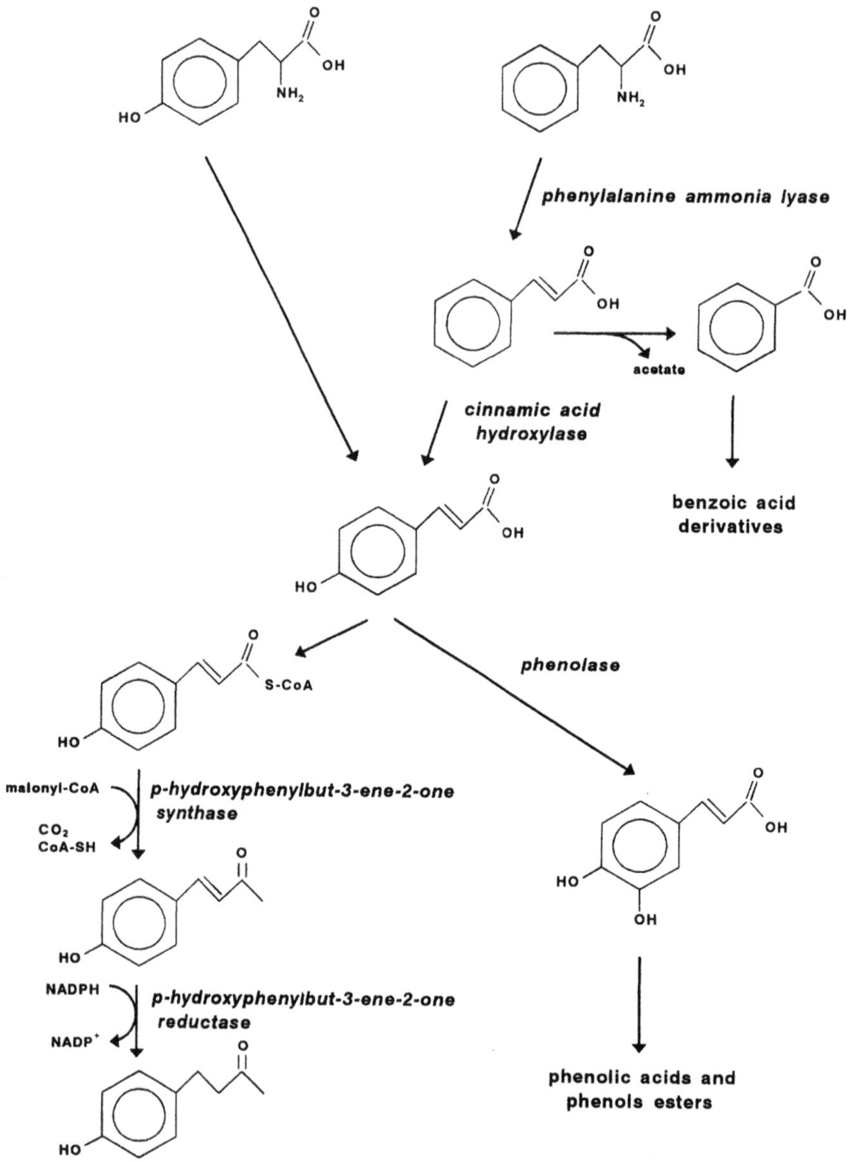

Fig. 7.4. Conversion of aromatic amino acids into volatile aroma compounds.

volatile component responsible for the typical aroma of bell peppers (Buttery *et al.*, 1969). A hypothesis has been postulated for the biosynthesis of these compounds based on chemical synthetic preparation and the correspondence between the alkyl moiety and valine, isoleucine and leucine, respectively. This proposed biosynthetic pathway, shown in Fig. 7.5 (Murray *et al.*, 1970;

Nursten and Sheen, 1974; Murray and Whitfield, 1975), involves the amidation and subsequent condensation with glyoxal of the amino acids listed above. No evidence for the enzymatic activities involved in this pathway have been found.

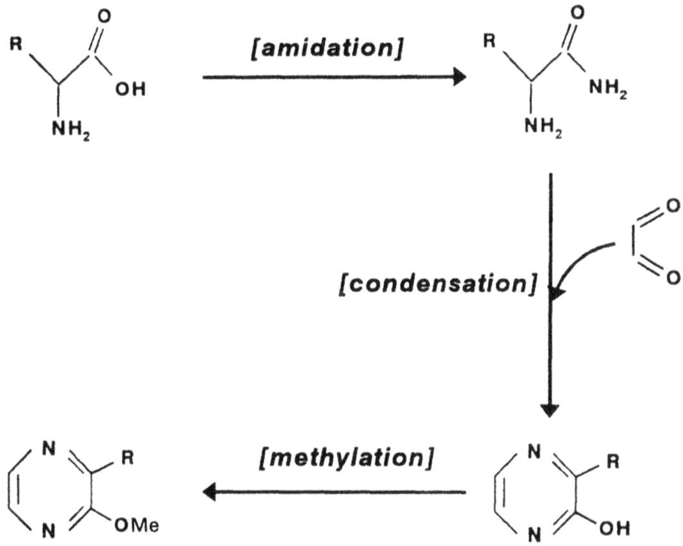

Fig. 7.5. Proposed biosynthesis of 3-alkyl-2-methoxypyrazines (from Murray *et al.*, 1970).

Little is also known about the biogenetic origin of some other minor compounds derived from amino acids important in the aroma of some fruits. Kazeniac and Hall (1970) suggested that methionine is involved in the biosynthesis of 2-isobutylthiazole in tomato. However, Schutte (1974) hypothesized that this important compound in tomato aroma is formed from the reaction of cysteamine with 3-methylbutanal (from leucine). On the other hand, 4-methoxyallylbenzene, identified as a component contributing to the spice-like note in the aroma of a number of apple cultivars, is believed to be formed from phenylalanine (Williams *et al.*, 1977).

Nitrogen- and sulfur-containing compounds from amino acids

As mentioned above, amino acids can act as substrates to produce numerous amino acid derived compounds containing nitrogen or sulfur. These compounds decisively determine the aroma of many vegetables after enzymatic degradation upon cell disruption. The biogenesis and catabolism of two of the most important classes of these compounds are described below.

S-alk(en)yl-cysteine sulfoxides The interest in investigating antibiotic substances present in onion, and especially in garlic, influenced a number of laboratories to initiate qualitative and quantitative analysis, and studies on enzymatic pathways involved in the biosynthesis and degradation of these compounds. The most outstanding characteristic of the chemical composition of *Allium* plants is the unusually large amount of organically bound sulfur (1–5% dry wt)(Whitaker, 1976). Sulfur labelling experiments demostrated that one class of these secondary metabolites, the *S*-alk(en)yl-cysteine sulfoxides, are the precursors to the characteristic aroma of *Allium* spp. plants. Four cysteine sulfoxides are found to occur in *Allium*, (+)-*S*-methyl-, (+)-*S*-propyl-, *(E)*-(+)-*S*-(1-propenyl)-, and *S*-allyl-cysteine sulfoxides. These compounds have been found to be present free and in conjugated forms as τ-glutamyl peptides, in a 1 : 1 ratio in dormant onion bulbs (Mattikala and Virtanen, 1967; Granroth, 1968). This proportion is highly important to onion aroma since the free forms are natural substrates for alliinase (cysteine sulfoxide lyase) giving rise to aroma volatiles, whereas the conjugated forms are not, and a change in the ratio could lead to changes in onion aroma potential. Early extensive studies on the biosynthesis of cysteine sulfoxides in onion, garlic, and chives (Suzuki *et al.*, 1962; Sugii *et al.*, 1963, 1964; Granroth, 1970) suggested that valine, cysteine, and serine were involved, and proposed cysteine as the precursor of *(E)*-(+)-*S*-(1-propenyl)-cysteine sulfoxide, and serine for the rest. Later, Lancaster and Shaw (1989) have shed some light on our knowledge of this pathway in experiments with ^{35}S-sulfate labelling. They proposed a biosynthetic pathway (Fig. 7.6) where τ-glutamyl cysteine and glutathione are the starting compounds, and τ-glutamyl peptides are the inmediate precursors of cysteine sulfoxides. The *S*-2-carboxypropyl moiety of the τ-glutamyl peptide and glutathione derivatives would come from the consecutive transamination, decarboxylation and dehydrogenation of valine to give rise to methacrylic acid (Granroth, 1970), in a way quite similar to that proposed for amino acids acting as direct precursors. Again, identification of intermediates have been stressed but little work have been done in terms of the enzymatic systems involved, except for some studies on τ-glutamyl transpeptidase (Kean and Hare, 1980; McCallion and Lancaster, 1984; Lancaster *et al.*, 1989). This enzyme seems to be partially associated to membrane, perhaps from peroxisomes, but most of its activity, together with τ-glutamyl cysteine sulfoxides and reaction products, is found in the soluble cytoplasmatic fraction.

The key enzymatic step in relation to the generation of aroma in *Allium* and *Brassica* species is carried out by alliinase (alliin alkyl-sulfenate lyase), also called alliin lyase, alkylcysteine lyase, cysteine sulfoxide lyase, or C-S lyase. This enzyme has been extensively studied, particularly in *Allium* plants (Stoll and Seebeck, 1947, 1948, 1951; Tobkin and Mazelis, 1979; Kazaryan and Goryachenkova, 1978; Nock and Mazelis, 1985), and in *Brassica* plants

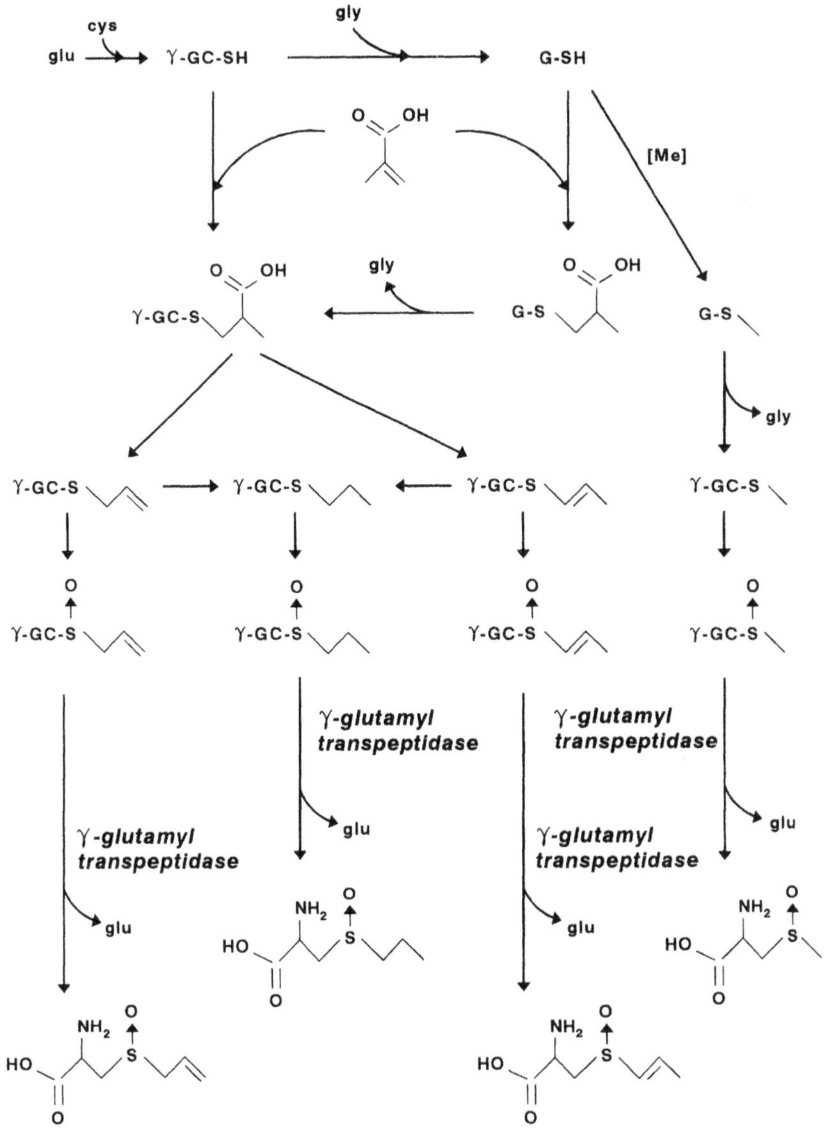

Fig. 7.6. Proposed pathway for the biosynthesis of S-alk(en)yl-cysteine sulfoxides in *Allium* plants (adapted from Lancaster and Shaw, (1989). γ-GC-SH, γ-glutamyl cysteine; G-SH, glutathione.

(Mazelis, 1963; Hamamoto and Mazelis, 1986). Alliinases catalyse the splitting of *S*-alk(en)yl-cysteine sulfoxides to produce dialk(en)yl thiosulfinates as primary products as shown below:

$$2\ \mathrm{R\text{-}\overset{\overset{O}{\uparrow}}{S}\text{-}CH_2\text{-}\underset{\underset{NH_3^+}{|}}{CH}\text{-}COO^-} \rightarrow \mathrm{R\text{-}\overset{\overset{O}{\uparrow}}{S}\text{-}S\text{-}R} + 2\ \mathrm{CH_3\text{-}CO\text{-}COO^-} + 2\ \mathrm{NH_4^+}$$

These thiosulfinates are quite unstable and rapidly undergo spontaneous non-enzymatic reactions to form a large number of sulfurous compounds characteristic of the headspace of the crushed plants mentioned above (Fenwick and Hanley, 1985; Carson, 1987; Lancaster and Boland, 1990; Block, 1992; Chin and Lindsay, 1994).

Alliinases have been purified and cDNA clones obtained from garlic, onion, and shallot (Kazaryan and Goryachenkova, 1978; Tobkin and Mazelis, 1979; Nock and Mazelis, 1985, 1987; Van Damme *et al.*, 1992; Rabinkov *et al.*, 1994). They are glycoproteins containing 5.8–6% carbohydrate by weight, with a molecular weight of about 51.5 kDa. *Allium* alliinases seem to be associated into dimers, trimers, and tetramers to form a functional holoenzyme (Lancaster and Boland, 1990; Rabinkov *et al.*, 1994). They are compartmentalized in the vacuole whereas their substrates, *S*-alk(en)yl-cysteine sulfoxides, are located in the cytoplasm (Lancaster and Collin, 1981). Cell rupture by means of tissue homogenization, crushing, or even chewing puts in contact enzyme and substrate to produce the characteristic aroma of *Allium* plants. Alliinase activities in *Brassica* plants have also been purified and characterized showing a striking similarity with those from *Allium* plants, except for a wider substrate specificity acting on other non-cysteine sulfoxide compounds, such as L-cystine.

Glucosinolates Glucosinolates are sulfur-containing compounds that yield breakdown products that contribute to the characteristic flavor of Cruciferae crops. Some of these products can also have undesirable effects in animal feedstuffs owning to their pungency and goitrogenic activity (Fenwick *et al.*, 1983*a*,*b*; Greer, 1962; McLeod, 1976).

The catabolism and toxicology of glucosinolates have been well documented (Larsen, 1981; Underhill, 1980), but relatively little is known about the biosynthesis of these sulfur-containing secondary plant products. A biosynthetic pathway for glucosinolates starting from amino acids has been elucidated by means of isotopic labelling studies (Benn, 1962; Underhill *et al.*, 1962, 1973; Wetter and Chisholm, 1968). Figure 7.7 shows this pathway by which amino acids might be converted to the corresponding glucosinolates with aldoximes, thiohydroximates, and desulfoglucosinolates acting as intermediates. From those studies, a conclusion immediately emerges: C-1 of the amino acid is lost during conversion, while the aminic nitrogen is preserved. However, confirmation of this pathway at the enzymatic level has not been completely achieved.

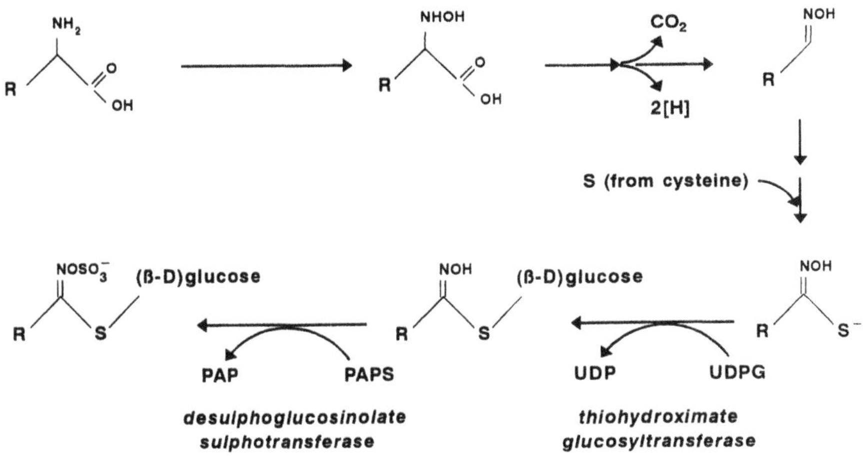

Fig. 7.7. Possible pathway for glucosinolate formation.

Decarboxylation of the *N*-hydroxyamino acid to aldoximes could be carried out in a single enzymatic step (Mahadevan, 1973), as has been proposed for the biosynthesis of cyanogenic glycosides. In this sense, both pathways would share the two first oxidative steps (Kjær, 1981). There are no conclusive studies showing how sulfur is introduced either, although cysteine is proposed as the sulfur carrier and a C-S lyase enzymatic activity is probably implied.

Matsuo and Underhill (1971) and Jain *et al.* (1990) partially purified a thiohydroximate glucosyltransferase, which catalyses the glucosylation step using uridine-5′-diphosphate glucose (UDPG) as glucose donor. The final step in this pathway seems to involve the sulfation of desulfoglucosinolates with 3′-phosphoadenosine-5′-phosphosulfate (PAPS) by a sulfotransferase. This enzymatic activity has been demonstrated and characterized in cell-free extracts of cress (*Lepidium sativum*) seedlings (Glendening and Poulton, 1988), but it was not possible to separate it chromatographically or electrophoretically from the glucosyltransferase enzymatic activity from *Brassica juncea* extracts (Jain *et al.*, 1990).

Plants containing glucosinolates also contain enzymes degrading these compounds. These enzymes, responsible for the aroma of many members of the order Capparales including a large number of Cruciferae, are usually called myrosinases. As may be inferred from their systematic name, thioglucoside glucohydrolase, myrosinases catalyse the hydrolysis of the thioglucosidic linkage in glucosinolates (Fig. 7.8). In most cases, a non-enzymatic intramolecular rearrangement of the released aglucone occurs producing isothionate and sulfate. This represents the major degradation path under conditions normally prevailing in crushed or injured plant tissues. However, depending on the sidechain R and the special conditions during the enzymatic process, fragmentation to nitriles with the liberation of sulfur (Daxenbichler *et al.*, 1977), or rare rearrangement to thiocyanates (Benn, 1977), may occur. The former is favoured by low pH or the presence of ferrous ions, whereas the latter might be enzymatically catalysed (Benn, 1977; Lüthy and Benn, 1977).

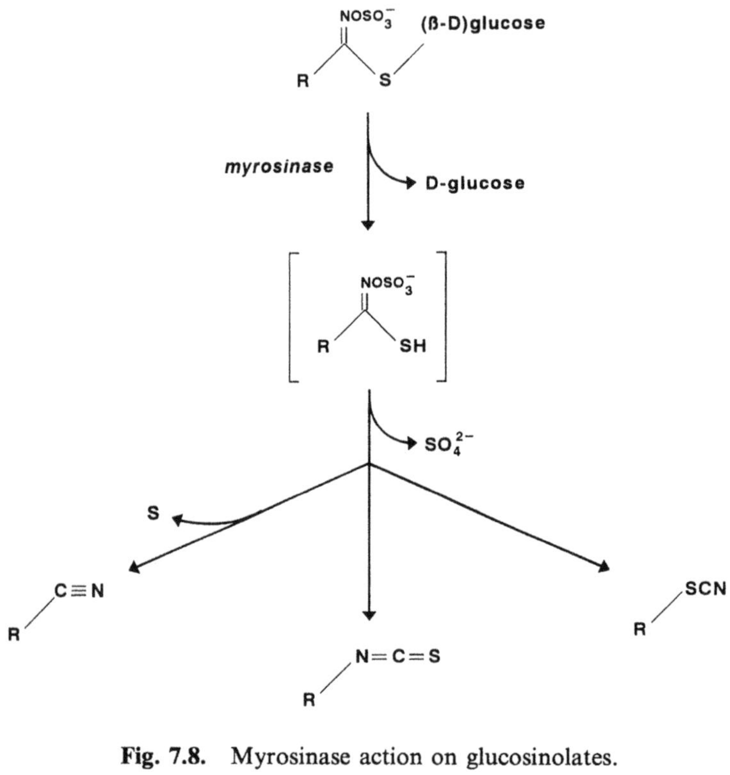

Fig. 7.8. Myrosinase action on glucosinolates.

Myrosinases have been extensively purified from various plants, and the presence of isoenzymes has been stablished in a number of cases. Recently, a cDNA clone for myrosinase from rape (*Brassica napus*) has been isolated and sequenced (Falk *et al.*, 1992). Generally speaking, myrosinases are glycoproteins containing sulfydryl groups which are essential for the catalytic activity (Björkman, 1976), with a moleclar weight of 125 000–150 000 Da and composed of two or four monomers. Myrosinase isoenzymes are characterized by their different requirements for ascorbic acid. This compound activates myrosinase catalysis probably through an allosteric effect, but its role is not clear since it can promote either activation, inhibition, or may have no effect on myrosinase activity (Ettlinger *et al.*, 1961; Ohtsuru and Hata, 1973; Björkman and Lönnerdal, 1973; Björkman, 1976). Wilkinson *et al.* (1984) have demonstrated that ascorbate concentration necessary to promote maximal activity varied with species.

It is generally assumed that myrosinases are present in special cells, so-called myrosin cells, separated from glucosinolates. Thus, aroma development becomes conditional on rupture of the cell structure. Most evidence indicates that these cells contain large amounts of myrosinases, although the latter may also be found in tissues lacking myrosin cells (Phelan *et al.*, 1984; Pocock *et al.*, 1987).

Carbohydrate metabolism

Although one may consider all plant aromas to come from carbohydrates, as they are the primary compounds produced by photosynthesis, only a few classes of aroma components in fruits and vegetables originate directly from carbohydrate metabolism. Among them, terpenes, considered to be formed from either carbohydrates or lipids, are biosynthesized through the isoprenoid pathway, while furanones seems to be produced from intermediates of the pentose phosphate cycle.

Isoprenoid pathway

Terpenes of fruit and vegetable aromas, mainly monoterpenes, are a class of volatile compounds that constitute the characteristic components of many essential oils, and contribute to, and in some cases determine, the aroma of many fruits. Volatile terpenes are just part of a large family of natural compounds, the isoprenoids, found in higher plants, and also in lower plants, fungi, animals and prokaryotes. They have been assigned many physiological functions including plant defence and hormonal or electron transport.

It is generally assumed that the early steps in the biosynthesis of volatile terpenes leading to the formation of isopentenyl diphosphate (IPP) (Fig. 7.9) are the same as those involved in the formation of other isoprenoid compounds. Mevalonic acid (MVA) is considered to be the first specific terpene precursor. The biosynthesis of MVA from acetate has not been demonstrated in higher plants, although radioactive isotope experiments leave no question as to the participation of acetate as a primary precursor in terpene synthesis. These experiments are consistent with the biosynthesis of one molecule of MVA from three of acetate (Goodwin, 1971) through an enzymatic sequence including acetyl-CoA acetyltransferase, hydroxymethylglutaryl-CoA synthetase, and hydroxymethylglutaryl-CoA reductase in a two-step reaction requiring NADPH (Kleinig, 1989). However, Ruzicka's (1953) hypothetical 'active isoprene', the basic C_5 unit from which all isoprenoid compounds are built, is IPP. This compound is formed from MVA by a series of reactions shown in Fig. 7.9. The first step is the synthesis of mevalonic acid 5-phosphate (MVAP) by the action of mevalonate kinase, an enzyme demonstrated in many plant systems (Spurgeon and Porter, 1980). MVAP undergoes a second phosphorylation with ATP catalysed by 5-phosphomevalonate kinase to give rise to mevalonic acid diphosphate (MVAPP). Finally, the last step in the biosynthesis of IPP is the decarboxylation and dehydration of MVAPP by MVAPP decarboxylase. These enzymatic activities were demonstrated to operate in orange extracts (Potty and Bruemmer, 1970*a*).

Although the synthesis of acetyl-CoA within plastids via glycolysis seems to be well established (Liedvogel and Bäuerle, 1986), the origin and compartmentation of acetyl-CoA in the plant cell are not completely understood. This compound apparently cannot permeate through membranes, so it has also to be formed in the cytoplasm. For this reason, a hypothesis has been put forward (Kaethner and ap Rees, 1985; Liedvogel, 1986) by which ATP citrate lyase could account for the production of acetyl-CoA needed for terpene biosynthesis in the cytoplasm.

The enzymatic action of prenyltransferases, forming monoterpenes, requires one molecule of IPP to be in the dimethylallyl diphosphate (DMAPP) form. This isomerization is carried out by isopentenyl diphosphate isomerase, an enzyme that has been purified and characterized in tomato (Spurgeon *et al.*, 1984), and red pepper chromoplasts (Dogbo and Camara, 1987), and seems to have a molecular weight between 28 and 37 kDa.

In the isoprenoid pathway, prenyltransferases catalyse the condensation of two isoprenoid diphosphates producing geranyl diphosphate (GPP), when DMAPP and IPP are substrates, or farnesyl diphosphate (FPP), when GPP and IPP are substrates. Prenyltransferase activity was partially purified from the flavedo of *Citrus sinensis* (de la Fuente *et al.*, 1981). Later, contrary to previous published results, Suga and Endo (1991) detected the presence of two different prenyltransferases, GPP synthase and FPP synthase in geranium leaves, requiring Mg^{2+} or Mn^{2+} as cofactors.

GPP is the direct precursor of monoterpenes. The biosynthesis of these volatile compounds is comparatively well understood, and it seems to comprise several independent or combined enzymatic reactions including hydrolysis, cyclations, and oxidoreductions (Schreier, 1986; Croteau, 1987). Most of the numerous naturally occurring monoterpenes are oxygenated

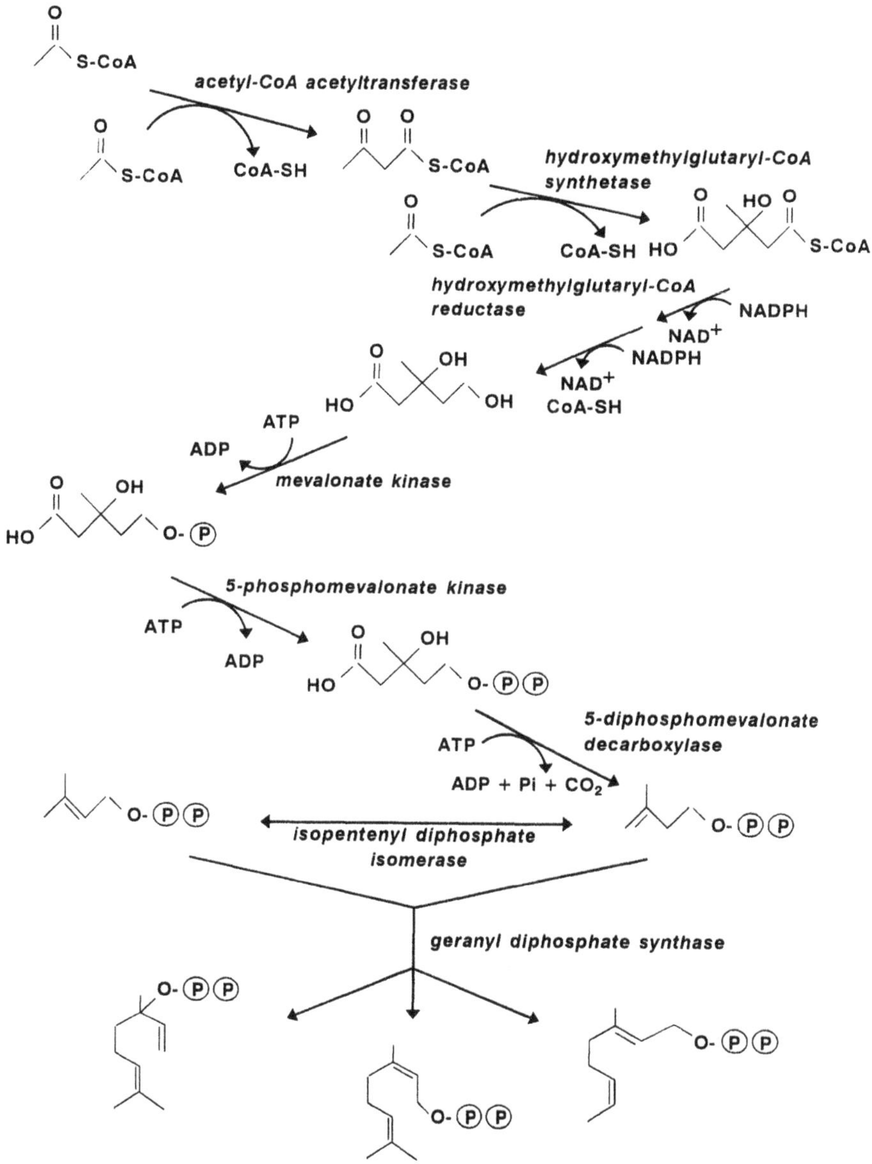

Fig. 7.9. Hypothetical conversion of acetyl-CoA into mevalonic acid and formation of monoterpenes.

compounds that are of particular significance because of the aromatic properties they impart to essential oils. The key step in the biosynthesis of most monoterpenes is the cyclation of GPP, although this enzymatic reaction rarely produces oxygenated products. Cyclases are located in plasts, and they are, except for some exceptions, soluble proteins with molecular weights of 50–100 kDa, requiring Mg^{2+} or Mn^{2+} as cofactors. Non-oxygenated monoterpenes are generally oxygenated in a series of steps which seems to involve cytochrome P450 systems and molecular oxygen (Croteau, 1987; Karp and Croteau, 1988).

Furanone biosynthesis

Furaneol (2,5-dimethyl-4-hydroxy-3(2H)-furanone) and its methylether derivative mesifurane (2,5-dimethyl-4-methoxy-3(2H)-furanone) are important aroma compounds which have been identified in many fruits, e.g. pineapple (Rodin *et al.*, 1965), raspberry (Honkanen *et al.*, 1980), mango (Pickenhagen *et al.*, 1981), arctic bramble (Kallio *et al.*, 1984), grapefruit (Lee and Nagy, 1987), tomato (Buttery *et al.*, 1994), or strawberry (Sanz *et al.*, 1995). Both furanones, having a pleasant sweet aroma, have been described as being the major character-impact compounds of fruits such as pineapple (Rodin *et al.*, 1965) or arctic bramble (Kallio *et al.*, 1984).

The interest in furanones during the last two decades has not only been due to their demonstrated contribution to many fruit aromas: furaneol has been also detected in several heat-processed foods, e.g. beef broth, roasted almonds, roasted coffee, wheat bread crust or popcorn (Schieberle, 1992; Kroh, 1994).

Despite the identification and importance of furanones to fruit aromas, little is known about their biosynthesis. After some experiments with labelled fructose on arctic bramble homogenates, Kallio (1975) proposed a synthesis hypothesis (Fig. 7.10) by which furaneol is formed from fructose through a series of unknown reactions involving hydrogen transfer and dehydration. Mesifurane would then be synthesized from furaneol by enzymatic methylation. More recently, Sasaki *et al.* (1991) proposed a biosynthetic access to the furaneol homologue 2-ethyl-4-hydroxy-5-methyl-3(2H)-furanone in yeast, proving that the metabolic pentose phosphate cycle was an early stage in this synthesis, and pointing to a 6,7-dideoxyheptose as the precursor. Similarly, furaneol could be biosynthesized from an intermediate of this pentose phosphate cycle, fructose-6-phosphate being the best precursor candidate.

Ester formation

Volatile esters formed by esterification of alcohol and carboxylic acids constitute one of the largest and main group of volatile compounds identified in fruit aroma. Biogeneration of volatile alcohols and acids is generally well explained through the enzymatic routes previously described. However, little is known about the esterification reaction, the final step in the biogenetic pathway to volatile esters.

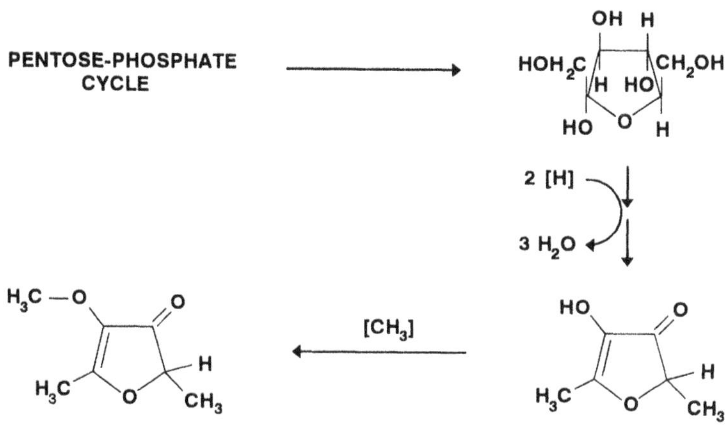

Fig. 7.10. Proposed biosynthetic origin and pathway for furanones (from Kallio, 1975 and Sasaki *et al.*, 1991).

Although the ester-forming capacity of different fruits has been proved (Yamashita *et al.*, 1975; Bartley *et al.*, 1985; Olías *et al.*, 1992), most studies have been carried out using whole fruits or tissue discs and did not involve deeper biochemical aspects. Esterification of added alcohols to banana pulp discs was investigated as early as 1970 by Myers *et al.* Further studies focused on aldehyde reduction to alcohol and subsequent ester formation in various fruit tissues (Yamashita *et al.*, 1976*a*,*b*, 1977; Ueda and Ogata, 1976). These early investigations showed that ester-forming activity is related to fruit ripening, since no activity was found in unripe fruits (Yamashita *et al.*, 1977), and also gave initial information about substrate specificity. Ester formation rate in banana pulp discs was found to be higher for straight-chain than for branched-chain acids (Ueda and Ogata, 1978), and a correlation between substrate specificity and aroma ester composition was observed in incubation experiments using intact strawberry fruits (Yamashita *et al.*, 1979).

Volatile esters are important contributors not only to the aroma of fruits but also to the flavor of fermented products. The mechanism of ester formation is better known in micro-organisms in which two different enzymes seems to be implicated: alcohol acyltransferase (AAT) and esterase. AAT catalyses the transfer of an acyl moiety of an acyl-CoA on to the corresponding alcohol, while esterase functions mainly by hydrolysing esters, although an ester-forming activity is also reported. Both enzymatic activities have also been described in fruits (Ueda and Ogata, 1976). Experiments using esterase inhibitors demonstrated that esterification of alcohols and acids in banana and strawberry is a coenzyme A-dependent reaction, and that esterase in fruits has only hydrolytic activity (Ueda and Ogata, 1977). The presence of the esterase hydrolytic activity could well account for not detecting ester formation in fruit homogenates, and explains the difficulty of AAT activity assays. In this sense, as measurement of the end-product ester is not accurate enough, detection of the acyl-CoA hydrolysis product, coenzyme A, is recommended (Fellman *et al.*, 1995).

The first attempt at AAT purification in fruits was carried out by Harada *et al.* (1985). The enzyme was localized in the soluble fraction of banana pulp cells, and described as an alcohol acetyltransferase. More recently, strawberry AAT has been purified and characterized in our laboratory (Pérez *et al.*, 1993). AAT enzymes already described in the literature appear to be a very heterogeneous group with a few common characteristics. Several studies have shown these enzymes to be associated with slowly sedimenting or floating structures of lipoprotein nature released during cell disruption. Most enzymes reported in micro-organisms are membrane associated (Yoshioka and Hashimoto, 1981; Malcorps and Dufour, 1992). Harada *et al.* (1985) observed a positive effect of Triton X-100 on banana AAT activity extraction, and our work on strawberry has proved that AAT is partially bound to membranes (Pérez *et al.*, 1993). Both, pH and temperature optima are quite similar in all AAT enzymes, pH range 7–8.5, and optimum temperature around 30 °C. Sulfydryl groups have been reported to be essential for the AAT activities of various sources. AAT activities appear to be closely linked to lipid metabolism in micro-organisms. Thus, the enzyme from *Sacharomyces cerevisiae* which catalyses the synthesis of ethyl and isoamyl acetate also acetylates 12-hydroxystearic acid, in what could be a prerequisite for incorporation of hydroxy fatty acids into cell-membrane phospholipids (Malcorps and Dufour, 1992). The relationship between AAT and lipid metabolism in fruits remains unknown.

Although availability of substrates has proved to be an important factor in ester biogeneration (Pérez *et al.*, 1992), it is clear that AAT specificity plays a key role in this process. Most enzymes so far studied have been described as alcohol acetyltransferases since their activities have not been tested with acyl-CoAs other than acetyl-CoA. The only AAT which was tested and able to use various acyl-CoAs, the enzyme from *Neurospora* sp., did not act on acetyl-CoA at all (Yamauchi *et al.*, 1989). Strawberry AAT was purified in our laboratory, showing an apparent molecular weight of 70 kDa, an optimum pH of 8.0, an optimum temperature of 35 °C, and was tested for its ability in using different acyl-CoAs and alcohols. Strawberry AAT was found to prefer acetyl-CoA and hexanol as substrates (Pérez *et al.*, 1993). Additional work carried out with different strawberry varieties showed clear differences among the AAT activity levels and maximum activity peaks during fruit ripening, although the same substrate preferences were observed for each of them (unpublished data). In a further study strawberry and banana enzyme specificities were compared, showing clear differences among acyl-CoA and alcohol specificities (Olías *et al.*, 1995). In both cases, results indicated a clear correlation between substrate specificity and volatile esters present in the aroma of each fruit, suggesting a determining role for AAT in aroma biogenesis.

Despite the importance of AAT as a key enzyme in aroma synthesis, many aspects such as its mechanism of action or physiological relevance are still not fully understood. In this sense, only preliminary studies on the effect of oxygen on AAT activity have been carried out (Fellman *et al.*, 1993; Ke *et al.*, 1994).

Conclusions

As we have seen throughout this brief review, biochemical knowledge about aroma formation in fruits and vegetables is still very limited. Apart from some specific biochemical pathways extensively studied mainly as a result of previous interests related to health, most of them require a deeper study for elucidation. Biotechnological applications in the field of aroma and fragances might make these biochemical studies possible.

In addition to the biochemical aspects covered in this paper it is worth mentioning the great strides that the study of glycoside derivatives has made in recent years. These compounds are now considered as direct aroma precursors because of the large number so far identified, and because the hypothetical physiological roles assigned to glycosides—as transport vehicles through membranes or as membrane protectors—do not always fit (Stahl-Biskup *et al.*, 1993).

References

Albrecht, W. Heidlas, J., Schwarz, M. and Tressl, R. (1992). In *Flavor precursors. Thermal and enzymatic conversions* (ed. R. Teranishi, G.R. Takeoka and M. Güntert), pp. 46–58. ACS Symposium Series 490, Washington, DC.

Bartley, I.M. and Hindley, S.J. (1980). *J. Exp. Bot.* **31**, 449–59.

Bartley, I.M, Stoker, P.G., Martin, A.D.E., Hatfield, S.G. and Knee, M. (1985). *J. Sci. Food Agric.* **36**, 567–74.

Benn, M.H. (1962). *Chem. Ind.* 1907.

Benn, M.H. (1977). *Pure Appl. Chem.* **49**, 197–210.

Berger, R.G. and Drawert, F. (1984). *J. Sci. Food Agric.* **35**, 1318–25.

Bicsak, T.A., Kann, L.R., Reiter, A. and Chase, T.Jr. (1982). *Arch. Biochem. Biophys.* **216**, 605–15.

Björkman, R. (1976). In *The biology and chemistry of the Cruciferae* (ed. J.G. Vaughan, A.J. McLeod and B.M.G. Jones), pp. 191–205. Academic Press, London.

Björkman, R. and Lönnerdal, B. (1973). *Biochim. Biophys. Acta* **327**, 121–31.

Block, E. (1992). *Angew. Chem. Int. Ed. Engl.* **31**, 1135–78.

Borejsza-Wysocki, N. and Hrazdina, G. (1994). *Phytochemistry* **35**, 623–8.

Brackmann, A., Streif, J. and Bangerth, F. (1993). *J. Amer. Soc. Hort. Sci.* **118**, 243–7.

Bramwell, A.F., Burrell, J.W.K. and Riezebos, G. (1969). *Tetrahedron Lett.* **37**, 3215–16.

Buttery, R.G., Seifert, R.M., Lundin, R.E., Guadagni, D.G. and Ling, L.C. (1969). *Chem. Ind.* **15**, 490–1.

Buttery, R.G., Takeoka, G.R., Krammer, G.E. and Ling, L.C. (1994). *Lebensm. Wiss. Technol.* **27**, 592–4.

Carson, J.F. (1987). *Food Rev. Int.* **3**, 71–103.

Cayrel, A., Crouzet, J., Chan, H.W.S. and Price, K.R. (1983). *Am. J. Enol. Vitic.* **34**, 77–82.

Chin, H.W. and Lindsay, R.C. (1994). *J. Agric. Food Chem.* **42**, 1529–36.

Conn, E.E. and Stumpf, P.K. (1963). In *Outlines of biochemistry*. Wiley, New York.

Croteau, R. (1987). *Chem. Rev.* **87**, 929–54.

Crouzet, J., Nicolas, M., Molina, I. and Valentin, G. (1984). In *Progress in flavour research* (ed. J. Adda), pp. 401–8. Elsevier, Amsterdam.

Daxenbichler, M.E., Van Etten, C.H. and Spencer, G.F. (1977). *J. Agric. Food Chem.* **25**, 121–4.

Dettweiler, G.R., Berger, R.G. and Drawert, F. (1990). *Dtsch. Lebensm. Rdsch.* **86**, 174–6.

Dogbo, O. and Camara, B. (1987). *Biochim. Biophys. Acta* **920**, 140–8.

Drawert, F. (1974). In *Proc. Colloq. Int. CNRS* **238** (ed. M. Ulrich), pp. 309–23. CNRS, Paris.

Drawert, F. (1975). In *Aroma research* (ed. H. Maarse and P.J. Groenen), pp. 13–39. Pudoc, Wageningen.

Drawert, F. and Berger, R.G. (1982). *Z. Naturforsch.* **37c**, 849–56.

Engel, K.H., Heidlas, J., Albrecht, W. and Tressl, R. (1989). In *Flavor chemistry: trends and development* (ed. R. Teranishi, R.G. Buttery and F. Shahidi), pp. 8–22. ACS Symposium Series **388**, New York.

Engel, K.H. and Tressl, R. (1983). *J. Agric. Food Chem.* **31**, 796–801.

Ettlinger, M.G., Dateo, G.P., Harrison, B.W., Marby, J. and Thompson, C.P. (1961). *Proc. Nat. Acad. Sci.* **47**, 1875–9.

Falk, A., Xue, J., Lenman, M. and Rask, L. (1992). *Plant Sci.* **83**, 181–6.

Fellman, J.K. and Mattheis, J.P. (1995). In *Fruit flavors. Biogenesis, characterization, and authentication* (ed. R.L. Rouseff and M.M. Leahy), pp. 149–62. ACS Symposium Series 596. Washington, DC.

Fellman, J.K., Mattinson, D.S., Bostick, B.C., Mattheis, J.P. and Patterson, M.E. (1993). *Postharvest Biol. Technol.* **3**, 201–14.

Fenwick, G.R. and Hanley, A.B. (1985). *CRC Crit. Rev. Food Sci. Nutr.* **22**, 273–7.

Fenwick, G.R., Griffiths, N.M. and Heanly, R.K. (1983*a*). *J. Sci. Food Agric.* **34**, 73–80.

Fenwick, G.R., Heaney, R.K. and Mollin, W.J. (1983*b*). *CRC Crit. Rev. Food Sci. Nutr.* **18**, 123–201.

Feys, M., de Mot, R., Naesens, W. and Tobback, P. (1982). *Z. Lebensm. Unter. Forsch.* **174**, 360–5.

Fuente, M., Pérez, L.M., Hashagen, U., Chayet, L., Rojas, C., Portilla, G. and Cori, O. (1981). *Phytochemistry* **20**, 1551–7.
Galliard, T. and Matthew, J.A. (1976). *Biochim. Biophys. Acta* **424**, 26–35.
Galliard, T. and Matthew, J.A. (1977). *Phytochemistry* **16**, 339–43.
Galliard, T. and Phillips, D.R. (1976). *Biochim. Biophys. Acta* **431**, 278–87.
Galliard, T., Phillips, D.R. and Reynolds, J. (1976*a*). *Biochim. Biophys. Acta* **441**, 181–92.
Galliard, T., Matthew, J.A., Fishwick, M.J. and Wright, A.J. (1976*b*). *Phytochemistry* **15**, 1647–50.
Galliard, T, Matthew, J.A., Wright, A.J. and Fishwick, M.J. (1977). *J. Sci. Food Agric.* **28**, 863–8.
Gardner, H.W. and Plattner, R.D. (1984). *Lipids* **19**, 294–9.
Gazeu-Reyjal, M. and Crouzet, J. (1976). *Phytochemistry* **15**, 1619–22.
Givan, C.V. (1980). In *The biochemistry of plants*, Vol. 5 (ed. P.K. Stumpf and E.E. Conn), pp. 329–57. Academic Press, New York.
Glendening, T.M. and Poulton, J.E. (1988). *Plant Physiol.* **86**, 319–21.
Goodwin, T.W. (1971). In *Carotenoids* (ed. O. Isler), pp. 577–636. Birkhaeuser, Basel.
Granroth, B. (1968). *Acta Chem. Scand.* **22**, 3333–5.
Granroth, B. (1970). *Ann. Acad. Sci. Fenn. Ser.* **154**, 1–71.
Greer, M.A. (1962). *Recent Prog. Horm. Res.* **18**, 187–219.
Hamamoto, A. and Mazelis, M. (1986). *Plant Physiol.* **80**, 702–6.
Harada, M., Ueda, Y. and Iwata, T. (1985). *Plant Cell Physiol.* **26**, 1067–74.
Hatanaka, A. (1993). *Phytochemistry* **34**, 1201–18.
Hatanaka, A., Kajiwara, T. and Sekiya, J. (1986). In *Biogeneration of aromas* (ed. T.H. Parliment and R. Croteau), pp. 167–75. ACS Symposium Series **317**, Washington, DC.
Heinz, D.E. and Jennings, W.G. (1966). *J. Food Sci.* **31**, 69–80.
Honkanen, E., Pyysalo, T. and Hirvi, T. (1980). *Z. Lebensm. Unter. Forsch.* **171**, 180–2.
Jadhav, S., Singh, B. and Salunkhe, D.K. (1972). *Plant Cell Physiol.* **13**, 449–59.
Jain, J.C., Groot-Wassink, J.W.D., Reed, D.W. and Underhill, E.W. (1990). *J. Plant Physiol.* **136**, 356–61.
Jennings, W.G. (1967). In *Chemistry and physiology of flavours* (ed. H.W. Schultz, E.A. Day and L.M. Libbey), pp. 419–430. AVI Publishing, Westport, CT.
Jennings, W.G. and Tressl, R. (1974). *Chem. Mikrobiol. Technol. Lebensm.* **3**, 52–5.
Jennings, W.G., Creveling R.K. and Heinz, D.E. (1964). *J. Food Sci.* **29**, 730–4.
Kaethner, T.M. and ap Rees, T. (1985). *Planta* **163**, 290–4.
Kallio, H. (1975). S.M. Acad. Diss., University of Turku, Finland.
Kallio, H., Lapvetelläinen, A., Hirvi, T. and Honkanen, E. (1984). In *Analysis of volatiles* (ed. P. Schreier), pp. 433–46. Walter de Gruyter, Berlin.
Karp, F. and Croteau, R. (1988). In *Bioflavour'87* (ed. P. Schreier), pp.173–98. Walter de Gruyter, Berlin.
Kazaryan, R.A. and Goryachenkova, E.V. (1978). *Biochemistry* **43**, 1502–8.
Kazeniac, S.J. and Hall, R.M. (1970). *J. Food Sci.* **35**, 519–30.
Ke, D., Zhou, L. and Kader, A.A. (1994). *J. Amer. Soc. Hort. Sci.* **119**, 971–5.
Kean, E.A. and Hare, E.R. (1980). *Phytochemistry* **19**, 199–203.
Kemp, T.R., Knavel, D.E. and Stoltz, L.P. (1974). *J. Agric. Food Chem.* **22**, 717–8.
Kim, I.S. and Grosch, W. (1978). *Z. Lebensm. Unter. Forsch.* **167**, 324–6.
Kim, I.S. and Grosch W. (1981). *J. Agric. Food Chem.* **29**, 1220–5.

Kjær, A. (1981). *Food Chem.* **6**, 223–34.
Kleinig, H. (1989). *Annu. Rev. Plant Physiol. Plant Mol. Biol.* **40**, 39–59.
Knee, M. and Hatfield, S.G.S. (1981). *J. Sci. Food Agric.* **32**, 593–600.
Kroh, L. (1994). *Food Chem.* **51**, 373–9.
Lancaster, J.E. and Boland, M.J. (1990). In *Flavor biochemistry* (ed. J.L. Brewster and H.E. Rabinowitch), pp. 33–72. CRC Press, Boca Raton, FL.
Lancaster, J.E. and Collin, H.A. (1981). *Plant Sci. Lett.* **22**, 169–76.
Lancaster, J.E. and Shaw, M.L. (1989). *Phytochemistry* **28**, 455–60.
Lancaster, J.E., Reynolds, P.H.S., Shaw, M.L., Dommisse, E.M. and Munro, J. (1989). *Phytochemistry* **28**, 461–4.
Larsen, P.O. (1981). In *The biochemistry of plants*, Vol. 7 (ed. E.E. Conn and P.K. Stumpf), pp. 502–23. Academic Press, New York.
Lee, H.S. and Nagy, S. (1987) *J. Food Sci.* **52**, 163–5.
Lidster, P.D., McRae, K.B. and Katherine, A.S. (1981). *J. Amer. Soc. Hort. Sci.* **106**, 159–62.
Liedvogel, B. (1986). *J. Plant Physiol.* **124**, 211–22.
Liedvogel, B. and Bäuerle, R. (1986). *Planta* **169**, 481–9.
Lüthy, J. and Benn, M.H. (1977). *Can J. Biochem.* **55**, 1028–31.
Maga, J.A. (1982). In *Food flavours* (ed. I.D. Morton and A.J. McLeod), pp. 283–323. Elsevier, Amsterdam.
Mahadevan, S. (1973). *Ann. Rev. Plant Physiol.* **24**, 69–88.
Malcorps, P. and Dufour, J.P. (1992). *Eur. J. Biochem.* **210**, 1015–22.
Matsuo, M. and Underhill, E.W. (1971). *Phytochemistry* **10**, 2279–86.
Mattikala, E.J. and Virtanen, A.I. (1967). *Acta Chem. Scand.* **21**, 2891–3.
Mazelis, M. (1963). *Phytochemistry* **2**, 15–22.
McCallion, B.J. and Lancaster, J.E. (1984). *Physiol. Plant.* **62**, 370–4.
McLeod, A.J. (1976). In *The biology and chemistry of the Cruciferae* (ed. J.G. Vaughan, A.J. McLeod and B.M.G. Jones), pp. 307–30. Academic Press, London.
Murray, K.E. and Whitfield, F.B. (1975). *J. Sci. Food Agric.* **26**, 973–86.
Murray, K.E., Shipton, J. and Whitfield, F.B. (1970). *Chem. Ind.* **27**, 897–8.
Myers, M.J., Issenberg, P. and Wick, E.L. (1969). *J. Food Sci.* **34**, 504–9.
Myers, M.J., Issenberg, P. and Wick, E.L. (1970). *Phytochemistry* **9**, 1693–1700.
Nock, L.P. and Mazelis, M. (1985). *Plant Physiol.* **77**, S-115.
Nock, L.P. and Mazelis, M. (1987). *Plant Physiol.* **85**, 1079–83.
Nursten, H.E. and Sheen, M.R. (1974). *J. Sci. Food Agric.* **25**, 643–63.
Ohtsuru, M. and Hata, T. (1973). *Agric. Biol. Chem.* **37**, 1975–2.
Olías, J.M., Sanz, L.C., Ríos, J.J. and Pérez, A.G. (1992). *J. Agric. Food Chem.* **40**, 266–70.
Olías, J.M., Pérez, A.G., Ríos, J.J. and Sanz, L.C. (1993). *J. Agric. Food Chem.* **41**, 2368–73.
Olías, J.M., Sanz, C., Ríos, J.J. and Pérez, A.G. (1995). In *Fruit flavors. Biogenesis, characterization, and authentication* (ed. R.L. Rouseff and M.M. Leahy), pp. 134–41. ACS Symposium Series 596. Washington, DC.
Paillard, N.M.M. (1979). *Phytochemistry* **18**, 1165–71.
Pérez, A.G., Ríos, J.J., Sanz, C. and Olías, J.M. (1992). *J. Agric. Food Chem.* **40**, 2232–5.
Pérez, A.G., Sanz, C. and Olías, J.M. (1993). *J. Agric. Food Chem.* **41**, 1462–6.
Phelan, J.R., Allen, A. and Vaughan, J.G. (1984). *J. Exp. Bot.* **35**, 1558–64.
Phillips, D.R. and Galliard, T. (1978). *Phytochemistry* **17**, 355–8.

Phillips, D.R., Matthew, J.A., Reynolds, J. and Fenwick, G.R. (1979). *Phytochemistry* **18**, 401–4.

Pickenhagen, W., Velluz, A., Passerat, J.P. and Ohloff, G. (1981). *J. Sci. Food Agric.* **32**, 1132–4.

Pocock, K., Heaney, R.K., Wilkinson, A.P., Beaumont, J.E., Vaughan, J.G. and Fenwick, G.R. (1987). *J. Sci. Food Agric.* **41**, 245–57.

Potty, V.H. and Bruemmer, J.H. (1970*a*). *Phytochemistry* **9**, 99–105.

Potty, V.H. and Bruemmer, J.H. (1970*b*). *Phytochemistry* **9**, 1001–7.

Pridham, J.B. (1965). *Ann. Rev. Plant Physiol.* **16**, 13–36.

Rabinkov, A, Ziao, Z.Z., Grafi, G., Galili, G. and Mirelman, D. (1994). *Appl. Biochem. Biotech.* **48**, 149–71.

Rech, J. and Crouzet, J. (1974). *Biochim. Biophys. Acta* **350**, 392–9.

Rhodes, M.J.C. (1973). *Phytochemistry* **12**, 307–14.

Rodin, J.O., Himel, C.M., Silverstein, R.M., Leeper, R.W. and Gortner, W.A. (1965). *J. Food Sci.* **30**, 280–5.

Ruzicka, L. (1953). *Experientia* **9**, 357–96.

Sanz, C., Richardson, D.G. and Pérez, A.G. (1995). In *Fruit flavors. Biogenesis, characterization, and authentication* (ed. R.L. Rouseff and M.M. Leahy), pp. 268–75. ACS Symposium Series **596**. Washington, DC.

Sasaki, M., Nunomura, N and Matsudo, T. (1991). *J. Agric. Food Chem.* **39**, 934–8.

Schieberle, P. (1992). In *Flavor precursors* (ed. R. Teranishi, G.R. Takeoka and M. Güntert), pp. 164–74. ACS Symposium Series **490**, Washington, DC.

Schreier, P. (1984). In *Cromatographic studies of biogenesis of plant volatiles*. Hüthig, Heidelberg.

Schreier, P. (1986). In *Developments in food flavours* (ed. G.G. Birch and M.G,. Lindley), pp. 89–106. Elsevier, Amsterdam.

Schreier, P. and Lorenz, G. (1981). In *Flavour'81* (ed. P. Schreier), pp. 495–507. Walter de Gruyter, Berlin.

Schreier, P. and Lorenz, G. (1982). *Z. Naturforsch* **37**, 165–73.

Schutte, L. (1974). *CRC Crit. Rev. Food Technol.* **4**, 457–505.

Sentheshanmuganathan, S. (1960). *Biochem. J.* **74**, 568–76.

Spurgeon, S.L. and Porter, J.W. (1980). In *The biochemistry of plants*, Vol. 4 (ed. P.K. Stumpf), pp. 419–83. Academic Press, New York.

Spurgeon, S.L., Sathyamoorthy, N. and Porte, J.W. (1984). *Arch. Biochem. Biophys.* **230**, 446–54.

Stahl-Biskup, E., Intert, F., Holthuijzen, J., Stengele, M. and Schulz, G. (1993). *Flavour Frag. J.* **8**, 61–80.

Stone, E.J., Hall, R.M. and Kazeniac, S.J. (1975). *J. Food Sci.* **40**, 1138–41.

Stoll, A. and Seebeck, E. (1947). *Experientia* **3**, 114–15.

Stoll, A. and Seebeck, E. (1948). *Helv. Chim. Acta* **31**, 189–210.

Stoll, A. and Seebeck, E. (1951). *Adv. Enzymol.* **11**, 377–400.

Streif, J. and Bangerth, F. (1988). *J. Hort. Sci.* **63**, 193–9.

Suga, T. and Endo, T. (1991). *Phytochemistry* **30**, 1757–61.

Sugii, M., Nagasawa, S. and Suzuki, T. (1963). *Chem. Pharm. Bull.* **11**, 135–6.

Sugii, M., Suzuki, T., Kakimoto, T. and Kato, J. (1964). *Bull. Inst. Chem. Res. Kyoto Univ.* **42**, 246–51.

Suzuki, T., Sugii, M. and Kakimoto, T. (1962). *Chem. Pharm. Bull.* **10**, 328–31.

Tobkin, H.E.Jr. and Mazelis, M. (1979). *Arch. Biochem. Biophys.* **193**, 150–7.

Tressl, R. and Albrecht, W. (1986). In *Biogeneration of aromas* (ed. T.H. Parliment and R. Croteau), pp. 114–33. ACS Symposium Series 317. Washington, DC.
Tressl, R. and Drawert, F. (1971). *Z. Naturforsch* **26b**, 774–9.
Tressl, R. and Drawert, F. (1973). *J. Agric. Food Chem.* **21**, 560–5.
Tressl, R., Emberger, R., Drawert, F. and Heimann, W. (1970). *Z. Naturforsch.* **25b**, 704–7.
Tressl, R., Holzer, M. and Apetz, M. (1977). *J. Agric. Food Chem.* **25**, 455–9.
Tressl, R., Engel, K.H., Albrecht, W. and Bille-Abdullah, H. (1985). In *Characterization and measurement of flavor compounds* (ed. D.D. Bills and C.J. Mussinan), pp. 43–60. ACS Symposium Series **289**. Washington, DC.
Tressl, R., Heidlas, J., Albrecht, W. and Engel, K.H. (1988). In *Bioflavour'87* (ed. P. Schreier), pp. 221–36. Walter de Gruyter, Berlin.
Ueda, Y. and Ogata, K. (1976). *Nippon Shokuhin Kogyo Gakkaishi* **23**, 288–94.
Ueda, Y. and Ogata, K. (1977). *Nippon Shokuhin Kogyo Gakkaishi* **24**, 624–30.
Ueda, Y. and Ogata, K. (1978). *Nippon Shokuhin Kogyo Gakkaishi* **25**, 440–5.
Underhill, E.W. (1980). In *Secondary plant products*, Vol. 8 (ed. E.A. Bell and B.V. Charlwood), pp. 493–506. Springer-Verlag, New York, NY.
Underhill, E.W., Chisholm, M.D. and Wetter, L.R. (1962). *Can. J. Biochem. Physiol.* **40**, 1505–14.
Underhill, E.W., Wetter, L.R. and Chisholm, M.D. (1973). *Biochem. Soc. Symp.* **38**, 303–26.
Van Damme, E.J.M., Smeets, K., Torrekens, S., Van Leuven, F. and Peumans, W.J. (1992). *Eur. J. Biochem.* **209**, 751–7.
Vaughan, P.F.T. and Butt, V.S. (1967). *Biochem. J.* **104**, 65p.
Vaughan, P.F.T. and Butt, V.S. (1969). *Biochem. J.* **113**, 109–15.
Vick, B.A. and Zimmerman, D.C. (1987). In *The biochemistry of plants*, Vol. 9 (ed. P.K. Stumpf and E.E. Conn), pp.53–90. Academic Press, New York.
Wardale, D.A. and Lambert, E.A. (1980). *Phytochemistry* **19**, 1013–16.
Wetter, L.R. and Chisholm, M.D. (1968). *Can. J. Biochem.* **46**, 931–5.
Whitaker, J.R. (1976). *Adv. Food Res.* **22**, 73–133.
Wilkinson, A.P., Rhodes, M.J.C. and Fenwick, R.G. (1984). *J. Sci. Food Agric.* **35**, 543–52.
Williams, A.A., Tucknott, O.G. and Lewis, M.J. (1977). *J. Sci. Food Agric.* **28**, 185–90.
Wurzenbergen, M. and Grosch, W. (1984). *Biochim. Biophys. Acta* **794**, 25–30.
Yamashita, I., Nemoto, Y. and Yoshikawa, S. (1975). *Agric. Biol. Chem.* **39**, 2303–7.
Yamashita, I., Nemoto, Y. and Yoshikawa, S. (1976*a*). *Phytochemistry* **15**, 1633–7.
Yamashita, I., Nemoto, Y. and Yoshikawa, S. (1976*b*). *Agric. Biol. Chem.* **40**, 2231–5.
Yamashita, I., Iino, K., Nemoto, Y. and Yoshikawa, S. (1977). *J. Agric. Food Chem.* **25**, 1165–8.
Yamashita, I, Iino, K. and Yoshikawa, S. (1978*a*). *Agric. Biol. Chem.* **42**, 1125–32.
Yamashita, I., Naito, S., Iino, K. and Yoshikawa, S. (1978*b*). *Nippon Shokuhin Kogyo Gakkaishi* **25**, 378–82.
Yamashita, I., Iino, K. and Yoshikawa, S. (1979). *Nippon Shokuhin Kogyo Gakkaishi* **26**, 256–9.
Yamauchi, H., Hasuo, T., Amachi, T., Akita, O., Hara, S. and Yoshikawa, K. (1989). *Agric. Biol. Chem.* **53**, 1551–6.
Yoshioka, K., and Hashimoto, N. (1981). *Agric. Biol. Chem.* **45**, 2183–90.
Yoshioka, H., Ueda, Y. and Iwata, T. (1982). *Nippon Shokuhin Kogyo Gakkaishi* **29**, 333–9.

Yu, M.H. and Spencer, M. (1969). *Phytochemistry* **8**, 1173–8.
Yu, M.H. and Spencer, M. (1970). *Phytochemistry* **9**, 341–3.
Yu, M.H., Olson, L.E. and Salunkhe, D.K. (1968*a*). *Phytochemistry* **7**, 555–60.
Yu, M.H., Olson, L.E. and Salunkhe, D.K. (1968*b*). *Phytochemistry* **7**, 561–5.
Yu, M.H., Salunkhe, D.K. and Olson, L.E. (1968*c*). *Plant Cell Physiol.* **9**, 633–8.

8. The psychology of food choice: the influence of attitudes and flavour perception

RICHARD SHEPHERD

Institute of Food Research, Earley Gate, Whiteknights Road, Reading RG6 6BZ, UK

Factors affecting food choice

The choice of foods by free-living people is a research issue of concern for many of those involved in the production and distribution of foods, and also for those concerned with nutrition and health education. Despite a great deal of knowledge gained on the impacts of diet on health and on specific diseases (Committee on Medical Aspects of Food Policy, 1994), relatively little is known about why people choose the foods that make up their diets or about how their choices might be influenced. Given recommendations, for example, to reduce fat in the diet or to increase the consumption of fruit and vegetables, it is then necessary to understand the determinants of food choice and what obstacles there might be to such changes. Although official recommendations have been in place in the UK since the report by COMA in 1984 (Committee on Medical Aspects of Food Policy, 1984) for a reduction in the energy in the diet derived from fat there has been relatively little change (Committee on Medical Aspects of Food Policy, 1994).

Food choice, like any complex human behaviour, will be influenced by many factors and these factors will not operate in isolation. A number of models seeking to delineate the effects of likely influences have been put forward (e.g. Pilgrim, 1957; Khan, 1981; Randall and Sanjur, 1981; Shepherd, 1985; reviewed by Shepherd, 1989). However, few of these models present any indication of likely mechanisms of action of the factors identified, nor do they quantify the relative importance of, or interactions between, factors. Likewise they do not allow any quantitative tests which are predictive of food choice. Thus many such models are really only catalogues of the likely influences but as such they can be useful in highlighting the variables worthy of study.

An example of such a model is shown in Fig. 8.1. The factors influencing food choice are categorized as those related to the food, to the person making the choice and to the external economic and social environment within which the choice is made.

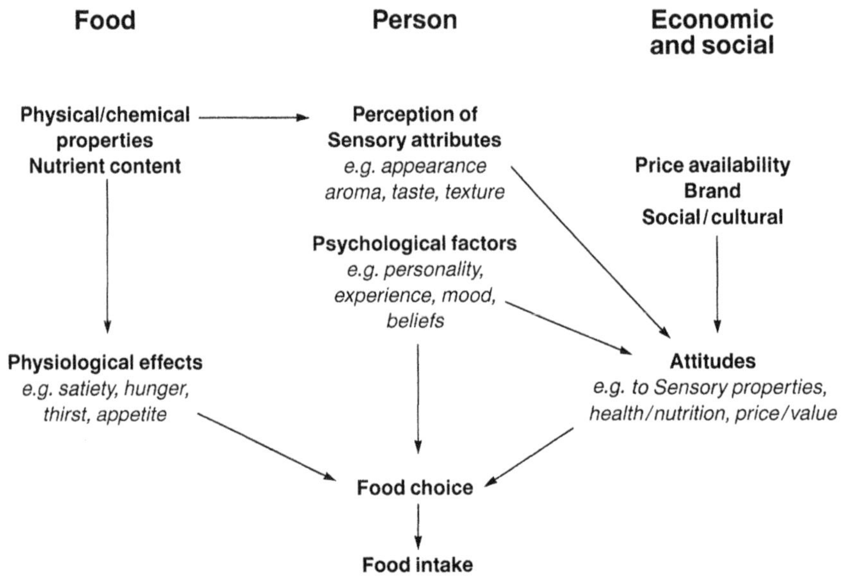

Fig. 8.1. Some factors affecting food choice and intake (from Shepherd, 1985).

There are certain variations in chemical and physical properties of the food which are perceived in terms of sensory attributes such as flavour, texture or appearance. However, many aspects of the chemical composition of foods are not perceived and therefore not every variation of chemical composition will influence choice via this route; on the other hand there will be some chemical variations which are perceived at very low concentrations and hence these may have a major impact on sensory responses to the food. These sensory attributes do not, in themselves, determine whether a person will choose a food—rather it is the person's liking for an attribute in a particular food which will determine choice. There are some common themes such as sweet foods being highly liked, which might be innate and is certainly shown clearly in newborn babies (Steiner, 1977), and bitter foods being unpalatable, which might reflect the tendency for toxins to have a bitter taste. Despite these commonalities there are many exceptions and the liking for attributes is often food-specific, e.g. a degree of bitterness in coffee or chocolate is highly liked. These preferences are most probably learned through experience (Cowart and Beauchamp, 1986). Within a particular culture there tends to be a great deal of agreement on the appropriateness of particular sensory attributes for particular foods, but there are nonetheless very substantial differences between individuals in their liking, which may reflect individual experiences.

Other chemical components in the foods, such as the amount of protein or carbohydrate, will have effects upon the person, either in the short or the longer term. Thus consuming a high-energy food will reduce hunger. The learning of the association between the sensory attributes of a food and its postingestional consequences would appear to be a major way in which preferences develop (Hill and Blundell, 1986). Certainly the association between ingestion and vomiting or sickness will lead to a very strong aversion to the food (Rozin, 1986).

In addition to physiological effects on food choice, psychological differences, such as personality, may also influence food choice. Various marketing and economic variables as well as social, cultural, religious or demographic factors will also influence food choice (Murcott, 1989; Shepherd, 1989). These are likely to operate through differences in the beliefs, values and attitudes held by people.

It is clear from all of these schemes that the area of understanding food choice may be attempted from a number of perspectives and hence a number of disciplines might be involved in the research. These would range from food science/technology and the chemistry and physics of food systems, through biochemistry, physiology and nutrition to psychology, sociology, anthropology, economics and marketing. Given these different orientations it is not surprising that these inputs have generally tended to be independent rather than coordinated. However, in order to arrive at an overall understanding of the reasons for food choice it will be necessary to adopt an interdisciplinary approach, taking account of inputs from these various sources.

Let us now consider attempts which have been made to quantify the relative importance of these different types of factors. This is difficult to achieve, given the disparate nature of the factors. One possible method might be to ask subjects directly to rate or rank the most important influences. This has been done by a number of authors; in one study husbands ranked taste as most important, with nutrition second, whereas wives placed nutrition highest with taste second (Schafer, 1978). Both agreed on cost and health occupying third and fourth positions. The problem with this type of procedure is that people may not be fully aware of the influences on their behaviour. Thus other approaches have been used to try to relate these factors to food choice.

Theory of planned behaviour

As outlined above, many of the influences on food choice are likely to be mediated by the beliefs and attitudes held by an individual. Beliefs about the nutritional quality and health effects of a food may be more important than the actual nutritional quality and health consequences in determining a person's choice. Likewise various marketing, economic, social, cultural, religious or demographic factors may act through the attitudes and beliefs held by the person. Thus the study of the relationship between choice and the beliefs and attitudes held by a person offers a possible route towards a better understanding of the influence of different factors on food choice.

The idea behind measuring attitudes is that they are expected to be causally related to behaviour. This is true both in the common use of the term 'attitude' and in the use of this term in the research literature in social psychology (Eagly and Chaiken, 1993). However, the empirical evidence for this link has not always been clear. In the nutrition literature, for example, many studies have attempted to measure the degree of association between attitudes and consumption of foods. Axelson *et al.* (1985) performed a meta-analysis of such studies and found evidence for a small (although statistically significant) correlation between attitudes and behaviour ($r=0.18$). Thus a superficial survey of this area might lead to the conclusion that attitudes are not related to behaviour to an important degree. The same type of finding in social psychology led to a crisis in attitude research in the late 1960s (Wicker, 1969), which resulted in the generation of a number of structured attitude models (e.g. Ajzen and Fishbein, 1980). One example of such a structured attitude model is the theory of reasoned action (TRA) (Ajzen and Fishbein, 1980) and its extension in the form of the theory of planned behaviour (TPB) (Ajzen, 1988). These models have been widely applied in social psychology and more recently have been applied to food choice issues.

The TRA seeks to explain rational behaviour which is under the control of the individual. In addition, however, the TPB seeks to be applicable to non-volitional behaviours, goals and outcomes which are not entirely under the control of the person. With volitional behaviours it is argued that intention to perform a behaviour is the best predictor of the behaviour. Intention, in turn, is predicted by two components: the person's own attitude (e.g. whether the person sees the behaviour as good, beneficial, pleasant, etc.) and perceived social pressure to behave in this way (termed the subjective norm). These relationships are shown schematically in Fig. 8.2. The TPB includes a component of perceived control, in addition to attitude and subjective norm in the prediction of behavioural intentions, and as a possible influence on the intention-behaviour link (see Fig. 8.2).

In turn, attitude is predicted from beliefs. Beliefs about expected outcomes of the behaviour are assessed and these responses are multiplied by the person's evaluations of whether they see these outcomes as good or bad. The sum of these belief-evaluations is used as the predictor of attitudes. The subjective norm is predicted by the sum of products of normative beliefs, which are perceived pressure from specific people or groups (e.g. doctors, family) and the person's motivation to comply with the wishes of these people or groups. In a similar fashion perceived control is determined by specific control beliefs.

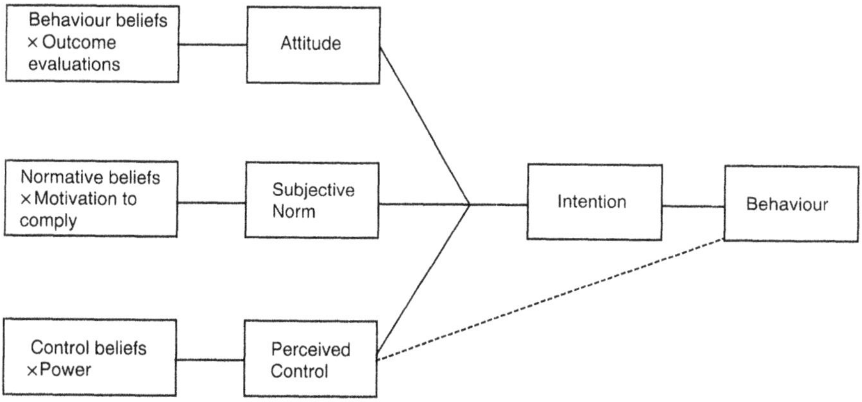

Fig. 8.2. Schematic representation of the theory of planned behaviour.

An important point in the original conception of the TRA is that influences other than beliefs, attitudes, social pressure and intention should act through these variables (Ajzen and Fishbein, 1980); this would also apply to the TPB, although here perceived control would form part of the model. Thus demographic variables, such as age or social class, should influence behaviour only through the model variables and not act as independent influences on behaviour.

The TRA has been widely applied to many issues in social psychology (Ajzen and Fishbein, 1980; Tesser and Shaffer, 1990), and more recently also successfully applied to a range of food choice issues (Axelson *et al.*, 1983; Shepherd and Stockley, 1985, 1987; Shepherd and Farleigh, 1986; Shepherd, 1988, 1989; Tuorila, 1987; Tuorila and Pangborn, 1988). Sheppard *et al.* (1988) carried out a meta-analysis of 87 studies using this model in the area of general consumer choice. They found an estimated correlation of 0.53 between intention and behaviour and a multiple correlation of 0.66 between attitude plus subjective norm against intention (Sheppard *et al.*, 1988). Thus this model has validity both in the study of general consumer choice and the study specifically of food choice.

The incorporation of perceived control has received some support in social psychological applications (e.g. Beale and Manstead, 1991), and in studies of weight loss (Schifter and Ajzen, 1985) and dietary health behaviours (Ajzen and Timko, 1986), although not in all applications (Fishbein and Stasson, 1990). In a study of biscuit and bread consumption (Sparks *et al.*, 1992), intentions to consume wholemeal bread were not influenced by perceived control but intentions to consume biscuits were. Thus, inclusion of perceived control may be important in predicting choices of some, although not all, foods. However, a failure to find a significant effect for perceived control does not invalidate the TPB because it would be argued that in those cases the behaviour is under the control of the person.

Relative importance of flavour: fat intake

This method can be used to get a more objective measure of the relative importance of flavour as against other variables influencing choice. This can be done by assessing the relative impact of particular beliefs on attitudes and intention. One study using this method investigated the factors influencing fat intake (Shepherd and Towler, 1992) and was a direct development from previous studies in this area (Shepherd and Stockley, 1985, 1987). The basic TRA was used to develop a questionnaire which was completed by 538 people.

The questionnaire investigated the consumption of four types of foods (meat, meat products, dairy products and fried foods), which together contribute about 70% of the total intake of dietary fat in the UK (Shepherd and Stockley, 1985). The subjective norm and normative beliefs were not included here since previous studies had indicated that these were not important components in relation to the consumption of these foods (Shepherd and Stockley, 1985, 1987).

Intention was assessed with one question for each food type on the person's intention to eat the food during the next week. There were three attitude items for each food type, worded, for example, 'My eating meat is good', with 7-point response scales ranging from 'strongly agree' to 'strongly disagree'. The other attitude questions related to judgements that eating the food was pleasant and beneficial.

Belief items were generated from short structured interviews with 34 people, who did not take part in the main study. Six beliefs were included for each food type. Three belief items were common to all food types: the food was seen as healthy, high in fat and tasted good. The remaining three items for each food type depended upon which beliefs were found to be important in the pre-interviews. There was a corresponding outcome evaluation item for each belief: for example 'Food which is high in fat is desirable', with response options from 'strongly agree' to 'strongly disagree', again on 7-point category scales.

The correlations between the components of the model are shown in Table 8.1. These were found to be reasonably high and were equivalent to previous findings using this model (Shepherd, 1989).

The relative importance of the different belief-evaluation combinations in determining attitudes was investigated using multiple regressions of the individual belief-evaluation scores against the attitude scores. The standardized regression coefficients (beta coefficients) from these analyses can be

Table 8.1. Correlation coefficients for four food groups between components of the model.

Food types	df	Beliefs versus attitude	Attitude versus intention	Intention versus behaviour
Meat	502	0.70	0.64	0.73
Meat products	501	0.63	0.59	0.78
Dairy products	515	0.58	0.40	0.57
Fried foods	509	0.59	0.61	0.77

All $p < 0.001$.

taken as a measure of the relative importance of the different beliefs in determining attitudes and hence intention. These coefficients are shown in Table 8.2. In most cases, the belief-evaluation score relating to taste had the highest coefficient from the regression with attitude, with foods perceived as healthy second highest. For dairy products the belief-evaluation score for healthiness was higher than the one for taste. The belief-evaluation score for fat content had a lower regression coefficient than the ones for taste or health for all of the food types. Some of the other belief-evaluation scores included were also significant in the regression equations: in particular those related to protein and vitamin content, but also those referring to additives in meat products and to the greasy taste of fried foods. Belief-evaluation scores related to expense and convenience were not significant contributors to these equations.

Thus this study demonstrates the usefulness of this type of model in relation to food choice. There were clear relationships between belief-evaluations, attitude and intention. In this study both the taste and health belief items were the most important, with other items like expense and convenience less so. This clearly demonstrates the importance of taste and flavour in the consumption of foods contributing significantly to fat intake.

Extensions of the TPB: self-identity and organic food consumption

Although the TRA and TPB have proved successful in many applications in the food choice area, there are a number of shortcomings in their conceptualization and implementation. This has led to a number of suggested modifications and extensions. One suggested modification to the TPB is that a person's self-identity may influence behaviour independently of his or her attitudes (Biddle *et al.*, 1987; Charng *et al.*, 1988). An example of evidence for such a modification is in the field of blood donation, where a person's

Table 8.2. Results of regressions of attitude against individual belief-evaluations. The standardized (β) coefficients and *t* values are shown

Belief	Regression coefficient (β)	*t*
Meat		(df=497)
Healthy	0.25	7.7***
Fat	0.12	4.1***
Taste	0.53	16.9***
Expense	0.04	1.2
Protein	0.12	4.1***
Vitamins	0.07	2.2*
Meat products		(df=498)
Healthy	0.27	7.8***
Fat	0.12	3.7***
Taste	0.51	15.6***
Expense	–0.01	–0.5
Convenience	–0.01	–0.3
Additives	0.16	4.8***
Dairy products		(df=512)
Healthy	0.42	12.2***
Fat	0.18	5.2***
Taste	0.31	9.2***
Protein	0.16	4.6***
Vitamins	0.09	2.8**
Calories	–0.03	–0.9
Fried foods		(df=507)
Healthy	0.16	4.6***
Fat	0.13	3.7***
Taste	0.51	14.0***
Convenience	0.01	0.3
Greasy	0.15	4.2***
Smell	–0.03	–0.9

* $p<0.05$; ** $p<0.01$; *** $p<0.001$

identification of him or herself as a blood donor is a predictor of intention to donate blood over and above the effects of the person's attitude towards blood donation (Charng *et al.*, 1988). These findings are surprising since if a person identifies him or herself as a blood donor this would be expected to be reflected in favourable attitudes towards blood donation, rather than giving an effect on intention independent of attitude. The inclusion of self-identity in the TPB was investigated in a study of the consumption of organic vegetables (Sparks and Shepherd, 1992). Two hundred and sixty-one subjects completed questionnaires which included measures of the components of the TPB, along with measures of self-identity.

There were five attitude items: 'eating organic vegetables is . . .' with response scales of 'extremely foolish—extremely wise', 'extremely bad—extremely good', 'extremely harmful—extremely beneficial', 'extremely un-enjoyable—extremely enjoyable', 'extremely unpleasant—extremely pleasant'. There was one subjective norm item in the form: 'Most people who are important to me think I should . . . I should not eat organic vegetables'.

Belief items were generated from a short pre-test questionnaire given to 75 people not involved in the main study. Seven beliefs were derived from this pre-test and were included in the main questionnaire: 'eating organic vegetables entails eating vegetables (i) that have a good flavour, (ii) that are healthy, (iii) that are expensive to purchase, (iv) that are easy to obtain in the shops, (v) that are associated with an 'alternative' lifestyle image, (vi) that have an attractive appearance, (vii) that are environmentally friendly'. The response scales were marked 'extremely unlikely' to 'extremely likely'. Outcome evaluations were assessed by asking subjects to rate each of the seven characteristics of vegetables from 'extremely bad' to 'extremely good'.

Perceived control was measured with three items: (i) 'How much control do you have over whether you do or do not eat organic vegetables', with the response scale end-points marked 'very little control' and 'complete control'; (ii) 'For me to eat organic vegetables is . . .', with response scale endpoints marked 'extremely difficult' and 'extremely easy'; (iii) 'If I wanted to, I could easily eat organic vegetables whenever I eat vegetables', with response scale endpoints marked 'extremely unlikely' and 'extremely likely'. Two measures of identification with green consumerism were constructed: 'I think of myself as a green consumer' and 'I think of myself as someone who is very concerned with green issues' (with responses from 'disagree very strongly' to 'agree very strongly').

Correlations between the components of the TPB were reasonably high, confirming the basic applicability of the model in this case (see Table 8.3). A multiple regression tested the hypothesis that 'green identity' would not add to the prediction of intentions over and above the contributions of the other components. The multiple correlation was 0.52 ($p < 0.001$) and, as expected, attitudes, subjective norm and perceived control all revealed independent effects but there was also a highly significant effect for 'green identity' (see Table 8.3).

One possible reason for 'green identity' having such an independent effect is that the people were using this as a proxy for past behaviour, i.e. inferring their self-identity response by thinking about how they have behaved in the past; past behaviour has been found to be a powerful predictor of both intention and future behaviour (Bentler and Speckart, 1979). However, adding responses on past behaviour into the above regression showed that the measure of self-identity was still an independent predictor of intention thus arguing against this interpretation of the effect.

The reason why self-identity should have such an independent effect is not entirely clear (Sparks and Shepherd, 1992), although it may reflect the inability of current measures of attitude adequately to assess various symbolic and emotional factors which might nonetheless influence intention and behaviour. Extensions of this kind to the basic model offer a potential means towards a better overall understanding of the different factors determining food choice.

Table 8.3. Simple correlations between intention to buy organic vegetables and model components, along with standardized (beta) regression coefficients from a multiple regression of intention against attitude, subjective norm, perceived control and self-identity.

Model component	Correlation coefficient	Regression coefficient (β)
Attitude	0.38***	0.21**
Subjective norm	0.30***	0.16*
Perceived control	0.27***	0.26***
Self-identity	0.37***	0.22**

* $p<0.05$; ** $p<0.01$; *** $p<0.001$.

Genetic engineering and ethical concerns

One aspect of behaviour not well covered by existing components of the TPB is that of moral or ethical concerns. In its basic form the model is purely utility based with behaviour leading to outcomes which are seen as beneficial or not. However, there are some behaviours where moral considerations might also have a significant impact, irrespective of beliefs about outcomes. Moral obligation has been investigated in studies on behaviours such as stealing or cheating, which have a very obvious moral component (e.g. Beck and Ajzen, 1991). In such cases it has been found to add significantly to the predictive power of the TPB. Although food choice is a less obvious area within which such factors might operate there are nonetheless particular instances where moral or ethical issues might be of some importance. These include those instances where decisions are made on behalf on others (Raats *et al.*, 1995) and the application of new techniques to food production, such as genetic engineering, which might be considered to have a significant moral dimension.

Genetic engineering is a technology which offers enormous potential in medicine, agriculture and food production. There are already a number of applications in pharmaceuticals but there are currently relatively few commercial examples in the field of food. One example is the use in cheesemaking of genetically engineered chymosin replacing rennet extracted from the stomachs of calves. The first genetically engineered complete food to be marketed is Calgene's Flavr-Savr tomato which was marketed in the USA in 1994. This has been genetically modified in order to extend shelf-life and, it is argued, improve the flavour through not having to pick and transport the fruit under-ripe. There are, however, many more possible applications where crops are being genetically engineered, often to provide some form of agricultural advantage such as pest resistance or herbicide resistance. The intention is that many of these crops will be used in food production in the future. Genetic engineering does not have the same limitations as conventional cross-breeding, which has been used for centuries to change the genetic makeup of animals and plants. Whereas cross-breeding relies on transferring genetic characteristics within a species, genetic engineering can be used to transfer genetic material across species boundaries and between very different types of organisms (e.g. from micro-organisms to animals, or from plants to animals).

Although some evidence suggests that the general public knows or understands little about the technology (Hamstra, 1991; Sparks *et al.*, 1994), the direction of public opinion is thought likely to be a major factor in influencing its future role in society (Advisory Council on Science and Technology, 1990). The application of this type of technology raises a number of issues of potential concern to the public, including issues of safety and what benefits are likely to derive from the application of the technology but it also raises moral and ethical concerns in the minds of many people. The Advisory Council on Science and Technology (1990) emphasized this in its recommendation that 'government departments need to be sufficiently alert to the ethical issues that are likely to cause concern in the future' (p. 28).

As part of an extensive series of studies examining public attitudes towards the application of genetic engineering to food production (Sparks *et al.*, 1994; Frewer *et al.*, 1994), two studies were conducted applying an extended form of the TPB to this issue.

In the first study (Sparks *et al.*, 1995), 334 members of the public completed a questionnaire designed following the basic structure of the TPB. Since public awareness and knowledge of the technology had been found to be very low (Sparks *et al.*, 1994), the questionnaire began with a simple short description of the technology. Instead of intention, two expectation questions were included, one related to eating food produced using genetic engineering and the other to supporting the use of the technology. In addition to the TPB questions, two questions measured perceived ethical obligation: 'I feel that I have an ethical obligation to avoid eating food produced by gene technology' and 'I feel that I have an ethical obligation to support the use of gene technology in food production' (both with responses from 'disagree very strongly' to 'agree very strongly'). One was framed positively and the other negatively so as not to suggest that ethical and moral issues can only be negative and also to determine whether ethical issues have the same impact when presented positively.

Ten possible outcomes of the application of genetic engineering to food production were used to assess beliefs and outcome evaluations. These were selected from a cross-section of the literature on genetic engineering (e.g. Pimentel *et al.*, 1989; Straughan, 1991). These are not necessarily the beliefs which are important to a random sample of the population but because people generally profess little knowledge about the technology, it was not considered feasible in this first study to assess beliefs in the standard way. The selected outcomes were

1. increased food production
2. ecological damage
3. cheaper food
4. animal and/or plant extinction
5. reduced use of pesticides
6. detrimental effects on animal welfare
7. improvements in food quality
8. unpredictable animal and/or plant mutations
9. economic growth
10. increased use of weedkillers.

Multiple regression analysis showed that people's expectations of (1) supporting the technology and (2) consuming food produced by the technology could be predicted from their attitudes (respectively: $\beta = 0.77$, $p < 0.001$ and $\beta = 0.23$, $p < 0.05$). These attitudes, in turn, could be predicted, not only from the perceived tangible advantages and disadvantages of the technology but also from a consideration of people's levels of ethical concern (Table 8.4). The positive beta coefficient for ethical obligation for supporting the technology and negative coefficient for eating the foods is because of the opposite wording of the question in the two cases (see above).

A second study (Sparks and Shepherd, in preparation) essentially replicated these findings but looked at specific applications of the technology. These were pork produced using genetically engineered growth hormone and genetically engineered tomatoes. In the previous study, the beliefs were generated from the literature rather than from a representative sample of the population and hence the beliefs included may not really have been important for the people taking part. This might then have underestimated the contribution from the belief component in the prediction of attitude, and hence possibly overestimated the contribution of ethical concerns. In this second study, a method was used to generate the beliefs directly from each person taking part (Towriss, 1984). In this case, each person generated his or her own set of beliefs and it was this set of beliefs that the person then received in the questionnaire rather than an overall set being derived from pre-interviews and then everyone receiving this same set of beliefs. The method used here would then be expected to give the highest estimate of the impact of belief-evaluations on attitude.

Table 8.4. Standardized regression coefficients (β) from multiple regressions predicting attitudes; in Study 1 attitudes towards eating foods produced using genetic engineering and supporting the use of the technology and in Study 2 the purchase of two foods produced using genetic engineering.

	Number of subjects	Belief-evaluations	Ethical obligation
Study 1			
Eating genetically engineered foods	284	0.46***	–0.41***
Supporting genetic engineering	284	0.33***	0.60***
Study 2			
Pork produced with genetically engineered growth hormone	60	0.30***	–0.51***
Genetically engineered tomatoes	61	0.56***	–0.37***

*** $p < 0.001$.

Following a similar pattern to that found in the previous study, it was apparent that respondents' perceived ethical obligation had a significant independent effect on both their attitudes towards the purchase of the pork and their attitudes towards purchasing the tomatoes (see Table 8.4). Intentions to purchase these products were very highly correlated with their attitudes: $r = 0.80$ ($p < 0.001$) in the case of pork, and $r = 0.70$ ($p < 0.001$) in the case of the tomatoes.

Thus ethical and moral concerns seem to play a major role in determining attitudes, and hence intention, when considering the application of novel technologies such as genetic engineering. It remains to be determined how widespread is the influence of ethical concerns on the acceptance of other technologies and on other food choice issues.

Conclusions

Food choice is influenced by a large range of potential factors. Many models put forward in this area involve merely listing the likely influences rather than offering a framework for empirical research and practical application. Although there is general agreement on the types of influences likely to be important, the integration of these factors into a coherent and quantitative model of food choice remains an area in need of development.

When directly asked what are the important influences on food choice, people place taste or flavour very high on their list of influences. These influences are also important when more indirect methods are used for assessing the relationships between beliefs, attitudes and intention.

The attempt to model food choice via an understanding of people's beliefs and attitudes requires a structured framework within which to measure and relate the variables of interest. One model from social psychology for achieving this is the TPB. This model appears to have widespread applicability and can be used to determine the relative importance of different factors in influencing food choice.

Various extensions of this model, including self-identity and moral obligation, offer a means for developing a clearer understanding of the factors influencing the choice of particular types of foods in particular contexts. In particular they pave the way for exploring the more emotional elements potentially important in food choice rather than simply addressing the rational cognitive issues which have been studied in the past.

Acknowledgements

The work reported here was funded by the Biotechnology and Biological Sciences Research Council and the UK Ministry of Agriculture, Fisheries and Food.

References

Advisory Council on Science and Technology (1990). *Developments in biotechnology*. HMSO, London.

Ajzen, I. (1988). *Attitudes, personality, and behavior*. Open University Press, Milton Keynes.

Ajzen, I. and Fishbein, M. (1980). *Understanding attitudes and predicting social behavior*. Prentice-Hall, Englewood Cliffs, NJ.

Ajzen, I. and Timko, C. (1986). *Basic Appl. Soc. Psychol.* **7**, 259–76.

Axelson, M. L., Brinberg, D. and Durand, J. H. (1983). *J. Nutr. Educ.* **15**, 94–8.

Axelson, M. L., Federline, T. L. and Brinberg, D. (1985). *J. Nutr. Educ.* **17**, 51–4.

Beale, D. A. and Manstead, A. S. R. (1991). *J. Appl. Soc. Psychol.* **21**, 409–31.

Beck, L. and Ajzen, I. (1991). *J. Res. Pers.* **25**, 285–301.

Bentler, P. M. and Speckart, G. (1979). *Psychol. Rev.* **86**, 452–64.

Biddle, B. J., Bank, B. J. and Slavinge, R. L. (1987). *Soc. Psychol. Quart.* **50**, 322–37.

Charng, H.-W., Piliavin, J. A. and Callero, P. L. (1988). *Soc. Psychol. Quart.* **51**, 303–17.

Committee on Medical Aspects of Food Policy (1984). *Diet and cardiovascular disease*. HMSO, London.

Committee on Medical Aspects of Food Policy (1994). *Nutritional aspects of cardiovascular disease*. HMSO, London.

Cowart B. J. and Beauchamp G. K. (1986). In *Interaction of the chemical senses with nutrition* (ed. M. R. Kare and J. G. Brand), pp. 25–44. Academic Press, Orlando.

Eagly, A. H. and Chaiken, S. (1993). *The psychology of attitudes*. Harcourt, Brace, Jovanovich, San Diego.

Fishbein, M. and Stasson, M. (1990). *J. Appl. Soc. Psychol.* **20**, 173–98.

Frewer, L. J., Shepherd, R. and Sparks, P. (1994). *J. Food Safety* **14**, 19–40.

Hamstra, A. (1991). *Biotechnology in foodstuffs: Towards a model of consumer acceptance*. Institute for Consumer Research, The Hague.

Hill, A. J. and Blundell J. E. (1986). *Nutr. Behav.* **3**, 133–44.

Khan, M. A. (1981). *CRC Crit. Rev. Food Sci. Nutr.* **15**, 129–53.

Murcott, A. (1989). In *World review of nutrition and dietetics*, **55** (ed. G. H. Bourne), pp. 1–40. Karger, Basel.

Pilgrim, F. J. (1957). *Am. J. Clin. Nutr.* 5, 171–5.

Pimentel, D., Hunter, M. S., LaGro, J. A., Efroymson, R. A., Landers, J. C., Mervis, F. T., McCarthy, C. A. and Boyd, A. E. (1989). *Biosci.* **39**, 606–14.

Raats, M. M., Shepherd, R. and Sparks, P. (1995). *J. Appl. Soc. Psychol.* **25**, 484–94.

Randall, E. and Sanjur, D. (1981). *Ecol. Food Nutr.* **11**, 151–61.

Rozin P. (1986). *Learn. Motiv.* **17**, 180–9.

Schafer, R. B. (1978). *J. Am. Dietet. Assoc.* **72**, 138–43.

Schifter, D. E. and Ajzen, I. (1985). *J. Pers. Soc. Psychol.* **49**, 843–51.

Shepherd, R. (1985). *Nutr. Food Sci.* **96**, 10–11.

Shepherd, R. (1988). *J. Hum. Nutr. Dietet.* **1**, 421–8.

Shepherd, R (1989). In *Handbook of the psychophysiology of human eating* (ed. R. Shepherd), pp. 3–24. Wiley, Chichester.

Shepherd, R. and Farleigh, C. A. (1986). *Hum. Nutr.: Appl. Nutr.* **40A**, 195–208.

Shepherd, R. and Stockley, L. (1985). *Hum. Nutr.: Appl. Nutr.* **39A**, 431–42.

Shepherd, R. and Stockley, L. (1987). *J. Am. Dietet. Assoc.* **87**, 615–9.

Shepherd, R. and Towler, G. (1992). *J. Hum. Nutr. Dietet.* 5, 387–97.

Sheppard, B. H., Hartwick, J. and Warshaw, P. R. (1988). *J. Consum. Res.* **15**, 325–43.

Sparks, P. and Shepherd, R. (1992). *Soc. Psychol. Quart.* **55**, 388–99.

Sparks, P. and Shepherd, R. (in preparation). The moral dimension of attitudes towards genetic engineering in food production.

Sparks, P., Hedderley, D. and Shepherd, R. (1992). *Europ. J. Soc. Psychol.* **22**, 55–71.

Sparks, P., Shepherd, R. and Frewer, L. J. (1994). *Agric. Hum. Val.* **11**, 19–28.

Sparks, P., Shepherd, R. and Frewer, L. J. (1995). *Basic Appl. Soc. Psychol.* **16**, 267–85.

Steiner J. E. (1977). In *Taste and development—The genesis of sweet preference* (ed. J. M. Weiffenbach), pp. 173–88. NIH-DHEW, Bethesda.

Straughan, R. (1991). *Brit. Food J.* **92**(7), 13–26.
Tesser, A. and Shaffer, D. R. (1990). *Ann. Rev. Psychol.* **41**, 479–523.
Towriss, J. G. (1984). *J. Market Res. Soc.* **26**, 63–75.
Tuorila, H. (1987). *Appetite* **8**, 1–14.
Tuorila, H. and Pangborn, R. M. (1988). In *Food acceptability* (ed. D. M. H. Thomson), pp. 267–79. Elsevier Applied Science, London.
Wicker, A. W. (1969). *J. Soc. Issues* **25**, 41–78.

9. Coumarins in fruit and vegetables

ALICJA M. ZOBEL

Department of Chemistry, Trent University, Peterborough, Ontario, Canada K9J 7B8

Introduction

Coumarins (benzopyrones) are very common secondary metabolites that have been found in all botanical families investigated. Some products of coumarins are even found in animals, as in the case of beaver scent glands, which contain anabolites of ellagic acid (Dean, 1963). Since the beginning of the 19th century, when Vogel (1820) reported isolation of coumarin, until 1991, about 1500 coumarins had been identified (Murray *et al.*, 1982; Murray, 1991), at the rate of about 20 per year. The term coumarin is derived from *Coumarouna odorata* Aubl. (Guillemette, 1835), the name of the tonka bean tree, from which coumarin was first isolated. The structure of coumarin was independently elucidated about 50 years later independently by Fittig (1868) and Strecker (1868).

Coumarin, 5,6-benzopyrone, can be formally considered to originate from fusion of benzene and α-pyrone, and the systematic name is 2-H-1-benzopyran-2-one. Psoralen, the simplest linear furanocoumarin, is 7-H-furo [3,2-γ][1]benzopyran-7-one.

Since, as is seen above, the chemical names of coumarins are quite complex, they are given trivial names. These can sometimes be confusing, because closely similar names often refer to structurally unrelated compounds, and related compounds sometimes have very different names (Murray *et al.*, 1982).

Occurrence

Some coumarins can occur in all parts of the plant; for instance, fraxoside and isofraxoside are found in all parts of the *Campanula* plant in similar concentrations (Plouvier, 1970). In other plants different coumarins can be characteristic of various plant organs. Table 9.1 lists some fruit and vegetable tissues containing coumarins which are eaten by humans and livestock. Even within the same organ specific tissues can show distinctive profiles of coumarins (Zobel and Brown, 1989*b*). *Psoralea subcaulis* seeds contain

Table 9.1. Some fruit and vegetable tissues containing coumarins which are eaten by humans and livestock (modified from Murray *et al.*, 1982)

Family	Species	Name of coumarin	Part of plant	Investigating author
Ascomycetae	*Aspergillus* sp.	Aflatoxin B1, G1		Hartley *et al.*, 1963
		B3, M1		Heathcote and Dutton, 1969
	Sacchonomyces sp.	Coumarin		VanSumere, 1960
Basidiomycetae	*Puccinia* sp	Coumarin		VanSumere *et al.*, 1957
		Aesculetin		
		Daphnetin		
		Scopoletin		
		Umbelliferone		
Gramineae	*Zea mays*	Coumarin	Root	VanSumere *et al.*, 1958
		Aesculetin	Leaves	Krupnikova and Dranik, 1973
		Scopoletin	Leaves and roots	
		Umbelliferone	Leaves and roots	
Orchidacea	*Vanilla planifolia*	Coumarin	Leaves	
Paimae	*Phoenix dactylifera*	Coumarin	Fruit	Murray *et al.*, 1982
Chenopodiaceae	*Beta vulgaris*	Coumarin	Seed	VanSumere, 1960
Compositae	*Cichorium intybus*	Aesculetin	Fruit	Hee, 1965
		Cichorin	Fruit	Nietzki, 1876
		Scopoletin	Fruit	Dem'yanenko and Dranik, 1971
		Umbelliferone	Fruit	
	Helianthus annuus	Coumarin	Stem	Neumann, 1962
		Scopoletin	Stem	Murray *et al.*, 1982
Cruciferae	*Raphanus sativus*	Coumarin	Leaves	Harborne and Corner, 1961
Labiatae	*Mentha piperita*	Aesculetin	Leaves	Hermann, 1959
Leguminosae	*Glycine* sp.	Scopoletin		Martin, 1963

Family	Species	Name of coumarin	Part of plant	Investigating author
	Medicago sp.	Coumestrol	Leaves	Simon, 1967
	Medicago sativa	Aesculetin	Leaves	Reppel and Wagenbreth, 1958
		Coumestrol	Leaves	Murray *et al.*, 1982
		Lucernol	Leaves	Spencer *et al.*, 1966
		Medicagol	Leaves	Livingston *et al.*, 1965
	Melilotus alba	Aesculetin	Leaves	Reppel and Wagenbreth, I 956
		Coumarin	Roots and Leaves	Bourquelot and Herissey, 1920
		Dicoumerol	Leaves	Campbell and Link, 1941
		Herniarin	Leaves	Reppel and Wagenbreth, 1958
		Scopoletin	Roots and Leaves	Reppel and Wagenbreth, 1958
		Umbelliferone	Roots and Leaves	Reppel and Wagenbreth, 1958
	Phaseolus vulgaris	Scopoletin	Fruit	Murray *et al.*, 1982
	Pisum sativum	Daphnetin	Leaves and roots	Reppel and Wagenbreth, 1958
		Scopoletin	Leaves and roots	Rappel and Wagenbreth, 1958
		Umbelliferone	Leaves and roots	Reppel and Wagenbreth, 1958
Moraceae	*Ficus* sp.	Bergapten	Fruit	El-Gamal *et al.*, 1975
		Psoralen	Fruit	El-Gamai *et al.*, 1975
		Scopoletin	Fruit	El-Gamal *et al.*, 1975
	Morus alba	Scopoletin	Root	Uno, 1972
		Umbelliferone	Root	Uno, 1972
Papaveraceae	*Papaver somniferum*	Aesculetin		Kosheleva and Nikonov, 1970
Rosaceae	*Fragaria orientalis*	lmperatorin	Leaves	lvanitskaya and Kuchinskaya, 1977
		lsoimperatorin	Leaves	Ivanitskaya and Kuchinskaya, 1977
		Umbelliferone	Leaves	Ivanitskaya and Kuchinskaya, 1977
Rutaceae	*Citrus aurantifolia*	Bergamotin	Fruit	Stanley and Vannier, 1967
		Bergapten	Fruit	Caldwell and Jones, 1945

Table 9.1. *contd*

Family	Species	Name of coumarin	Part of plant	Investigating author
		Bergaptol	Fruit	Caldwell and Jones, 1945
		Imperatorin	Fruit	Stanley and Vannier, 1967
		Isoimperatorin	Fruit	Stanley and Vannier, 1967
		Isopimpinellin	Fruit	Caldwell and Jones, 1945
		Limethin	Fruit	Caldwell and Jones, 1945
		Phellopterin	Fruit	Stanley and Vannier, 1967
	Citrus aurantium	Aurapten	Leaves	Komatsu *et al.*, 1932
		Auraptenal	Fruit	Stanley *et al.*, 1965
		Bergapten		Stanley and Vannier, 1967
		Bergaptol	Fruit	Kariyone and Matsuno, 1953
		Isoimperatorin	Fruit	Kariyone and Matsuna, 1953
		Limetlin	Fruit	Venturella *et al.*, 1966
		Meranzin		Bohme and Pictsch, 1939
		Scoparone	Fruit	Lu *et al.*, 1977
		Umbelliferone	Fruit	Komatsu *et al.*, 1932
Scrophulariaceae	*Digitalis putpurea*	Scopoletin	Leaves	Hermann, 1959
Solanaceae	*Solanum tuberosum*	Aesculin		VanSumere, 1960
		Coumarin		VanSumere, 1960
		Scopoletin		VanSumere, 1960
		Umbelliferone		VanSumere, 1960
Umbelliferae	*Ammi majus*	Alloimperatorin	Fruit	Abu-Mustafa *et al.*, 1975
		Ammirin	Fruit	Abu-Mustafa *et al.*, 1975
		Bergapten	Fruit	Fahmy and Abu-Shady, 1948
		Imperatorin	Fruit	Fahmy and Abu-Shady, 1948

Family	Species	Name of coumarin	Part of plant	Investigating author
		Isobergapten	Fruit	Blazek and Stary, 1965
		Isoimperatoritin	Fruit and stem	Abu-Mustafa *et al.*, 1968
		Isopimpinellin	Fruit and leaves	Nikonov, 1965
		Majurin	Fruit and leaves	Murray *et al.*, 1982
		Marmesin	Fruit	Abu-Mustafa *et al.*, 1958
		Marmesinin	Fruit	Starkowsky and Badran, 1958
		(+)-Oxypeucedanin	Leaves	Ivie, 1978
		Pabulenol	Leaves	Ivie, 1978
		Saxalin	Leaves	Ivie, 1978
		Umbelliferone	Leaves	Crowden *et al.*, 1969
		Umbelliprenin	Fruit	Abu-Mustafa *et al.*, 1971
		Xanthotoxin	Fruit and leaves	Fahmy and Abu-Shady, 1948
	Anethum graveolens	Aesculetin	Fruit	Dranik and Prokopenko, 1969
		Bergapten	Fruit	Dranik and Prokopenko, 1969
		Graveolone	Leaves	Aplin and Page, 1967
		Scopoletin	Fruit and leaves	Aplin and Page, 1967
		Umbelliferone	Fruit	Dranik and Prokopenko, 1969
		Umbelliprenin	Fruit	Dranik and Prokopenko, 1969
	Apium graveolens	Apigravin	Leaves and stem	Garg *et al.*, 1979
		Apiumetin	Leaves	Garg *et al.*, 1978
		Apiumoside	Leaves	Garg *et al.*, 1979
		Bergapten	Stem	Innocenti *et al.*, 1976
		Celerin	Stem	Murray *et al.*, 1982
		Cefereoside	Stem	Murray *et al.*, 1982
		8-Hydroxybergapten	Stem	Garg *et al.*,1979
		Isoimperatorin	Stem	Garg *et al.*,1979

Table 9.1. *contd*

Family	Species	Name of coumarin	Part of plant	Investigating author
		Isopimpinellin	Stem	Garg *et al.*,1979
		Osthenol	Stem	Garg *et al.*,1979
		Rutaretin	Stem	Garg *et al.*,1978
		Seselin	Stem	Garg *et al.*,1979
		Umbelliferone	Stem	Garg *et al.*,1979
	Coriandrum sativum	Bergapten	Leaves	Kartnig, 1966
		Scopoletin	Leaves	Kartnig, 1966
		Umbelliferone	Leaves	Kartnig, 1966
		Umbelliprenin	Leaves	Kartnig, 1966
	Daucus carota	Coumarin	Seed	Seifert *et al.*, 1968
		Peucedanin	Leaves	Lescova and Ananichev, 1969
		Umbelliferone	Leaves	Crowden *et al.*, 1969
	Heracleum mantegazzianum	Angelicin	Fruit and leaves	Beyrich, 1968
		Herachenol	Stem	Kato *et al.*, 1978
		Isopimpinellin	Stem	lnnocenti *et al.*, 1976
		Xanthotoxin	Stem	Innocenti *et al.*, 1976
		Apterin	Roots	Ficher and Baerheim-Svendsen, 1976
		Bergapten	Leaves and fruit	Beyrich, 1968
		Columbianetin	Roots	Murray *et al.*, 1982
		lmperatorin	Fruit	Beytich, 1968
		Isobergapten	Roots	Komissarenko *et al.*, 1965
		Isopimpinellin	Roots and Leaves	Beyrich, 1968
		Osthol	Roots	Giorgobiani *et al.*, 1969
		Phellopterin	Fruit	Beyrich, 1968

Family	Species	Name of coumarin	Part of plant	Investigating author
		Pimpinellin	Fruit	Komissaranka *et al.*, 1967
		Psoralen	Leaves	Molho *et al.*, 1971
		Sphondin	Roots	Komissarenko *et al.*, 1965
		Umbellifue	Roots	Komissarenko *et al.*, 1965
		Xanthotoxin	Roots and Fruit	Beyrich, 1968
	Levisticum officinale	Bergapten	Roots	Kowalska and Skrzypezakowa, 1964
		Psoralen	Roots	Karisen *et al.*, 1968
		Umbelliferone	Roots	Horhammer *et al.*, 1966
	Pastinaca sativa	Apterin	Roots	Fisher and Baerheim-Svendsen, 1976
		Bergapten	Fruit and leaves	Soine *et al*, 1956
		Coumarin	Fruit	Casparis and Manella, 1944
		Imperatorin	Fruit and Stem	Soine *et alt*, 1956
		Isobergapten	Fruit	Gusak and Safina,1976
		Isopimpinellin	Fruit and stem	Beyrich, 1965
		Osthot	Fruit	Maksyutina, 1967
		Pimpinellin	Fruit	Gusak and Safina,1976
		Sphondin	Fruit and stem	Maksyutina, 1965
		Umbelliferae	Fruit	Gusak and Safina,1976
		Xanthotoxin	Fruit and stem	Fahmy *et al*, 1956
		Xanthotoxol	Fruit	Maksyutina and Kolesnikov, 1959
	Petroselinum crispum	Bergapten		Innocenti *et al.*, 1976
		Coumarin		Casparis and Manella, 1944
		Herachenol	Stem	Kato *et al.*, 1978
		Isopimpinellin	Stem	Innocenti *et al.*, 1976
		Xanthotoxin	Stem	Innocenti *et al.*, 1976

psoralen in the seed coat but not in the embryo itself (Baskin *et al.*, 1967). Interestingly, the embryo of the *Heracleum* seed lacks coumarins, but during development of the seedling, synthesis begins (Carbonnier-Jareau *et al.*, 1978). During imbibition of barley embryos the concentration of coumarin decreased (VanSumere *et al.*, 1972), but in *Coumarouna odorata* it increased (Valio, 1973).

Fruits and flowers are very often the plant parts richest in coumarins. Fruits of *Angelica* contained coumarins as 2.2% of the dry weight (Sokolova and Nikonov, 1969), reaching up to 7 mg/g fresh weight (Zobel and Brown, 1991*b*). In the case of *Ruta graveolens* we found the opposite phenomenon, as leaves contained up to 2 mg/g fresh weight but fruits and seeds had several times less (Zobel and Brown, 1989*a*). Carbonnier and Molho (1978) suggested that in seeds obliged to germinate under harsh conditions in septic soil coumarins may play the role of protective agents of embryos against microbial attack. They suggested as well a role as agents inducing dormancy (Baskin *et al.*, 1967; Berrie *et al.*, 1968). Thus, often, fruits and roots are the organs with the highest content of coumarins (Wierzchowska-Renke, 1975; Carbonnier and Molho, 1978). *Heracleum* leaves contain concentrations twice as high as stems (Carbonnier and Molho, 1978), but the opposite holds for *Peucedanum* (Christensen and Wissing, 1972). Sometimes the difference can be intervarietal in one species, as in the case of rice seedling roots (Sadhu, 1975). Varga *et al.* (1978) found in *R. graveolens* roots high levels of rutaretin and low levels of isorutaretin, the reverse of that found in tissue cultures of this species. Steck and Bailey (1969) isolated different furanocoumarins from roots and leaves of *Angelica angelica*. From roots of *A. dahurica* Yen (1967) isolated six coumarins, but only byakangelicin from its fruits. In *Mammea africana* the bark contains dihydroxycoumarins, but in the xylem xanthones predominate (Crichton and Waterman, 1978).

During fruit ripening coumarins do not stay qualitatively and quantitatively constant (Bose and Mookerjee, 1944; Méndez and Castro-Poceiro, 1983; Lokar and Delber, 1988). In *Poncirus trifoliata* there are qualitative changes in the coumarins, unripened fruit contains predominantly simple coumarins, but ripened ones contain furanocoumarins (Guiotto *et al.*, 1977). *Citrus paradisi* contained the same kind of coumarins both in the peel and in the interior of the fruit, but in different proportions (Maier and Metzler, 1967).

Both qualitative and quantitative changes were observed during the vegetative period, and Innocenti *et al.* (1976) found that in *Apium graveolens* and *Petroselinum sativum* levels of furanocoumarins were high in spring and autumn, but decreased during summer. A range of coumarins in very high concentrations was found in immature fruits of *Ammi majus*, but in the ripened fruits all the concentrations were much lower (Balbaa *et al.*, 1972). The opposite occurred in *P. trifoliata* (Murray *et al.*, 1982).

Quantitative changes to furanocoumarins were observed in *Heracleum lanatum* depending on temperature (Zobel and Brown, 1990). They were several-fold higher than in the control when the plant growth temperature was about 36 °C for several days, and up to 100 times as high after two freezing nights.

Even daily changes in the content of leaf coumarins have been observed, Schlosser-Szigat (1964) found that leaves of *Melilotus* contained the highest amounts at noon, and that the longer the days the higher the concentration in the leaves. In *Levisticum officinale* all the organs contained similar amounts of coumarins, but the lowest content was in leaf petioles at the end of the vegetative period (Dauksha, 1968).

Geographical localization may influence concentrations of coumarins. In several plant species Komissarenko and Zoz (1976) found that psoralen was highest at a latitude of 45–55 N. But contrary examples exist, where there are no differences due to geographical location, as in the case of psoralens in *Prangos*, occurring in identical concentrations in plants from Bulgaria, Russia and Italy (Bubeva-Ivanova *et al.*, 1973). More influence is exerted by soil quality, with increased concentrations of coumarins in *Mammea africana* growing in poor soil (Crichton and Waterman, 1978). Similarly, scopoletin concentration was enhanced when sunflower plants grew on boron-deficient soils (Watanabe *et al.*, 1964). These findings support a hypothesis (Janzen, 1974) that on low-nutrient soils genetic regulation can result in high concentrations of secondary metabolites in plants, which would deter herbivores and pathogens under such conditions of greater vulnerability.

Detoxification and degradative metabolism

After coumarins have been taken up by animal or microbial cells, and if they are toxic to these cells (Ivie, 1978*a*,*b*, 1987*a*,*b*; Ivie *et al.*, 1981,) they undergo modification. The first detoxification experiments were those of Sieburg in 1921, showing that umbelliferone, daphnetin and aesculetin injected into a rabbit were later excreted in the urine as conjugated forms. Detoxification in animals involves hydroxylation of the benzene ring followed by the conjugation reaction. Hydroxylation is most often in positions 3,7 and 8, followed by 5, but rarely in position 4 (Mead *et al.*, 1958; Feuer *et al.*, 1966). However, in humans treated for prostate cancer with coumarin and cimetidine (Marshall, 1989; Marshall *et al.*, 1990; Maucher *et al.*, 1993), umbelliferone was excreted; hence up to 80% of the substitution was at position 7 (Schilling *et al.*, 1969). A gene for 7-hydroxylation has even been found (Yamano *et al.*, 1990). Substitution at position 7 was mediated by enzymes of a rat microsomal fraction, with activity dependent on the sex, and activity in the female three times that of the male (Creaven *et al.*, 1965). *In vitro*, cytochrome P450 was shown to participate in hydroxylation (Kratz and Staudinger, 1967).

Hydroxycoumarins in several plants are further conjugated as ethereal sulfates or glucuronides (Mead *et al.*, 1958; Furuya, 1958a). Further degradation of hydroxycoumarins can involve opening of the lactone ring followed by *cis–trans* inversion to produce the stable forms, 2′-hydroxycinnamic acid and melilotic acid, as well as 2′-hydroxyphenylacetic acid, which undergoes a parallel decarboxylation during catabolism (Booth *et al.*, 1959).

Kaighen and Williams (1961) found 2′-hydroxycinnamic acid to be the predominant form, but a minor amount of 2′-hydroxyphenyllactic acid was also formed; the possible effect of intestinal microflora on the orally administered coumarins was not considered. Demethylation of scoparone to scopoletin, isoscopoletin and aesculetin was described in rabbits (Furuya, 1958b) and mice (Zutshi *et al.*, 1978).

Degradation in micro-organisms

Hydroxylation of coumarin to umbelliferone also occurs in micro-organisms (Smith and Rosazza, 1974). But, *in vitro*, rat cecal extracts produced melilotic acid and dihydrocoumarin, but not umbelliferone, 2′-hydroxycinnamic acid or 2′-hydroxyphenylacetic acid (Scheline, 1968). Thus microflora in the gut degrade coumarins differently from other cells within the body (Indahl and Scheline, 1971). Soil bacteria can open the lactone ring by oxidizing it with mixed function oxidases, followed by 2′,3′ ring cleavage and catabolism to succinate and pyruvate (Levy and Frost, 1966). It is not known whether such reactions can occur in higher organisms.

Degradation in higher plants

Coumarin can be further metabolized within the plant in which it is synthesized. Most experiments have dealt with exogenous coumarin. When [^{14}C]coumarin was administered the products were melilotic acid and its glucoside (Kosuge and Conn, 1959). Such glucosides can be further converted to *cis*-2′-glucosyloxycinnamic acid (Stoker and Bellis, 1962). Harborne and Corner (1961) found that simple glucosylation was the fate of administered umbelliferone, scopoletin and aesculetin. In experiments with *Nicotiana tabacum* seedlings, Innerarity *et al.* (1972) found that scopolin was converted to scopoletin, with some residual glucoside.

Dicoumarol affects blood coagulation, but little is known of its degradative reactions. Burke and Parkins (1979) showed that rat hepatic microsomes metabolized it by mixed function oxidases to 7-hydroxydicoumarol (Lee and Conway, 1976).

Aflatoxins such as AfB1 can be converted into more toxic derivatives; some, such as AfP1, are very potent carcinogens. Demethylation appears usually to be the first step (Wogan *et al.*, 1967). Oral administration of AfB1 resulted in excretion of AfM1 (Dalezios *et al.*, 1973), and human liver produced significant levels of it (Roebuck and Wogan, 1977). Unexplained but interesting results were obtained by Vaught *et al.* (1977), who found that AfB1 can be metabolized by rat liver nuclei as well as by microsomes. Dietetary deficiencies can have significant influence on human detoxification and metabolism of aflatoxins, e.g. deficiency of vitamin A may increase carcinogenicity of AfB1 (Campbell and Hayers, 1976).

Other coumarins also undergo detoxification with demethylation as the first step, as in the case of xanthotoxin to xanthotoxol. Collins *et al.* (1974) found further detoxification as a result of opening of the furan ring, followed by the pyrone ring. Detoxification of furanocoumarins by insects may have a significant effect on ecology (Berenbaum and Zangerl, 1993).

Biosynthesis

Microbes not only contain simple coumarins (see Table 9.1) but can also synthesize more complex coumarins such as novobiocin, chartreusin and coumarmycin, pointing to biosynthetic diversity. One example is the fact that *Streptomyces* developed the route to coumarin by oxidative cyclization directly from tyrosine (Brown, 1963).

Since 1960 there has been increased interest in the reactivity of natural coumarins and their biosynthesis, starting with the work of Späth and Kuffner (1936), summarized in the review papers of Brown (1970, 1979, 1981). In higher plants different coumarins have different biosynthetic pathways dependent on the structures. Four types of coumarins could be envisioned:

1. simple coumarins, unsubstituted or with substitution in the benzene ring
2. furanocoumarins with the furan ring fused at the 6,7 positions (psoralens), or at 7,8 (angelicins)
3. pyranocoumarins, with a pyran ring fused at 6,7 or at 7,8
4. coumarins substituted in the pyran ring (e.g. 4-hydroxycoumarins).

Though (1) and (4) appear similar, they differ in biological action and biosynthesis. For review articles on biosynthesis see Kosuge and Gilchrist (1977), Brown (1981) and Murray *et al.* (1982).

The benzopyrone nucleus is said to be formed from 2′-hydroxycinnamic acid by lactonization of 2′-hydroxy and carboxyl functions; thus the simple coumarin nucleus is phenylpropanoid, having a C_6 benzene ring linked to a

C_3 aliphatic chain (Neish, 1964). Phenylpropanoids are most commonly derived in plants by the shikimate–chorismate pathway, and microbes can use the same pathway, but the coumarin nucleus of fungal aflatoxin can be entirely derived from acetate-polyketide intermediates. Coumarin itself exists in plants mostly in a bound form as the glucoside (Roberts and Link, 1937; Kosuge, 1961). Coumarin glucosides are compartmentalized in the cell vacuole (Zobel, 1993), while the glucosidases which can liberate free coumarin are within the cell wall (Oba *et al.*, 1981). Only mechanical disruption of the tonoplast and plasma membrane allows these two compartments to fuse and free coumarin to be released. We found free coumarin to be so cytotoxic and to alter the ultrastructure of mitochondria and endomembranes so drastically (Podbielkowska *et al.*, 1994, 1995) that storing free coumarin in the vacuole under hydrophilic conditions in an aqueous environment could be dangerous to the cells producing it. Bound coumarin is *cis*-2′-glucosyloxycinnamic acid (coumarinyl glucoside), and after hydrolysis with removal of glucose it spontaneously lactonizes to free coumarin.

Coumestans of leguminous plants have a coumarin nucleus derived in part from acetate, and in part from shikimate, and are biosynthetically isoflavonols (Brown, 1963). Furanocoumarins originate from umbelliferone and mevalonate. Other simple coumarins can arise by conversion of cinnamic acid to umbelliferone which then undergoes further substitution (Murray *et al.*, 1982).

Biological action

From the great range of effects produced by aflatoxins, and the diversity of species susceptible to them, it could be concluded that the mechanisms of action are numerous and complex (Wogan, 1977). Not surprisingly, therefore, data obtained by different researchers can differ greatly. AfB1 was found to inhibit synthesis of RNA in rat liver, and to a smaller extent human liver and kidney (Scaife, 1971). Activity of some enzymes could be increased (Brown and Abrams, 1965), but it lowered the activity of rat liver hydroxylase (Durtoo *et al.*, 1968). Prasanna *et al.* (1975) found increased activity of cyclic AMP. The percentage of saturated fatty acids increased (Veen, 1967), but activation of ATPases was depressed (Bababunmi and Bassir, 1969).

Other coumarins can stimulate some enzymes in animals (Hanson and Deleo, 1989) but inhibit others (Woo *et al.*, 1983; Laskin and Lee, 1991; Cai *et al.*, 1993). Bergapten and xanthotoxin inhibited microsomal enzymes (Letteron *et al.*, 1986) and inactivated cytochrome P450 through covalent modification of the apoprotein (Mays *et al.*, 1990). Aesculetin inhibited 5-lipoxygenase (Neicki *et al.*, 1993) and fatty acid production was reduced (Kimura *et al.*, 1985). In bacteria inhibition of DNA gyrase was observed by

Drlica and Coughlin (1989), as was topoisomerase II in eucaryotes (Hsiek and Brutlag, 1980; Nakasu and Kikuchi, 1985).

Dicoumarol and other 4-hydroxycoumarins play a role as anticoagulants (Scheel, 1978). Knobloch *et al.* (1952) explained that the existence of two tautomeric forms in equilibrium can account for the anticoagulant effect of such coumarins, e.g. 4-hydroxycoumarin and 2-hydroxychromone. Their suggestion could be supported by the fact that Chmielewska and Cieslak (1958) obtained two isomeric demethyl ethers of dicoumarol, but not much was achieved in elucidating the mechanism of dicoumarol's action. It was found to be antagonistic to vitamin K (Suttie and Jackson, 1977), and inhibited prothrombin synthesis (Ren *et al.*, 1977).

Since the work of Kuske (1942) the best-known instance of biological activity has been that of the induction of photophytodermatitis by furanocoumarins (Atherton *et al.*, 1988; Meffert *et al.*, 1982; Dall'Acqua and Rodighiero, 1985; Swain and Downum, 1991). Ivie (1986*a*) stated that furanocoumarins have the widest documented spectrum of biological action, both light-dependent (Cadet *et al.*, 1990) and -independent (Averbeck, 1988, 1989; Gast *et al.*, 1988). The light-dependent mechanism of this effect has been well established since the work of Musajo *et al.* (1953) followed by that of Dall'Acqua and coworkers (1969, 1978). Contact dermatitis was investigated by Pathak and associates (1974, 1986), but the full explanation for the 'contact' feature of furanocoumarin-containing plants was forthcoming only after demonstration that a high percentage of these coumarins is deposited on the plant surface. Several medicinal plants known to cause contact dermatitis (Zobel and Brown, 1991*d*) contain substantial concentrations of furanocoumarins on their surface, accounting for the fact that even brief touching may be enough to produce the lesion without ingestion of the plant. Such examples have been observed among people handling umbelliferous plants (e.g. celery, parsnip) and their fruits (Scheel *et al.*, 1963; Ataga *et al.*, 1993). Psoralen has been reported to be the most active of the various furanocoumarins (Pathak *et al.*, 1961). The suggestion of strong photoreactivity on the part of angelicin, which forms only a monoadduct, has been advanced by Potapenko (1986). Mechanisms of reaction are still under investigation (Potapenko and Sukhorukov 1984*a,b*). New photoactive coumarins are continually being discovered (Ceska *et al.*, 1988; Kashman *et al.*, 1993). For humans they can be used as anticancer agents in clinical treatment (Wattenberg *et al.*, 1979; Marshall, 1989; Egan *et al.*, 1990; Nair *et al.*, 1991; Kahn *et al.*, 1994). These compounds can be antimutagens, as Wall *et al.* (1990) found for imperatorin. But on the other hand they can exhibit genotoxicity (Averbeck *et al.*, 1988, 1992; Walther *et al.*, 1988; Gasparro *et al.*, 1990*a,b*; Reardon *et al.*, 1992) mediated by the singlet oxygen reaction (Beijersbergen *et al.*, 1989), and be mutagens even in the dark (Bridges and Mottershead, 1977; Billardon *et al.*, 1984; Billardon and Moustacchi, 1986; Moustacchi, 1988).

Coumarins in higher plants can play a protective role because they are antibacterial (Ashwood-Smith and Grant, 1977; Drlica and Synder, 1978; Towers, 1984), antifungal (Mikkelson *et al.*, 1961; Camm *et al.*, 1976; Beier and Oertli, 1985), antiviral (Musajo *et al.*, 1965; Kashman *et al.*, 1993) act against bacteriophage (Kittler *et al.*, 1986), and inhibit growth of the cells (Kupidlowska *et al.*, 1994). They might thus prevent consumption of a plant by generalist herbivores (Berenbaum, 1978; Diawara *et al.*, 1993). By absorbing ultraviolet radiation coumarins can form a protective shield against such radiation (Zobel and Brown, 1993*a*; Zobel *et al.*, 1994*a*). The binding site of furanocoumarins after radiation was established from the studies of Rodighiero *et al.* (1961), who showed binding of furanocoumarins, but not simple coumarins, to DNA. Dall'Acqua *et al.* (1971) demonstrated the formation of monoadducts, and, after further radiation, cross-linkages between the two strands of the double helix (Zarebska, 1994*a,b*). A few years later Dall'Acqua *et al.* (1976) found that 3,4 photoadducts do not absorb 365 nm radiation, and thus, unlike 4′,5′ photoadducts, are unable to cross-link DNA. Reaction with RNA involves either singlet or triplet states of linear furanocoumarins (Pathak *et al.*, 1961).

But phototoxicity of psoralen may not be directly involved in intercalation with DNA (Laskin *et al.*, 1985, 1986). In bacteria inhibition of DNA synthesis is connected with DNA relaxation. In eucaryotic cells as well coumarins cause DNA relaxation (Mattern and Painter, 1979), but they block eucaryotic DNA replication (Mattern and Scudiero, 1981), a process in which topoisomerase participates. The overall reaction is inhibition of DNA replication (Drlica and Coughlin, 1989).

Reaction with protein and lipids, causing changes in the membrane, has recently been investigated (Midden, 1988; Bordin, 1991; Caffieri *et al.*, 1991, 93; Dall'Acqua and Martelli, 1991; Zarebska, 1994*a,b*). The reaction mechanisms of furanocoumarins in the dark were investigated, suggesting that even without light furanocoumarins react with DNA (Laskin *et al.*, 1985).

Histochemical localization of coumarins is important because their localization is closely correlated with their role in the plant. We currently envision five major roles:

- antimicrobial activity
- deterrence of feeding
- autoinhibition of germination
- shielding against ultraviolet radiation
- transduction of ultraviolet into visible (usable) light.

Histochemical localization is possible because they autofluoresce (Zobel and March, 1993); hence they can be located in the epi-light microscope (Zobel and Brown, 1989*a*), and by the technique of tissue isolation (Zobel and Brown, 1989*b*, 1991*b*; Zobel *et al.*, 1991*a*). The first proof of the existence of furanocoumarins on the plant surface came from experiments using high temperature to melt epicuticular wax and remove compounds included in it. Zobel and Brown (1988) used almost-boiling water for brief dipping of leaves instead of long-time washing at room temperature with organic solvents which can be suitable for analysis of waxes but caused leakage from epidermal cells (Tulloch and Hoffman, 1982).

Several coumarins have been identified as co-occurring with psoralen, xanthotoxin and bergapten on the surface of several plant species (Zobel *et al.*, 1991*b*), including citrus fruits (Plomley and Zobel, unpublished). On fruits of Rutaceae, Umbelliferae and Leguminosae (Zobel and Brown, 1991*c*) psoralens may be responsible for photophytodermatitis on people handling them. On the surface of leaves there are deposits of crystal- like structure (Fig. 9.1a, arrow), which are removed with our almost-boiling water technique (Fig. 9.1b). Such an epicuticular wax layer can be broken in some areas, uncovering the smooth cuticular surface below (Fig. 9.1c, asterisk). On the cross-section the thickness of the surface deposits can be greater than that of the outer cell wall itself (Fig. 9.1b, arrow).

In the isolated tissues of fruits of the above-mentioned three families it was proved that location and concentration can be different for different species, sometimes depending on their habitats. In successive dippings of *Psoralea* fruit tissue, seed coat and the embryo itself, Zobel *et al.* (1991*c*) found psoralens to be localized in different concentrations on the surface of the fruit, in the interior of the fruit tissue, on the surface of the seed coat, in the interior of the seed coat tissue, on the surface of the embryo and in the embryo itself (Fig. 9.1e, arrows). In *Angelica archangelica* (Zobel and Brown, 1991*b*) the largest deposits were in the seed coat (Fig. 9.1f, SC), and some were visible on the surface (arrow), while fruit tissue contained lower concentrations of psoralen, bergapten and xanthotoxin. Some of the glands in the fruit tissue showed no fluorescence (Fig. 9.1f, asterisk), even though, on higher magnification, they contained some unidentified crystals (Fig. 9.1g). In *Sium* species, whose fruits germinate in water, deposits of furanocoumarins were different. There were some on the surface (Fig. 9.1h, arrow) and a high concentration in the embryo (E), but no fluorescence was visible either in fruit tissue or seed coat (asterisk).

Large concentrations of furanocoumarins on the surface of embryos and within dry fruit and seed tissues may be responsible for autoinhibition of germination (Zobel and Brown, 1991*a*,*b*,*c*). These compounds are antimitotics (Gawron and Glowniak, 1987; Gawron *et al.*, 1989; Podbielkowska *et al.*, 1994, 1995; Kupidlowska, 1995*a*,*b*) in concentrations of 50–200 ppm and, at this concentration they inhibit the uptake of oxygen (Kupidlowska, 1995*a*; Podbielkowska *et al.*, 1995, 1996), and they retard enzymes responsible for the energy status of the cell. With such properties, furanocoumarins

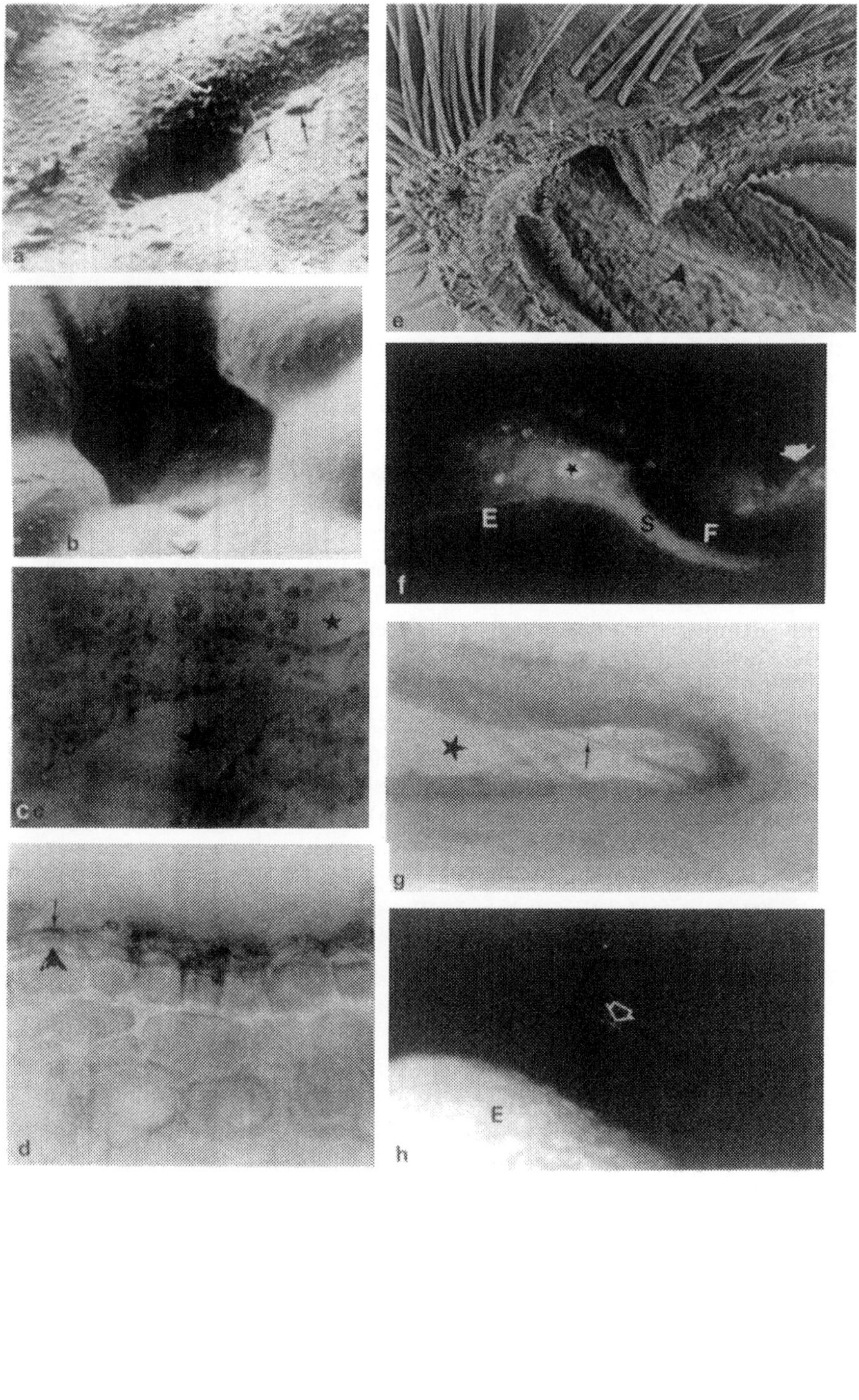
a
b
c
d
e
f
E
S
F
g
h
E

can at least partially account for embryo dormancy during autumn and winter, when the fruits are in the soil. Following the spring thaw and frequent rains the concentrations of coumarins are lowered, terminating dormancy. Additionally, we found (Nxumalo and Zobel, 1995) that 5–20 ppm of xanthotoxin actually stimulated oxygen uptake; thus we suggest that with diminishing concentrations of furanocoumarins in the spring the germination of the embryo could actually be stimulated (Keightley and Zobel, unpublished). The ecological significance of furanocoumarins as germination inhibitors was mentioned by Rodighiero *et al.* (1954), Sarma and Barooah (1982), Oster *et al.* (1992) and Cappelletti *et al.* (1992).

Relatively large umbelliferous fruits of *Ferula* contain a 10–20-fold range of concentrations among different individuals. This finding might explain the great time difference and difficulty of germination characteristic of these fruits, but it is a very useful ecological factor in enabling germination to take place over a long period of time. For us consumers this variation in concentration can explain the observation that some of these seeds can be poisonous to stock in winter, but then become edible in spring. This can be true for garden vegetables and spices too, as we observed seasonal changes in concentrations of furanocoumarins (Zobel and Brown, 1990, 1991*a*), and changes dependent on environmental conditions (Chandbury *et al.*, 1985; Ashwood-Smith *et al.*, 1985; Berenbaum and Nitao, 1985; Ceska *et al.*, 1986; Dercks, *et al.*, 1990; Lee and Berenbaum, 1992).

Surface localization of furanocoumarins has further ecological implications. Under stress conditions their surface concentrations change, e.g. with ultraviolet radiation (Zobel and Brown, 1993*a*; Zobel *et al.*, 1994*a*) and soil

---

Fig. 9.1. (a) Surface of *R. graveolens* leaf. Crystal-like structures and other deposits visible in the epicuticular wax. (b) Wax removed from the surface. Scanning electron micrograph. (c) Broken epicuticular wax reveals the smooth epicuticle (asterisk) under it. (d), Cross-section of the leaf petiole of *R. graveolens*. Thick layer over the cuticle (arrow) of the thickness of the cell wall below (arrowhead). (e) Scanning electron micrograph of *Psoralea bituminosa* fruit. Location of furanocoumarins is on the fruit surface (arrow) in the fruit tissue (asterisk), on the seed coat surface (double arrow), in the seed coat tissue (double asterisk), and on the surface of the embryo (arrowhead). (f) Cross-section of the fruit of *A. archangelica*. Location of autofluorescence after 366 nm radiation under the epi-light microscope. Surface deposit of furanocoumarins (arrow) on the fruit, not visible fluorescence inside fruit tissue (F) nor in gland in fruit tissue (asterisk). Fluorescent compounds visible in seed coat (S); some fluorescence in cell wall of embryo (E). (g) Magnification of (f), including gland (asterisk). No fluorescence in gland, but crystals in interior. (h) Cross-section of *Sium* fruit in epi-light microscope after ultraviolet radiation. Slight fluorescence on surface (white arrow), none within fruit or seed coat tissue. Intense fluorescence in interior of embryo cells (E).

mineral content (Zobel *et al.*, 1993*a,b*; Zobel *et al.*, 1995) as well as with acid and salt spray (Zobel *et al.*, 1991*c*). Surface deposition of these compounds may be responsible for communication of the plant with its environment (Zobel and Brown, 1995), and it seems that extrusion is a genetically controlled natural characteristic of most and probably all cells, as even callus cells extrude these compounds to the surface (Zobel and Brown, 1993*b*).

Matern's group, working on elicitor-stimulated biosynthesis of furanocoumarins by suspension culture cells explained extrusion of furanocoumarins from the cells as a means for getting rid of these toxic compounds (Matern *et al.*, 1988). We agree to some extent with this point of view, after seeing drastic structural changes in the electron microscope due to the effect of coumarins (Casley-Smith *et al.*, 1973; Podbielkowska *et al.*, 1995, 1996; Kupidlowska *et al.*, 1994, 1995*b*). But, on the other hand, their localization within intercellular spaces (Zobel, 1995) suggests their protective role against bacteria penetrating into the parenchyma tissue through the stoma and the labyrinth of intercellular passages.

As concentrations of coumarins may increase under stress conditions, they may act as phytoalexins (Natarajan *et al.*, 1981; Nover *et al.*, 1989; Klocke *et al.*, 1989; Lee and Berenbaum, 1992; Zangerl, 1990; Berenbaum, 1991). Many examples prove that coumarins increase as a result of infection, parasitism, ultraviolet radiation (Zangerl and Berenbaum, 1987), extreme temperatures (Zobel and Brown, 1990), air pollution and other environmental conditions. We found that some furanocoumarins may be phytoalexins for only certain species, as in the case of isopimpinellin in *R. graveolens*, but not celery; conversely, xanthotoxin is a phytoalexin for both species infected by red spider (Zobel *et al.*, 1994*b*). The role of coumarins as phytoalexins has attracted wide attention in recent years (Beier *et al.*, 1983; Zangerl *et al.*, 1991; Rodov *et al.*, 1992).

Coumarin and several of its derivatives have been used as anticancer drugs clinically (Jernis and Smith, 1981; Gruenert *et al.*, 1985; Marshall, 1989; Apseloff *et al.*, 1990; Marshall *et al.*, 1990; Raghunathan *et al.*, 1991; Tokura *et al.*, 1991), as they stop mitosis (Meyer *et al.*, 1985; Gawron and Glowniak, 1987; Podbielkowska *et al.*, 1994; Kupidlowska *et al.*, 1995*b*). We found that psoralen and xanthotoxin cause chromosomal aberrations (Schimmer and Abel, 1986) and questioned their usage in this context (Keightley *et al.*, 1995; Podbielkowska *et al.*, 1995). Coumarin and 4-hydroxycoumarin do not cause such chromosomal aberrations (Podbielkowska *et al.*, 1994); thus their usage for prostate cancer treatment may be considered safe.

To me, one of the most interesting ideas related to the physiological role of coumarins at the molecular level is their possible function *in vivo* as scavengers of free radicals (Halliwell and Gutteridge, 1989; Payá *et al.*, 1992*a,b*) and as antioxidants, which up to now has been proved only *in vitro*

(Payá *et al.*, 1993). The surface would be a very convenient location for their deposition. Ultraviolet radiation by itself does not totally destroy the epidermal and mesophyll cells of *R. graveolens* (Zobel and Kuras, 1996), but when it reacts with coumarins it can form potentially damaging free radicals. Surface localization of ultraviolet-absorbing coumarins (and in that respect of any other compounds, as found by Zobel and Lynch (1995), would not only prevent ultraviolet from penetrating into the cells, but free radicals would also be excluded from the cytoplasm.

Absorption of high-energy radiation can have several results, after molecules are excited their temperature can rise and heat is released, or electrons can be emitted, or longer wavelengths of radiation can be released. We found that autofluorescence, and thus transduction of ultraviolet into visible light, is a very important phenomenon for the plant. Our investigations with *R. graveolens* plants (Zobel and Brown, 1993*a*) showed that they not only survived four weeks under 366 nm monochromatic radiation, but still produced four times the furanocoumarin concentration of plants kept under white light. Such plants remained healthy under 254 nm radiation for 2 weeks, producing more coumarins than the control plants, and one additional coumarin, isopimpinellin, was detectable (Zobel *et al.*, 1994*a*). The supply of energy for these phenomena could have come from transduction of ultraviolet into usable wavelengths, and experiments on sustaining assimilation under such conditions of energy transduction are in progress (unpublished studies). Quantitative and qualitative differences of furanocoumarins located in two distinct compartments (surface of a plant and its interior) should be further investigated as the roles they play there may be different.

Structures

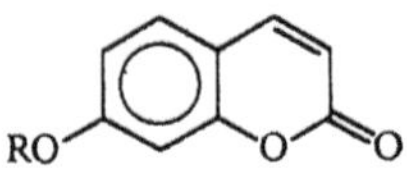

R = OH : Umbelliferone
R = CH_3 : Herniarin

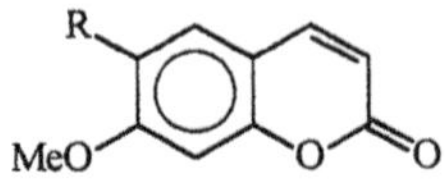

R = 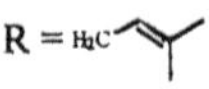 : Suberosin

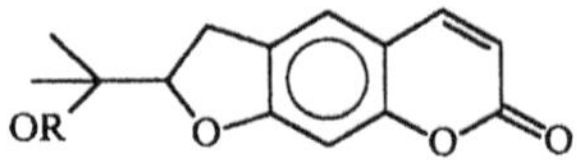

R = H : Marmesin
R = β-D-glucosyl : Marmesinin

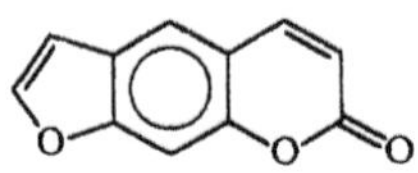

Psoralen

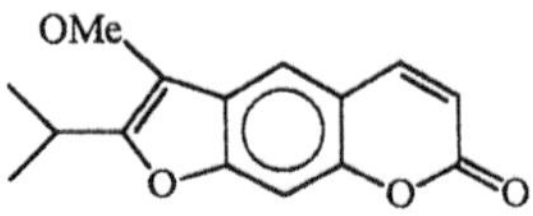

Peucedanin

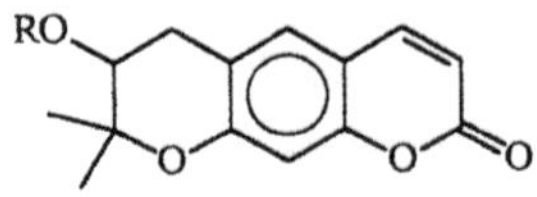

R = H : Aegelinol

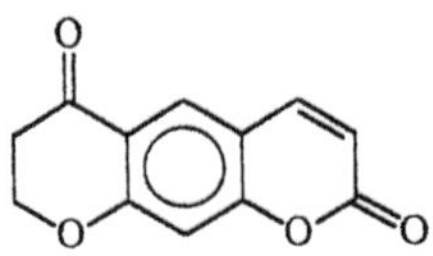

Graveolone

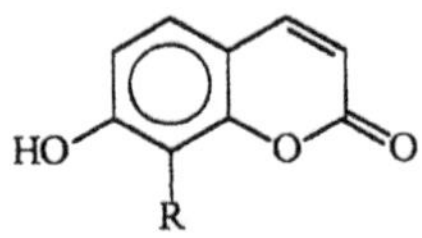

R = 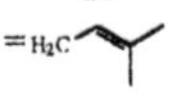 : Osthenol

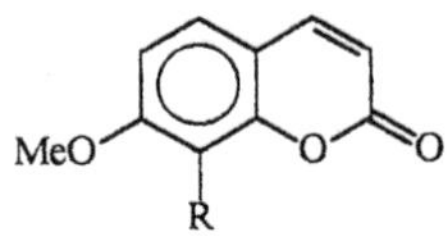

R = 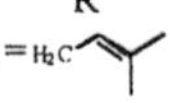 : Osthol

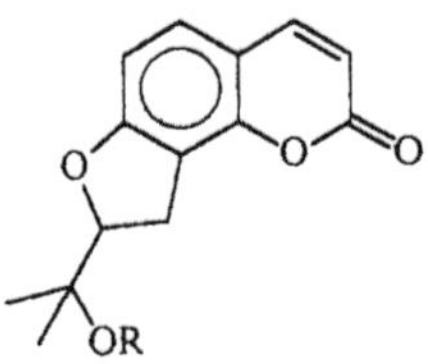

R = H : Columbianetin

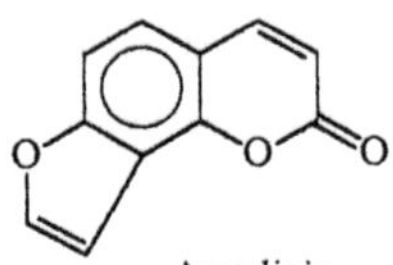

Angelicin

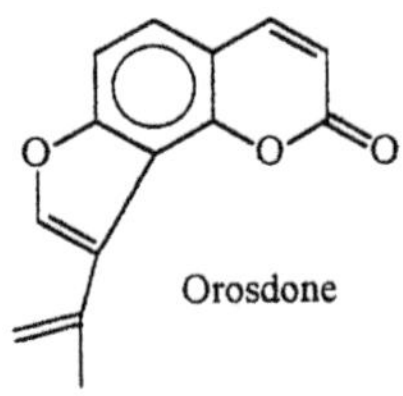

Orosdone

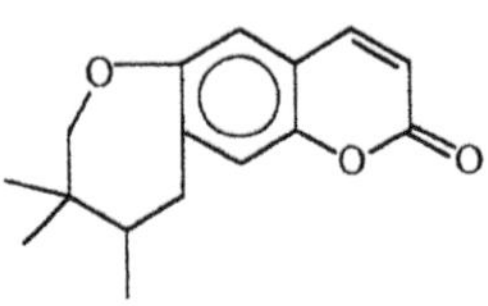

R = H : Lomatin

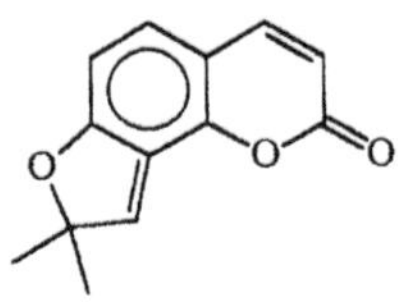

Seselin

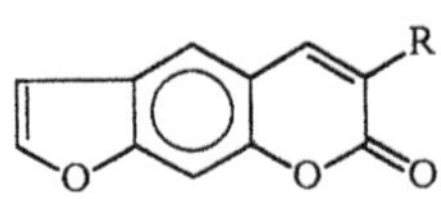

Chalepensin

R = 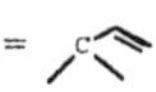 : Chalepensin

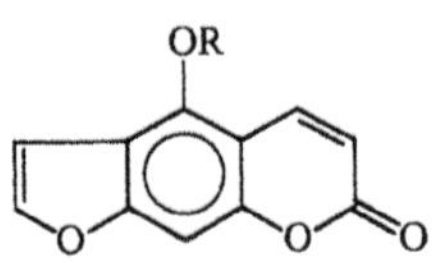

Bergapten

R = 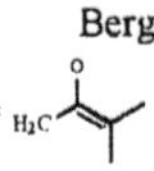 : Isooxypeucedanin

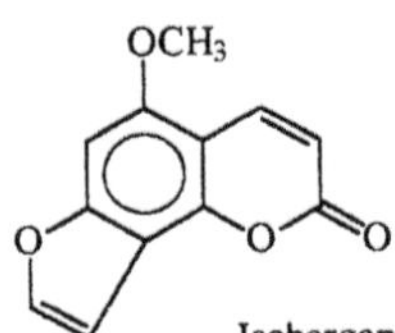

Isobergapten

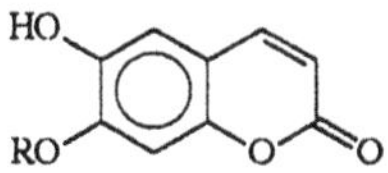

R = H : Aesculetin
R = CH_3 : Isoscopoletin
R = B-D glucosyl : Cichorin

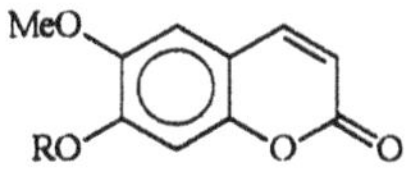

R = H : Scopoletin
R = CH_3 : Hydrangetin
R = B-glucosyl : Scopolin

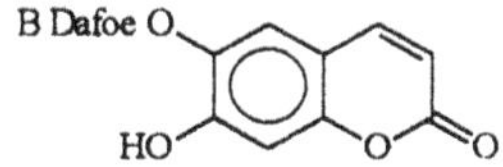

Aesculin

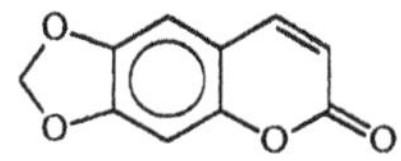

Ayapin

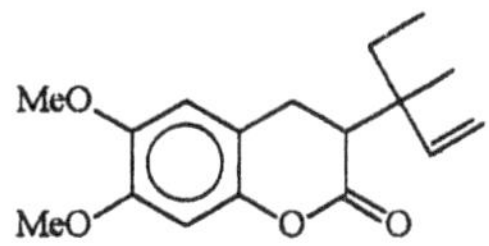

Rutaculin

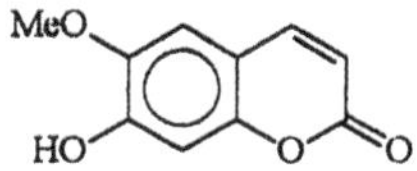

Obliquetin

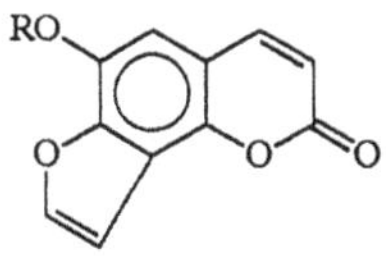

R = H : Heratomol
R = CH_3 : Sphondin
R = H_2C : B-D glucosyl cichorin

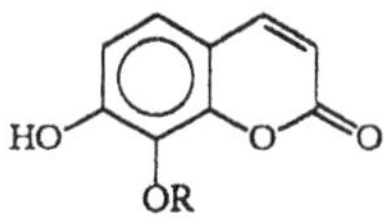

R = H : Daphnetin
R = CH_3 : Hydragetin

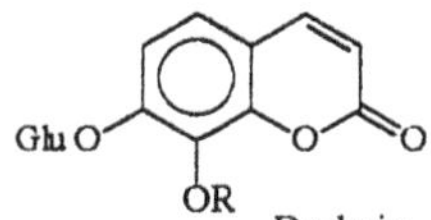

Daphnin

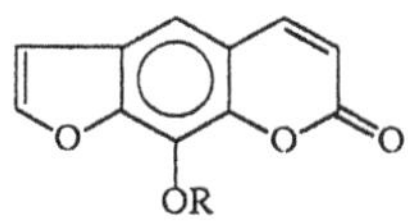

R = H : Xanthotoxol
R = CH_3 : Xanthotoxin
R = H_2C : Heratomin

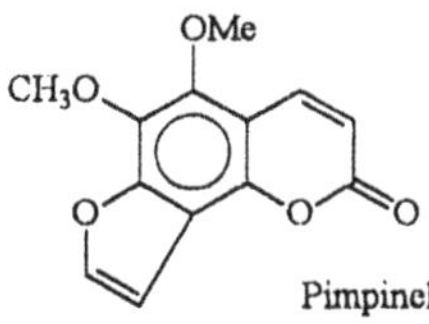

Pimpinellin

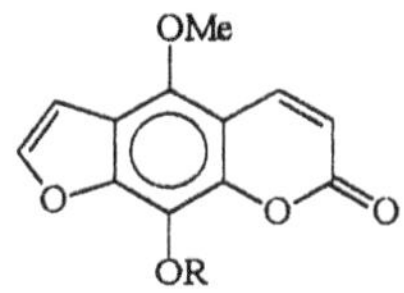

R = CH_3 : Isopimpinellin
R = H_2C : Phellopterin

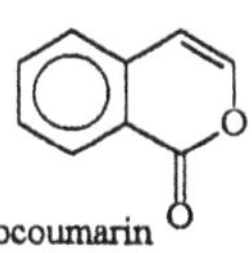

Isocoumarin

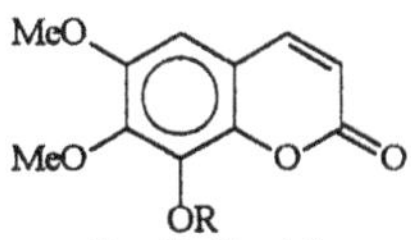

R = H : Fraxidin

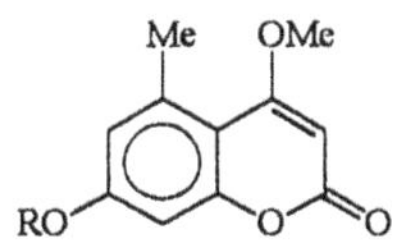

R = CH_3 : Siderin

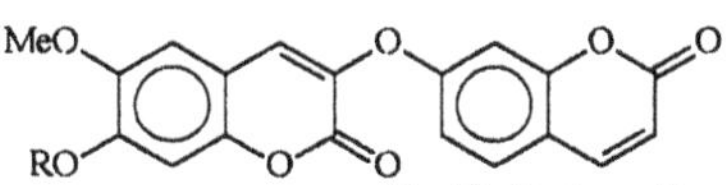

R = H : Daphnoretin

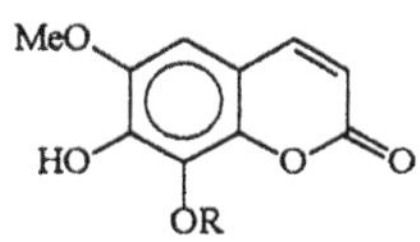

R = H : Heratomol
R = CH_3 : Isoflaxidin
R = Glu : Fraxin

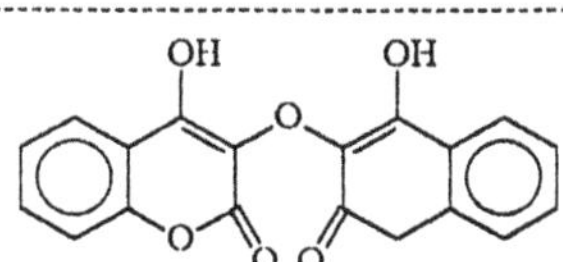

Dicoumarol lactone of

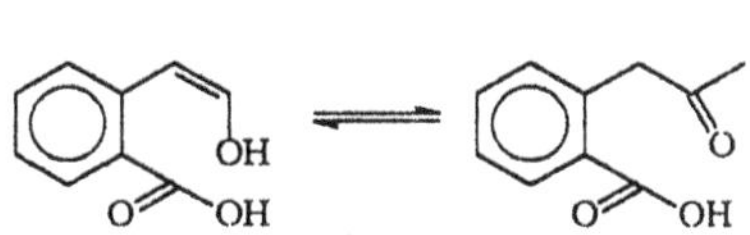

References

Abel, G. and Schimmer, O. (1981). *Planta Med.* **42**, 333–43.

Abel, G. and Schimmer, O. (1986). *Mutat. Res.* **169**, 51–5.

Abel, G., Erdelmeier, C., Neier, B. and Sticher, O. (1985). *Planta Med.* **46**, 250–2.

Abu-Mustafa, E.A., Badran, N., Fayez, M.B.E. and Starkowsky, N.A. (1958). *Nature* **182**, 54–7.

Abu-Mustafa, E.A., El-Bay, F.K.A. and Fayez, M.B.E. (1968). *Recl. Trav. Chim. Pays-Bas.* **87**, 925–29.

Abu-Mustafa, E.A., El-Bay, F.K.A. and Fayez, M.B.E. (1971*a*). *J. Pharm. Sci.* **60**, 788–92.

Abu-Mustafa, E.A., El-Bay, F.K.A. and Fayez, M.B.E. (1971*b*). *Tetrahedron Lett.* 1657–9.

Abu-Mustafa, E.A., El-Bay, F.K.A. and Fayez, M.B.E. (1975). *Naturwissenschaften* **62**, 39–45.

Aplin, R.T. and Page, C.B. (1967). *J. Chem. Soc. C.*, 2593–7.

Apseloff, G., Sheppard, D., Chambers, M., Nawoot, S., Mays, D. and Gerber, N. (1990). *Drug Metabol. Dispos.* **18**, 298–303.

Ashwood-Smith, M.J. and Grant, E. (1977). *Experientia* **33**, 387–8.

Ashwood-Smith, M., Ceska, M. and Chandhury, S. (1985). *British Med. J.* **290**, 1249–53.

Ataga, A., Epton, A.S. and Frost, R. (1993). *Physiol. Mol. Plant Pathol.* **42**, 161–168.

Atherton, D., Carbott, F., Glover, M. and Hawk, J. (1988). *Brit. J. Dermatol.* **118**, 791–5.

Averbeck, D. (1988). In *Photosensitization.* (ed. G. Moreno). Springer-Verlag, Berlin.

Averbeck, D. (1989). *Photochem. Photobiol.* **50**, 859–82.

Averbeck, D., Chandra, P. and Biswas, R. (1975). *Radical Environ. Biophys.* **12**, 252–7.

Averbeck, D., Papadopoulo, D. and Moustacchi, E. (1988). *Cancer Res.* **48**, 2015–20.

Averbeck, D., Dardalhon, M., Magana-Schwencke, N., Meira, L., Meniel, V., Boiteux, S. and Sage, E. (1992). *J. Photochem. Photobiol. B, Biol.* **14**, 47–63.

Bababunmi, E.A. and Bassir, O. (1969). *Br. J. Pharmacol.* **37**, 497–9.

Balbaa, S.I., Hilal, S.H. and Haggag, M.Y. (1972). *Planta Med.* **22**, 122–8.

Baskin, J.M., Ludlow, C.J., Harris, T.M. and Wolf, F.T. (1967). *Phytochemistry* **6**, 1209–13.

Beier, R.C and Oertli, E.H. (1985). *Phytochemistry* **22**, 2595–7.

Beier, R.E., Ivie, G.W. and Oertli, E.H. (1983). In *Xenobiotics in foods and feeds* (ed. J.W. Finley and D.E. Schwass). ACS symposium series 234. American chem. Soc., Washington DC.

Beijersbergen van Heneqouven, G., Wijn, E., Schoonderwoerd, S. and Dall'Acqua, F. (1989). *Photochem. Photobiol. B, Biol.* **3**, 631–85.

Berenbaum, M.R. (1978). *Science* **201**, 532–4.

Berenbaum, M.R. and Nitao, J.J. (1985). *J. Chem. Ecol.* **11**, 1349–58.

Berenbaum, M.R. and Zangerl, A. (1993). *Oecologia* **95**, 370–5.

Berenbaum, M.R., Nitao, J.J. and Zangerl, A.R. (1991). *J. Chem. Ecol.* **17**, 207–15.

Berrie, A., Parker, W., Knights, B. and Hendrie, M. (1968). *Phytochemistry* **7**, 567–73.

Bevilacqua, R., Baccichetti, F. and Musajo, L. (1973). *Ital. J. Biochem.* **22**, 160–6.

Beyrich, T. (1965). *Pharmazie* **20**, 655–9.

Beyrich, T. (1968). *Pharmazie* **23**, 336–42.

Bhattacharyya, P., Chatterjee, D., Chakrabarti, A. and Chakraborty, D. (1979). *Indian J. Chem.* **17B**, 111–12.

Birkofer, E., Loper, G., Hanson, C., Graham, J., Witt, S. and Spencer, R. (1967). *Crop Sci.* **7**, 259–67.

Billardon, B. and Moustacchi, E. (1986). *Mutat. Res.* **174**, 241–6.

Billardon, C., Levy, S. and Moustacchi, E. (1984). *Mutat. Res.* **138**, 63–70.

Blarek, Z. and Stary, F. (1965). *Mitt. Dtsch. Pharm. Ges.* **35**, 117–21.

Bohme, H. and Pietsch, G. (1939). *Ber. Dtsch. Chem. Ges.* **72B**, 773–6.

Booth, A., Masri, M., Robbins, D., Emerson, O., Jones, F. and De Eds, F. (1959). *J. Biol. Chem.* **234**, 946–9.

Bordin, F. (1991). *Photochem. Photobiol.* **1**, 127–134.

Bose, P.K. and Mookerjee, A. (1994). *J. Indian Chem. Soc.* **21**, 181–9.

Bourquelot, E. and Herissey, H. (1920). *Hebd. Seances Acad. Sci.* **170**, 1545–52.

Bridges, B.A. and Mottershead, R. (1977). *Mutat. Res.* **44**, 305–312.

Brown, J.M.M. and Abrams, L. (1965). *Onderstepoort J. Vet. Res.* **32**, 119–21.

Brown, S. A. (1963). *Lloydia* **26**, 211–20.

Brown, S.A. (1970). *Phytochemistry* **9**, 2471–3.

Brown, S.A. (1979). *Planta Med.* **36**, 299–302.

Brown, S.A. (1981). In *The biochemistry of plants, a comprehensive treatise*, Vol. 7 (ed. P.K. Stumpf and E.E. Conn). Academic Press, New York..

Bubeva-Ivanova, L., Zheleva, A.B., Mitrev, A., Mincheva, V. and Savchev, P. (1973). *Chem. Abstr.* **79**, 89461.

Burke, B.A. and Parkins, H. (1979). *Phytochemistry* **18**, 1073–7.

Cadet, J., Vigny and P. Midden, W. (1990). *J. Photochem. Photobiol., B, Biology* **6**, 197–206.

Caffieri, S., Zarebska, Z. and Dall'Acqua, F. (1991). *Med. Biol. Environ.* **19**, 45–50.

Caffieri, S., Zarebska, Z. and Dall'Acqua, F. (1993). In *Frontiers of photobiology* (ed. A. Shima, M. Ichahashi, Y. Fujiwara and H. Takeve) Elsevier, Amsterdam.

Cai, Y., Bennett, D., Nair, R., Ceska, O., Ashwood-Smith, M. and DiGiovanni, J. (1993). *Chem. Res. Toxicol.* **6**, 872–878.

Caldwell, A.G. and Jones, E.R.H. (1945). *J. Chem. Soc.* 540–3.

Camm, E.L., Wat, C.K. and Towers, G.H.N. (1976). *Can. J. Bot.* **54**, 2562–6.

Campbell, H.A. and Link, K.P. (1941). *J. Biol. Chem.* **138**, 21–3.

Campbell, T.C. and Hayes, J.R. (1976). *Toxicol. Appl. Pharmacol.* **35**, 199–211.

Cappelletti, E.M., Innocenti, G. and Caporale, G. (1992). *J. Chem. Ecol.* **18**, 155–163.

Carbonnier, J. and Molho, D. (1978). *Bull. Mus. Natl. Hist. Nat. Sci. Phys.-Chim.* **19**, 17–32.

Carbonnier-Jarreau, M., Carbonnier, J., Farille, M. and Molho, D. (1978). *Bull. Mus. Natl. Hist. Nat. Sci. Phys.-Chim.* **19**, 3–15.

Casley-Smith, J.R., Foldi-Borasok, E. and Foldi, M. (1973). *Br. J. Experiment. Pathol.* **53**, 1–5.

Casparis, P. and Manella, E. (1944). *Pharm. Acta Helv.* **19**, 158–62.

Ceska, O., Chaudhary, S., Warrington, P.J., Poulton, G. and Ashwood-Smith, M.J. (1986). *Experientia* **43**, 130–2.

Ceska, O., Chaudhary, S., Warrington, P.J., Ashwood-Smith, M.J., Bushnell, G. and Buldon, G. (1988). *Phytochemistry* **27**, 2083–7.

Chaudhary, S., Ceska, O., Warrington, P.J. and Ashwood-Smith, M.J. (1986). *J. Agr. Food Chem.* **33**, 1152–7.

Chmielewska, I. and Cieslak, J. 1958. *Tetrahedron* **4**, 135–8.

Collins, J.F., Donnelly, W.J., Grudon, M.F. and James, K.J.J. (1974). *Chem. Soc., Perkin Trans. I*, 2177–81.

Christensen, F. and Wissing, F. (1972). *Biochem. Pharmacol.* **21**, 975–9.

Creaven, P., Parke, D. and Williams, R. (1965). *Biochem. J.* **96**, 390–7.

Crichton, E. and Waterman, P. (1978). *Phytochemistry* **17**, 1783–6.

Crowden, R., Harborne, J. and Heywood, V. (1969). *Phytochemistry* **8**, 1963–9.

Dalezios, J., Hsieh, D. and Wogan, G. (1973). *Food Cosmet. Toxicol.* **11**, 605–9.

Dall'Acqua, F. and Caffieri, S. (1988). *Photomed. Photobiol.* **10**, 1–46.

Dall'Acqua, F. and Martelli, P. (1991). *J. Photochem. Photobiol. B, Biol.* **8**, 234–54.

Dall'Acqua, F., Marciani, S. and Rodighiero, G. (1969). *Z. Naturforsch.* **24b**, 307–14.

Dall'Acqua, F., Marciani, S., Ciavatta, L. and Rodighiero, G. (1971). *Naturforsch* **26B**, 561–5.

Dall'Acqua, F., Marciani, S., Zambon, F. and Rodighiero, G. (1976). *Chim. Ind. (Milan)* **58**, 733–9.

Dall'Acqua, F., Terbojevich, M., Marciani, S., Vadaldi, D. and Recher, M. (1978). *Chem. Biol. Interact.* **21**, 103–15.

Dall'Acqua, F. and Rodighieno, G. (1985). In *Primary photoprocesses in biology and medicine* (ed. R. Rensasson, G. Yori, F. Land, T. Truscott). Plenum Press, New York.

Dauksha, W. (1968). *Aktul. Vop. Farm.* **23**, 100–5.

Dean, F.M. (1963). *Naturally occurring Oxygen ring compounds.* Butterworth, London.

Dem'yanenko, V.G. and Dranik, L.I. (1971). *Khim. Prir. Soedin.* **7**, 115–20.

Dercks, W., Trumble, J., Winter, C. (1990). *J. Chem. Ecol.* **16**, 443–57.

Diawara, M., Trumble, J. and Quiros, C.F. (1993). *J. Agric. Food Chem.* **41**, 819–24.

Dranik, L.J. and Prokopenko, A.P. (1969). *Khim. Prir. Soedin.* **5**, 437–43.

Drlica, K. and Coughlin, S. (1989). *Pharmac. Ther.* **44**, 107–21.

Drlica, K. and Snyder, M. (1978). *J. Molec. Biol.* **120**, 145–54.

Egan, D., O'Kennedy, R., Moran, E., Cox, D., Prosser, E. and Thornes, R. (1990). *Drug Metabolism Rev.* **22**, 503–529.

El-Gamal, M.H.A., El-Bay, F.K., El-Tawil, B.A.H. and Gadalla, K.Z. (1975). *Egypt. J. Chem.* **18**, 767–74.

Fahmy, I.R. and Abu-Shady, H. (1948). *Q. J. Pharm. Pharmacol.* **21**, 499–507.

Fahmy, I.R., Saber, A.H. and Kadir, E.A.E. (1956). *J. Pharm. Pharmacol.* **8**, 653–9.

Feuer, G., Golberg, L. and Gibson, K.I. (1966). *Food Cosmet. Toxicol.* **4**, 157–64.

Fischer, F.C. and Baerheim-Svendsen, A. (1976). *Phytochemistry* **15**; 1079–84.

Fittig, R. (1868). *Z. Chem.* **4**, 595–610.

Furuya, T. (1958*a*). *Chem. Pharm. Bull.* **6**, 696–701.

Furuya, T. (1958*b*). *Chem. Pharm. Bull.* **6**, 701–4.

Garg, S.K., Gupta, S.R. and Sharma, N.D. (1978). *Phytochemistry* **17**, 2135–6.

Garg, S.K., Gupta, S.R. and Sharma, N.D. (1979). *Phytochemistry* **18**, 1580–3.

Garg, S.K., Gupta, S.R. and Sharma, N.D. (1980). *Planta Med.* **38**, 186–91.

Gasparro, F., Bevilacqua, P., Goldminz, D. and Edelson, R. (1990*a*). In *DNA damage and repair in human tissue* (ed. B. Sutherland and A. Woodhead). Plenum, New York.

Gasparro, F., Dall'Amico, R., O'Malley, M., Heald, P. and Edelson, R. (1990*b*). *Photochem. Photobiol.* 52, 315–321.

Gast, W., Rytter, M., Walther, T., Schaller, J., Hofmann, C. and Haustein, U. (1988). *Studia Biophys.* **124**, 173–8.

Gawron, A. and Glowniak, K. (1987). *Planta Med.* **53**, 526–9.

Gawron, A., Gorski, G. and Glowniak, K. (1989). *J. Pharm. Pharmacol.* **42**, 194–5.

Giorgobiani, E.D., Komissarenko, N.F. and Kemertelidze, E.P. (1969). *Chem. Abstr.***71**, 67947.

Gruenert, D, Ashwood-Smith, M., Mitchell R. and Cleaver, J. (1985). *Cancer Res.* **45**, 5394–8.

Guillemette, A. (1835). *Jutus Liebigs Ann. Chem.* **14**, 324–40.

Guiotto, A., Rodighiero, P., Pastorini, G. and Celon, E. (1977). *Phytochemistry* **16**, 1257–62.

Gurtoo, H.L., Campbell, T.C., Webb, R.E. and Plowman, K.M. (1968). *Biochem. Biophys. Res. Commun.* **31**, 588–94.

Gusak, L.E. and Safina, L.K. (1976). *Tr. Bot. Inst. Akad. Nauk Kaz. SSR* **35**, 145–58.

Halliwell, B. and Gutteridge, J. (1989). *Free radicals in biology and medicine*, 2nd edn. Clarendon Press, Oxford.

Hannuksela, M., Stenback, F. and Lahti, A. (1986). *Arch. Derm. Res.* **278**, 347–351.

Hanson, D.L. and Deleo, V.A. (1989). *Photochem. Photobiol.* **49**, 423–30.

Harborne, J.B. and Corner, J.J. (1961). *Biochem. J.* **81**, 242–8.

Hartley, R.D., Nesbitt, B.F. and O'Kelly, J. (1963). *Nature* **198**, 1056–9.

Heathcote, J.G. and Dutton, M.F. (1969). *Tetrahedron* **25**, 1497–505.

van Hee, L. (1965). *Ann. Gembloux* **71**, 211–20.

Hermann, K. (1959). *Arch. Pharm. Ber. Dtsch. Pharm. Ges.* **292**, 322–31.

Horhammer, L., Wagner, H. and Kraemer-Heydweiller, D. (1966). *Chem Abstr.* **65**, 2065.

Hsiek, D. and Brutlag, D. (1980). *Cell* **21**, 115–125.

Indahl, S.R. and Scheline, R.R. (1971). *Xenobiotica* **1**, 13–21.

Innerarity, L.T., Smith, E.C. and Wender, S.H. (1972). *Tob. Sci.* **16**, 86–90.

Innocenti, G., Dall'Acqua, F. and Caporale, G. (1976). *Planta Med.* **29**, 165–70.

Ivanitskaya, N.I. and Kuchinskaya, N.S. (1977). *Vopr. Farm. Dal'nem Vostoke* **2**, 184–93.

Ivie, G.W. (1978a). *J. Agric. Food Chem.* **26**, 1394–410.

Ivie, G.W. (1978b). In *Effects of poisonous plants on livestock* (ed. R.E. Keller, K.R. van Kamper and L.F. James). Academic Press, New York.

Ivie, G.W. (1987*a*). *Rev. Latinoam. Quim.* **18**, 1–6.

Ivie, G.W. (1987*b*). In *Light-activitated pesticides* (ed. J.R. Heitz and K.R. Downum). pp. 217–30. American Chemical Society **339**, Washington, DC.

Ivie, G.W., Holt, D.L. and Ivey, M.C. (1981). *Science* **213**, 909–10.

Iwanowska, A., Tykarska, T., Kuras, M. and Zobel, A.M. (1994). *Ann. Bot.* **74**, 313–20

Janzen, D. (1974). *Biotropica* **6**, 69–74.

Jernis, P. and Smith, A. (1981). *J. Org. Chem.* **46**, 577–9.

Kahn, J., Preis, P., Waldman, F. and Tseng Jr, A. (1994). *J. Cancer Res., Clin. Oncol.* **120**, 19–26.

Kaighen, M. and Williams, R. (1961). *J. Med. Pharm. Chem.* **3**, 25–31.

Kariyone, T. and Matsuno, T. (1953). *Pharm. Bull.* **1**, 119–27.

Karlsen, J., Boomsma, L. and Baerheim-Svendsen, A.(1968). *J. Chromatogr.* **42**, 550–6.

Kartnig, T. (1966). *Chem Abstr.* **71**, 276.

Kashman, Y., Gustafson, K.R., Fuller, R.W., Cardellina, J.H., Mcmahon, J.B., Currens, M.J., Buckheit, R.W., Hughes, S.H., Cragg, G.M. and Boyd, M.R. (1993). *J. Med. Chem.* **36**, 1110–19.

Kato, T., Kobayashi, M., Sasaki, N., Kitahara, Y. and Takahashi, N. (1978). *Phytochemistry* **17**, 158–63.

Keightley, A.M., Dobrzynska, M., Podbielkowska, M. Renke, K. and Zobel, A.M. (1996). *Int. J. Pharm.* **34**, 105–14.

Kerekjarto, B., Kratz, F. and Staudinger, H. (1964). *Biochem. Z.* **339**, 460–6.

Kimura, Y., Okuda, H., Arichi, S., Baba, K. and Kozawa, M. (1985). *Biochim. Biophys. Acta* **834**, 224–229.

Kittler, L., Midden, W. and Wang, S. (1986). *Studia Biophys.* **114**, 139–48.

Klocke, J., Balandrin,M., Barnby, M. and Yamasaki, R. (1989). In *Insecticides of plant origin,* Vol. **341.** American Chemical Society, Washington, DC.

Knobloch, E., Kakac, B. and Macha, F. (1952). *Chem. Listy* **46**, 416–18.

Knox, C., Land E. and Truscott, T. (1986). *Photochem. Photobiol.* **43**, 359–363.

Komatsu, S., Tanaka, S., Ozawa, S., Kubo, R., Ono, Y. and Matsuda, Z. (1932). *Chem. Abstr.* **26**, 717.

Komissarenko, N.F. and Zoz, I.G. (1976). *Rastit. Resur.* **12**, 336–40.

Komissarenko, N.F., Zoz, I.G., Chernobai, V. and Kolesnikov, D. (1961). *Biokhimiya* **26**, 980–90.

Komissarenko, N.F., Chernobai, I.G. and Kolesnikov, D.G. (1965). *Tr. Bot. Inst. Akad. Nauk SSSR, Ser. 5*, 12–58.

Kosheleva, L.I. and Nikonov, G.K. (1970). *Tr. Vses. S'ezda Farm., 1st 1967*, 245–8.

Kosuge, T. (1961). *Arch. Biochem. Biophys.* **95**, 211–16.

Kosuge, T. and Conn, E. (1959). *J. Biol. Chem.* **234**, 2133–40.

Kosuge, T. and Gilchrist, D. (1977). In *Mycotoxic fungi, mycotoxins, mycotoxicoses*, Vol. 1 (ed. T.D. Wyllie and L. G. Morehouse). Marcel Dekker, New York.

Kowalska, M. and Skrzypczalkowa, L. (1964). *Diss. Pharm.* **16**, 255–60.

Krasnovsky, A. (1982). In *Kinetics of conversion* (ed. A. Krasnovsky). Nauka, Leningrad.

Kratz, F. and Staudinger, H. (1967). *Hoppe-Seyer's Z. Physiol. Chem.* **348**, 568–77.

Krupnikova, T. and Dranik, L. (1973). *Fiziol.-Biokhim. Osn. Vzaimodeistviya Rast. Fitotsenozakh.* **64**, 1100–5.

Kupidlowska, E., Kowalec, M., Sulkowski, G. and Zobel, A.M. (1994). *Ann. Bot.* 73, 525–30.

Kupidlowska, E., Dobrzynska, K., Parys, E. and Zobel, A.M. (1995*a*). *J. Chem. Ecol.* 20, 2471–80.

Kupidlowska, E., Bieniak, B., Ruchirawat, A. and Zobel, A.M. (1995*b*). *Phytomedicine* **2**, 63–9.

Kuske, H. (1942). *Dermatologica* **82**, 273–80.

Laskin, J. and Lee, E. (1991). *Biochem. Pharmacol.* **41**, 125–32.

Laskin, J., Lee, E., Yurkow, E., Laskin, D. and Gallo, M. (1985). *Proc. Natl. Acad. Sci. USA* 82, 6158–6162.

Laskin, J., Lee, E., Laskin, D. and Gallo, M. (1986). *Proc. Natl. Acad. Sci. USA* **83**, 8211–15.

Lee, F.H. and Conway, W.D. (1976). *Pharmacol. Res. Commun.* **8**, 469–72.

Lee, K. and Berenbaum, M.R. (1992). *Biochem. Systemat. Ecol.* **20**, 197–207.

Lescova, E.V. and Ananichev, A.V. (1969). *Rastit. Resur.* **5**, 565–71.

Letteron, P., Descatoire, V., Larrey, D., Tinel, M., Geneve, J. and Pessayre, D. (1986). *J. Pharmacol. Exptl. Ther.* **238**, 685–92.

Levy, C.C. and Frost, P.J. (1966). *J. Biol. Chem.* **241**, 997–1000.

Lischka, G. and Decker, E. (1978). *Arch. Derm. Res.* **261**, 267–72.

Livingston, A.L., Witt, S.C., Lundin, R.E. and Bickoff, E.M. (1965). *J. Org. Chem.* **30**, 2353–6.
Lokar, L.R.C. and Delber, S. (1988). *Phytochemistry* **27**, 1073–7.
Lu, F.J., Wu, T.S. and Chang, S.I. (1977). *T' ai-wan Yao Hsueh Tsa Chih* **29**, 1.
Maier, V.P. and Metzler, D.M. (1967). *Phytochemistry* **6**, 1127–31.
Maksyutina, N.P. (1965). *Khim. Prir. Soedin.* **1**, 133–9.
Maksyutina, N.P. (1967). *Khim. Prir. Soedin.* **3**, 213–14.
Maksyutina, N.P. and Kolesnikov, D.G. (1959). *Chem. Abstr.* **53**, 16292.
Marshall, M.E. (1989). *J. Clin. Oncol.* **7**, 402–4.
Marshall, M.E., Butler, K. and Hermansen, D. (1990). *Prostate* **17**(2), 95–9.
Martin, P. (1963). *Z. Bot.* **51**, 300–11.
Matern, U., Shasser, H. Wendorff, H. and Hamerski, D. (1988). In *Cell culture and somatic cell genetics in plants,* Vol. 5 (ed. I. Vasil and F. Coustable). Academic Press, New York.
Mattern, M.R. and Painter, R.B. (1979). *Biochim. Biophys. Acta* **563**, 306–9.
Mattern, M.R., and Scudiero, D.A. (1981). *Biochim. Biophys. Acta* **653**, 248–58.
Maucher, A., Kager, M. and Vonangerer, E. (1993). *J. Cancer Res. Clin. Oncol.* **119**, 150–4.
Mays, D.C, Hilliard, J.B., Wong, D.D., Chambers, M.A., Park, S.S., Gelboin, H.V. and Gerber, N. (1990). *J. Pharmacol. Exptl. Ther.* **254**, 720–731.
Mead, J.A.R., Smith, J.N. and Williams, R.T. (1958). *Biochem. J.* **68**, 67–9.
Meffert, H., Bohm, P., Roder, B. and Sonnichsen, N. (1982). *Dermatol. Monatsschr.* **168**, 387–93.
Méndez, J. and Castro-Poceiro, J. (1983). *Phytochemistry* **22**, 2599–601.
Meyer, B., Wall, M., Wani, M. and Taylor, H. (1985). *J. Nat. Prod.* **48**, 952–6.
Midden, W. (1988). In *Psoralen DNA photobiology* (ed. F. Gasparro), CRC Press, Boca Raton, Florida.
Mikkelson, V.E., Fowlks, E.W. and Griffith, D.G. (1961). *Arch. Phys. Med. Rehabil.* **42**, 609–14.
Moreno, G., Salet, C., Kohen, C. and Kohen, E. (1982). *Biochim. Biophys. Acta.* **721**, 109–11.
Moustacchi, E. (1988). *Arch. Toxicol., Suppl.* **12**, 26–34.
Murray, R.D.H. (1991). In *Progress in chemistry of organic natural products* (ed. W. Herz, G.W. Kirby, W. Steglich and C. Tamm). Springer-Verlag, Berlin.
Murray, R.D.H., Méndez, J. and Brown, S.A. (1982). *The natural coumarins, occurrence, chemistry, and biochemistry.* Wiley, Toronto.
Musajo, L., Rodighiero, C. and Caporale, G. (1953). *Chim. Ind. (Milan)* **35**, 13–22.
Musajo, L., Rodighiero, C., Colombo, G., Torlone, V. and Dall'Acqua, F. (1965). *Experientia* **21**, 24–7.
Nair, R.V., Fischer, E.P., Safe, S., Cortez, C.H., Harvey, R.G. and DiGiovanni, J. (1991). *Carcinogenesis* **12**, 65–69.
Nakasu, S. and Kikuchi, A. (1985). *EMBO J.* **4**, 2705–10.
Natarajan, A., Verdegaal-Irmerzeel, E., Ashwood-Smith, M. and Poulton, T. (1981). *Mutat. Res.* **84**, 113–24.
Neichi, T., Koshihari, Y. and Murota, S. (1983). *Biochim. Biophys. Acta* **753**, 130–2.
Neish, A. (1964). In *Biochemistry of phenolic compounds* (ed. J.B. Harborne), Academic Press, London.
Neumann, J. (1962). *Phyton (Buenos Aires)* **18**, 95–7.
Nietzki, R. (1876). *Arch. Pharm.* **208**, 327–31.

Nikonov, G.K. (1965). *Chem Abstr.* **62**, 14989.
Nover, L., Neumann, D. and Scharf, K. (1989). *Heat shock and other stress response systems of plants.* Springer Verlag, Berlin.
Nxumalo, F. and Zobel, A.M. Submitted to *Int. J. Pharm.*
Oba, K., Conn, E., Canut, H. and Boudet, A. (1981). *Plant Physiol.* **68**, 1359–63.
Omar, A., Weissman, U. and Krebs, A. (1971). *Dermatologica* **155**, 65–79.
Oster, U., Kardinal, C., Burghardt, H., Werner, B., Lottspeich, F. and Rudiger, W. (1992). *J. Plant Physiol.* **140**, 110–15.
Pathak, M.A., Allen, B., Ingram, D. and Fellman, J. (1961). *Biochim. Biophys. Acta* **54**, 506.
Pathak, M.A., Kramer, D. and Fitzpatrick, T.B. (1974). In *Sunlight and Man* (ed. J. Fitzpatrick). University of Tokyo Press, Tokyo.
Pathak, M., Zarebska, Z., Mihm, M., Jarzabek-Chorzelska, M., Chorzelski, T. and Jabionska, S. (1986). *J. Invest. Dermatol.* **86**, 308–315.
Payá, M., Halliwell, B. and Hoult, J. (1992*a*). *Free Rad. Res. Comms.* **17**(5), 293–8.
Payá, M., Halliwell, B. and Hoult, J. (1992*b*). *Biochem. Pharmacol.* 44, 205–14.
Payá, M., Ferrandiz, M.L., Miralles, F., Montesinos, C., Ubeda, A. and Alcaraz, M.J. (1993). *Arzneimittel-Forschung / Drug Research* **43**–1(6), 655–8.
Plouvier, V. (1970). *C. R. Hebd. Seances Acad. Sci.* **270D**, 1526–9.
Podbielkowska, M., Kupidlowska, E., Waleza, M., Dobrzynska, K., Louis, S.A. Keightly, A. and Zobel, A.M. (1994). *Int. J. Pharm.* **32**, 262–73.
Podbielkowska, M., Piwocka, M., Waszkowska, E., Waleza, M. and Zobel, A.M. (1995). *Int. J. Pharm.* **33**, 7–15.
Podbielkowska, M., Waleza, M., Dobrzynska. K. and Zobel, A.M. (1996). *Int. J. Pharm.*, **34**, 96–105.
Potapenko, A. (1986). *Studia Biophys.* **114**, 149–158.
Potapenko, A. and Sukhorukov, V. (1984*a*). *Experientia* **40**, 264–5.
Potapenko, A. and Sukhorukov, V. (1984*b*). *Studia Biophys.* **101**, 89–98.
Prasanna, H.R., Ramakrishna, V., Gupta, S.R., Viswanathan, L. and Venkitasubramanian, T.A. (1975). *Indian J. Biochem. Biophys.* **12**, 207–11.
Raghunathan, R., Nair, V., Safe, S., Cortez, C., Harvey, R. and DiGiovanni, J. (1991). *Carcinogenesis* **12**, 65–69.
Reardon, Y., Spielmann, P., Huang, Y., Sastry, S., Sancar and Hearst, J. (1992). *Nucleic Acid Res.* **19**, 1623–9.
Ree, K., Johnsen, A. and Hovig, T. (1981). *Acta Pathol. Microbiol. Scand. (Sect. A)* **89**, 81–90.
Ren, P., Stark, P.Y., Johnson, R.L. and Bell, R.G. (1977). *J. Pharmacol., Exp. Ther.* **201**, 541–5.
Reppel, L. and Wagenbreth, D. (1958). *Flora (Jena)* **146**, 212–21.
Rezmitz, H. and Urban, R. (1957). *Naturwissenschaften* **44**, 13–21.
Roberts, W.L. and Link, K.P. (1937). *J. Biol. Chem.* **119**, 269–74.
Rodighiero, G., Caporale, G. and Albiero, C. (1954). *Gazz. Chim. Ital.* **84**, 874–9.
Rodighiero, G., Caporale, G. and Dolcher, T. (1961). *Atti. Accad. Naz. Lincei, Rend., Cl. Sci. Fis., Mat. Nat.* **30**, 84–93.
Rodov, V., Shimskou, B., Kim, Y., Szapro, B. and Sttak, Y. (1992). *J. Amer. Soc. Hort. Sci.* **117**, 788–92.
Roebuck, B. and Wogan, G. (1977). *Cancer Res.* **37**, 1649–53.
Ronto, G., Fekete, A., Gaspar, S. and Modos, K. (1989). *J. Photochem. Photobiol. B*, **3**, 497–507.

Rousset, S., Nocentini, S., Revet, B. and Moustacchi, E. (1990). *Cancer Res.* 50, 2443–8.
Sadhu, M.K. (1975). *Indian J. Exp. Biol.* **13**, 577–80.
Sarma, C. and Barooah, A. (1982). *Indian Agr.* **26**, 75–81.
Sasaki, M., Meguro, F., Kumazawa, E., Fujita, H., Kakishima, H. and Sakata T. (1988). *Mutat. Res.* **197**, 51–8.
Sato, M. and Hasegawa, M. (1971). *Phytochemistry* **10**, 2367.
Scaife, J. (1971). *FEBS Lett.* **12**, 143.
Scheel, L.D. (1978). In *Mycotoxic fungi, mycotoxins, mycotoxicoses*, Vol. 2 (ed. T.D. Wyllie and L. G. Morehouse). Marcel Dekker, New York.
Scheel, L.D., Perone, V.B., Larkin, R.L. and Kupel, R.E. (1963). *Biochem.* 2, 1127.
Scheline, R. (1968). *Acta Pharmacol. Toxicol.* **26**, 325.
Schilling, W., Crampton, R. and Longland, R. (1969). *Nature* **221**, 664.
Schimmer, O. and Abel, G. (1986). *Mutat. Res.* **169**, 47–50.
Schlosser-Szigat, G. (1964). *Naturwissenschaften* 51, 556.
Seifert, R., Buttery, R. and Ling, L. (1968). *J. Sci. Food Agric.* **19**, 383.
Sieburg, E. (1921). *Biochem. Z.* 113, 176.
Simon, J. (1967). *Aust. J. Bot.* **15**, 83.
Smith, R. and Rosazza, J. (1974). *Arch Biochem. Biophys.* **161**, 551.
Soine T., Abu-Shady, H. and Di Gangi, F. (1956). *J. Am. Pharm. Assoc.* **45**, 426.
Sokolova, A. and Nikonov, G. (1969). *Khim. Prir. Soedin.* **5**, 318.
Späth, E. and Kuffner, F. (1936). *Chem Ber.* **69**, 75–114.
Spencer, R.R., Bickoff, E.M., Lundin, R.E. and Knuckles, B.E. (1966). *J. Agric. Food Chem.* **14**, 162.
Stadler, E. and Buser, H.R. (1984). *Experientia* **40**, 1157–9.
Stanley, W. and Jurd, L. (1971). *J. Agric. Food Chem.* **19**, 1106–10.
Stanley, W.L. and Vannier, S.H. (1967). *Phytochemistry* **6**, 585–9.
Stanley, W.L.,Waiss, A.C., Lundin, R.E. and Vannier, S.H. (1965). *Tetrahedron* **21**, 89–91.
Steck, W. and Bailey, B.K. (1969). *Can J. Chem.* **47**, 2425–8.
Starkowsky, N.A. and Badran, N. (1958). *J. Org. Chem.* **23**, 1818–20.
Stoker, J. and Bellis, D. (1962). *J. Biol. Chem.* **7**, 2303.
Strecker, A. (1868). In *Regnault-Strecker's Kurzes Lehrbuch der organischen Chemie*, Vol. 5. F. Viewig und Sohn, Brunswick.
Suttie, J. and Jackson, C. (1977). *Physiol. Rev.* **57**, 1–15.
Swain, L. and Downum, K. (1991). *Am. Chem. Soc.* **343**, 361–70.
Tinel, M., Belghiti, J., Descatoire, V., Amouyal, G., Letteron, P., Geneve, J., Larrey, D. and Passayre, D. (1987). *Biochem. Pharmacol.* **15**, 951–5.
Tokura Y., Jagi, J., Edelson, R. and Gasparro, F. (1991). *Photochem. Photobiol.* **53**, 517–23.
Towers, G.H.N. (1984). *Can. J. Bot.* **62**, 2900–11.
Tulloch, O. and Hoffman, O. (1982). *Phytochemistry* **21**, 1629–42.
Uno, T. (1972). *Chem. Abstr.* **76**, 56567.
Valio, I.F.M. (1973). *J. Exp. Bot.* **24**, 442–9.
VanSumere, C. (1960). In *Phenolics in plants in health and disease* (ed. J. B. Pridham). Pergamon, Oxford.
VanSumere, C.F., VanSumere-De Preter, C., Wining, L.C. and Ledingham, G.A. (1957). *Can. J. Microbiol.*, **3**, 847–59.

VanSumere, C.F., Hilderson, H. and Massart, L. (1958). *Naturwissenschaften* **12**, 292–7.

VanSumere, C.F., Cottenie, J. and DeGreef, J. and Kint, J. (1972). *Phytochemistry* **4**, 165–221.

Varga, E., Kuzovkina, I.N., Rozsa, Z. and Szendrei, K. (1978). *Chem. Abstr.* **89**, 176400.

Vaught, J.B., Klohs, W. and Gurtoo, H.L. (1977). *Life Sci.* **21**, 1497–506.

Veen, W.A.G. (1967). *Acta Physiol. Pharmacol. Neerl.* **14**, 448–57.

Venturella, P., Bellino, A. and Cusmano, S. (1966). *Gazz. Chim. Ital.* **96**, 475–81.

Vogel, A. (1820). *Gilbert's Ann. Phys.* **64**, 161–75.

Wall, M.F, Wani, M.C., Hughes, T.J. and Taylor, H. (1990). In *Antimutagenesis and anticarcinogenesis mechanisms II* (ed. Y. S. Kurda, D.M. Shankel and M.D.Walters). Plenum Press, New York.

Walther, T., Gast, W., Rytter, M. and Haustein, U. (1988). *Studia Biophys.* **124**, 179–82.

Wantanabe, R., Chorney, W., Skok, J. and Wender, S.H. (1964). *Phytochemistry* **3**, 391–5.

Wattenberg, L.W., Lam, L.K.T. and Fladmoe, A.V. (1979). *Cancer Res.* **39**, 1651–4.

Wierzchowska-Renke, K. (1975). *Chem. Abstr.* **82**, 1927.

Wogan, G.N. (1977). *Mycotoxins Hum. Anim. Health, Proc. Conf.* **29**.

Wogan, G.N., Edwards, G.S. and Shank, R.C. (1967). *Cancer Res.* **27**, 1729–31.

Woo, W., Shin, K. and Lee, C. (1983). *Biochem. Pharmacol.* **32**, 1800–3.

Yamano, S., Tatsuno, J. and Gonzalez, F.J. (1990). *Biochem.* **29**, 1322–9.

Yen, K.Y. (1967). *Chem. Abstr.* **73**, 32342.

Zangerl, A.R. (1990). *Ecology* **71**, 1926–32.

Zangerl, A.R. and Berenbaum, M.R. (1987). *Ecology* **68**, 516–20.

Zangerl, A.R., Berenbaum M.R. and Nitao, J. (1991). *Evolutionary Ecol.* **5**, 136–45.

Zarebska, Z. (1994*a*). *J. Photochem. Photobiol. B, Biol.* **23**, 101–9.

Zarebska, Z. (1994*b*). *J. Photochem. Photobiol.* **23**, 101–9.

Zobel A.M. (1993). Proceedings of Second Princess Chulabhorn Meeting, Thailand.

Zobel, A.M. (1996). *Ann. Bot.*, submitted.

Zobel, A.M. and Brown, S.A. (1988). *J. Nat. Prod.* 51, 941–6.

Zobel, A.M. and Brown, S.A. (1989*a*). *Can. J. Bot.* 67, 915–21.

Zobel, A.M. and Brown, S.A. (1989*b*). *Can. J. Bot.* **67**, 1456–9.

Zobel, A.M. and Brown, S.A. (1990). *J. Chem. Ecol.* **16**, 1623–34.

Zobel, A.M. and Brown, S.A. (1991*a*). *J.Chem. Ecol.* **17**, 1801–10.

Zobel, A.M. and Brown, S.A. (1991*b*). *Environment. Experiment. Bot.* **31**, 447–52.

Zobel, A.M. and Brown, S.A. (1991*c*). *Can. J. Bot.* **69**, 485–8.

Zobel, A.M. and Brown, S.A. (1991*d*). *J. Toxicol. Cutaneous and Ocular Toxicol.* **10**, 223–31.

Zobel, A.M. and Brown, S.A. (1993*a*). *J. Chem. Ecol.* **19**, 939–52.

Zobel, A.M. and Brown, S.A. (1993*b*). *Can. J. Bot.* **71**, 966–9.

Zobel, A.M. and Brown S.A. (1995). *Allelopathy Journal* **2**, 9–20.

Zobel, A.M. and Kuras, M. (1996). *Ann. Bot.*, submitted.

Zobel, A.M. and Louis, S.A. (1990). In *Coumarins, research and applications.* Proceedings Ital. Phytochemical Society Meeting, Padua, Italy.

Zobel, A.M. and Lynch J.M. (1995). *Proceedings of Meeting on Air Pollution, Athens, Greece. September 1994*, Vol. 1 pp. 210–13.

Zobel, A.M. and March, R.E. (1993). *Ann. Bot.* **71**, 251–5.

Zobel, A.M., Brown, S.A. and Nighswander, J.E. (1991*a*). *Ann. Bot.* **67**, 213–18.
Zobel, A.M., Wang, J., March, R.E. and Brown, S.A. (1991*b*). *J. Chem. Ecol.* 17, 1859–1870.
Zobel, A.M., Brown, S.A. and March, R.E. (1991*c*). *Can. J. Bot.* **69**, 1673–8.
Zobel, A.M., Brown, S.A., Kuras, M., Crang, R., Wierzbicka, M. and Nighswander, J.E. (1993*a*). *Heavy Metals in the Environment*, 1, 214–17.
Zobel, A.M., Sandstrom, T., Nighswander, J.E. and Dudka, S. (1993*b*). *Heavy Metals in the Environment* **1**, 210–13.
Zobel, A.M., Chen, Y. and Brown, S.A. (1994*a*). *Acta Hort.* **381**, 355–360.
Zobel, A.M., Crellin, J., Brown, S.A. and Glowniak, K. (1994*b*). *Acta Hort.* **381**, 510–76.
Zobel, A.M., Wierzchowska-Renke, K., Brown, S.A., Primeau, B., Lynch, J., Kuras, M., Crang, R. and Nighswander, J.E. (1995). Proceedings of Meeting on Biochemogeology of Soil, May 1995, in press.
Zutshi, U., Rao, P., Soni, A. and Atal, C. (1978). *Indian J. Exp. Biol.* **16**, 836–46.

10. Physical and physiological changes in minimally processed fruits and vegetables

MIKAL E. SALTVEIT

Mann Laboratory, Department of Vegetable Crops, University of California, Davis, CA 95616–8631, USA

Introduction

Overview

The cuts and abrasion that are used to prepare minimally processed fruits and vegetables produce both an immediate and a subsequent physical and physiological response. Some of the injurious manipulations during minimal processing involve the peeling of apples, the slicing of cucumbers, the abrading and cutting of carrots, and the shredding of lettuce and cabbage. Not only do mechanical shocks during these manipulations produce surface bruises and cracks, but the propagated hydraulic shock can also extend deep within the tissue and cause extensive internal bruising and fracturing. Liquids released from cut and damaged cells and water used during processing can coat the surface, block pores and significantly reduce gas diffusion into and out of the tissue. Later, as the surface liquid evaporates or is taken up by the tissue, gas diffusion increases beyond its initial value since the natural protective layers of the epidermis have been removed during processing. The lack of an intact epidermis and the presence of free water on the surface accelerate both water loss and possible infestation by microorganisms and contamination by pollutants. Loss of moisture can also render the surface debris remaining after processing translucent and impart a whitish discoloration to the abraded surface of the processed tissue.

Immediate physiological responses to wounding include the formation and propagation of a wound signal that elicits a wide range of physiological and biochemical responses in adjacent and distant tissues. The nature of the wound signal and the method of its propagation are not well understood. Wounding also stimulates the consumption of oxygen and the production of carbon dioxide and ethylene. Coupled with impeded gas diffusion, enhanced respiration can stimulate anaerobic respiration. A modified atmosphere package designed on the basis of quiescent, non-wounded rates of respiration will be unable to maintain the desired gaseous composition during the burst of respiration following wounding. Accelerated respiration can deplete food reserves so that taste and flavour are significantly altered. Increased concentrations of ethylene within the tissue can accelerate ripening, alter phenolic metabolism, and promote senescence. Induced wound healing processes are usually minimal at the low storage temperatures and short storage durations employed for processed tissue, but altered phenolic metabolism and lignin formation could alter the taste and texture of some tissues that are chilling sensitive and cannot be stored near 0 °C. For example, inhibition of lignification with the herbicide glyphosate, significantly reduced the toughening and fibre content of minimally processed asparagus spears stored at 2.5 °C (Saltveit, 1988).

A better understanding of the physical and physiological changes taking place in minimally processed fruits and vegetables will help in devising storage and handling procedures to minimize their impact on the quality and shelf-life of the processed product. This review of the physical and physiological changes in minimally processed fruit and vegetables will cover material published during the past few years. Recent reviews on this topic include those by Brecht (1995), Huxsoll *et al.* (1989), Rolle and Chism (1987) and Watada *et al.* (1990).

Lightly processing is wounding

By its very nature, the preparation of minimally processed fruits and vegetables entails some degree of wounding. Cutting divides the whole fruit or vegetable into smaller segments and usually shortens the storage life (Bolin and Huxsoll, 1991). This separation can be accomplished by slicing, chopping, dicing or shredding. Examples include sliced cucumbers, chopped nuts, diced carrots and shredded cabbage. The deleterious effect of these cutting types of wounds can often be minimized by using very sharp cutting tools so that the number of injured cells is limited to those actually cut. Blunt cutting instruments can induce injury to cells many layers removed from the actual cut because of the mechanical shock imparted to the tissue. The physical and physiological response of the tissue is often significantly altered by cutting the tissue into ever smaller segments. The direction of the cut (transverse, longitudinal or diagonal) can also significantly alter the response of the tissue (Brecht, 1995).

Abrasion is another technique used to minimally process fruits and vegetables and can be used by itself, or in conjunction with cutting to produce segments of tissue with altered surface topology. The surface blush on 'baby' carrots is an example of how processing techniques such as cutting and abrading root segments can affect the surface appearance. Removal of the natural epidermal surface has a profound effect on water loss, gas exchange and contamination from microbes and chemical pollutants. Peeling is analogous to surface abrasion in that it removes the epidermis, but it is different in that it usually involves the removal of large sections of the epidermis and adjacent tissue in single pieces. It can be done mechanically, thermally and/or enzymatically. Examples include peeled apples, tomatoes and citrus.

Wounding is a powerful morphogenic stimulus. A wound signal appears to be formed at the site of injury and propagate out into adjacent tissue. Cells near the wound site respond first and with the greatest intensity to the injury. The consumption of oxygen and the production of carbon dioxide and ethylene are stimulated by wounding, as is the synthesis of many metabolically active compounds. The wound signal and resulting physiological responses probably arose to thwart attacking pests, to coordinate wound healing processes, and to induce adaptive changes that wound make the plant less susceptible to subsequent attacks and injuries.

Minimally processed fruits and vegetables are becoming more important

While commercial food services are still the predominant users of minimally processed fruits and vegetables, individual consumers are becoming an increasingly important segment in this market. Factors which influence consumers to purchase minimally processed fruits and vegetables include their ease of use, their high quality and long shelf-life, and the lack of food preparation time in families where both parents work outside the home. Application of postharvest and food science technologies to maintain quality and extend shelf-life will undoubtedly increase the use of minimally processed fruits and vegetables by corporate, institutional and private consumers.

Physical effects

Immediate and local physical effects

Mechanical shock to tissue The mechanical forces used in cutting, abrading and peeling fruits and vegetables not only injure the cells immediately exposed to these manipulations, but also damage distant cells. Bruises, cracks, and fractures are common injuries in minimally processed tissue. Soft tissues tend to be more easily bruised than rigid tissues, while cracks and fractures are more common in rigid tissue. During rapid mechanical processing, a great deal of force can be imparted to segment the tissue and a hydraulic shock wave can be produced that propagates through the tissue and injures distant cells. In some tissues the reflected and concentrated shock wave can bruise tissue far removed from the site at which the force was applied. Slower processing speeds and sharper cutting tools would reduce the amount of force transferred to the tissue and the amount of resulting damage sustained.

Liquid on cut surface blocks pores Fluids released from cut and damaged cells during mechanical processing and water used during processing to wash or transport tissue within the processing facility can coat the surface of tissue segments and greatly retard gas diffusion (Fig. 10.1). Diffusion of gases through a liquid is around 10 000 times slower than through a similar thickness of gas. Therefore, a thin aqueous film on the surface and liquid in the intercellular spaces of the first few cell layers can greatly retard the movement of oxygen into the tissue and of carbon dioxide and ethylene out of the tissue. Coupled with the increased rates of respiration induced by wounding, impeded gas diffusion can result in sufficiently elevated levels of carbon dioxide and reduced levels of oxygen to encourage anaerobic respiration. Accumulation of elevated levels of ethylene within a water-coated tissue segment can also promote ripening and senescence of sensitive tissues. The hydroscopic nature of the cell wall and the small pore size in tissues makes the physical removal of the added and/or exuded surface moisture very difficult to accomplish.

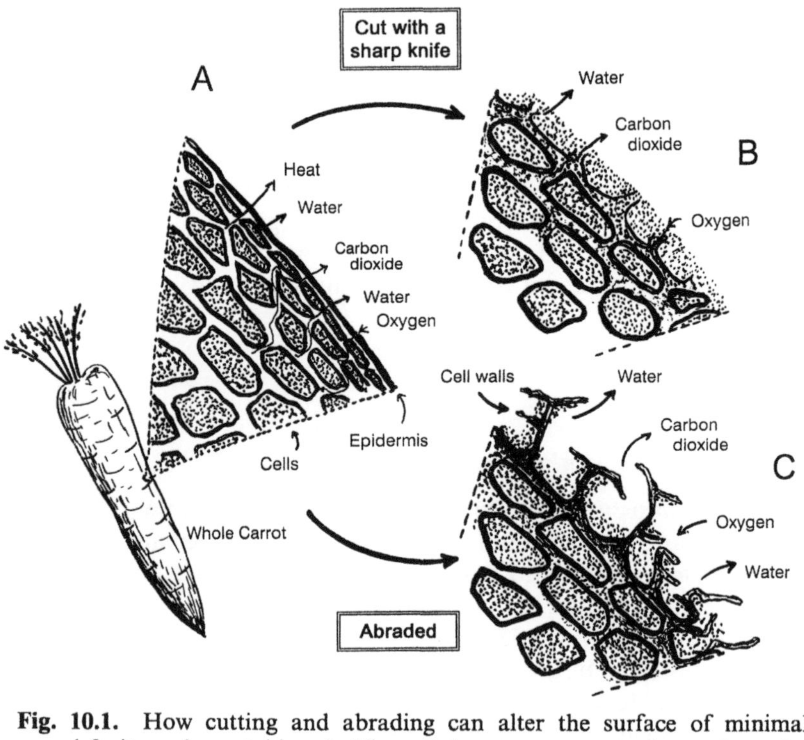

Fig. 10.1. How cutting and abrading can alter the surface of minimally processed fruits and vegetables: A. The cellular structure near the surface of a whole, uninjured carrot root. B. The immediate effect of cutting the epidermal layer from a carrot root. Water on the cut surface and in the intercellular spaces between the first few cell layers is represented by stipples. C. The abraded surface of a carrot root.

Water loss can be accelerated both by the removal of the natural epidermal barrier to gas diffusion and by the presence of free moisture on the surface of the processed segment. Unless thoroughly washed, surface moisture can also contain the cytoplasmic remains of ruptured cells; material rich in sugars and proteins. This surface solution would be in close proximity to sources of microbial contamination and would provide a nutritive substrate for microbial growth.

Entry of contaminants Minimal processing removes the natural epidermal barrier and exposes cells to both microbial infestation and to contamination by pollutants and chemicals in the wash water. The aqueous solutions used to wash the incoming whole plant material will accumulate spray residues and other chemicals and pollutants that have alighted on, or been incorporated into the plant during its growth. The water used to wash the product and to transport it along the processing line should be isolated from one another so that cross-contamination is minimized. The washing and transporting solutions should become increasingly cleaner as the product moves along the processing line.

Later and distant physical effects

Elimination of natural barriers Once the aqueous solutions on the surface have evaporated and/or been taken up by the tissue, the diffusion of gases can change from being greatly impeded to being greatly enhanced over that of uninjured tissue with an intact epidermal barrier. The internal concentration of oxygen can therefore increase and the concentration of carbon dioxide and ethylene can decrease. This enhanced gas diffusion also tends to maintain the rate of water loss above that of uninjured tissue; however, because the cut surface would be dryer after loss of the surface moisture, the rate of water loss should be lower than that from the initially wounded tissue. The possibility of contamination would remain a problem until wound healing processes can produce a new barrier, or until an artificial barrier can be erected by packaging or by the application of some surface coating. Edible coatings that provide such a barrier are a fascinating area of study (Krochta *et al.*, 1994).

Packaging which protects minimally processed tissue from desiccation and contamination can also, either through design or by accident, significantly alter the gas composition around the tissue. Packages can be specifically designed to create a modified atmosphere by relying on film permeability and rates of respiration of uninjured or aged tissue. However, in designing these packages, the transient elevation in the rate of respiration and ethylene production following wounding must be taken into account when modelling the dynamics of gas exchange to keep the specified gas composition within the desired range. Unacceptable high levels of carbon dioxide or ethylene, or low levels of oxygen, may occur because of the initial wound response in improperly designed packages.

Surface debris changes appearance As surface moisture is lost, the optical properties of the surface can also change, and what was a transparent mixture of water and cellular debris could become a whitish, semi-opaque layer (see Fig. 10.1). The amount of change in surface appearance caused by surface debris is minimal in tissue cut with sharp knifes and maximal in abraded tissue. During abrasion, portions of the resistant cell walls are left attached to the surface. Water on the surface holds these cell wall remnants close to the surface by surface tension and renders them almost invisible. As water is lost, the untethered portion of the attached particles swings away from the surface and forms a layer that imparts a whitish discolouration to the surface. This is a purely physical phenomenon and does not require organic synthesis or even the presence of living cells.

The development of a white blush on minimally processed carrots is a good example of the effect surface debris has on appearance. Wounding carrot root tissue induces enzymatic activity (e.g. PAL and POD) with the resultant accumulation of soluble phenolics and lignin formation in carrot sticks (Howard and Griffin, 1993). White discoloration corresponded with lignin formation in this tissue. However, it was later shown that white blush development in minimally processed carrots was mainly a physical, and not a physiological change (Avena-Bustillos *et al.*, 1993, 1994; Cisneros-Zevallos *et al.*, 1995, 1996). The development and loss of white blush on peeled and segmented carrot roots could be controlled by removing or adding back surface moisture.

Physiological effects

Immediate and local physiological effects

Wound responses Wounding elicits a myriad of physiological responses from fruit and vegetable tissues (Fig. 10.2). The initial responses to wounding, such as increased respiration and ethylene production, cause additional changes to the tissue as they alter ripening patterns and phenolic metabolism, and hinder cellular growth processes associated with wound healing. Storage technologies which interfere with some of these primary and secondary changes may leave others unaffected. An understanding of the numerous events altered by wounding and how they interact will help in the development and utilization of technologies to prolong quality and shelf-life of minimally processed fruits and vegetables.

Wound signal The observation that wounding not only effects the injured cells themselves, but also cells far removed from the site of injury, implies that some sort of signal arises at the site of injury and is propagated to distant cells. Although the exact nature of this wound signal is unknown, its mere identification and characterization could have a significant impact on the management of minimally processed plant tissue. Elimination or attenuation of the signal could prevent many of the deleterious physiological changes from occurring in minimally processed fruits and vegetables.

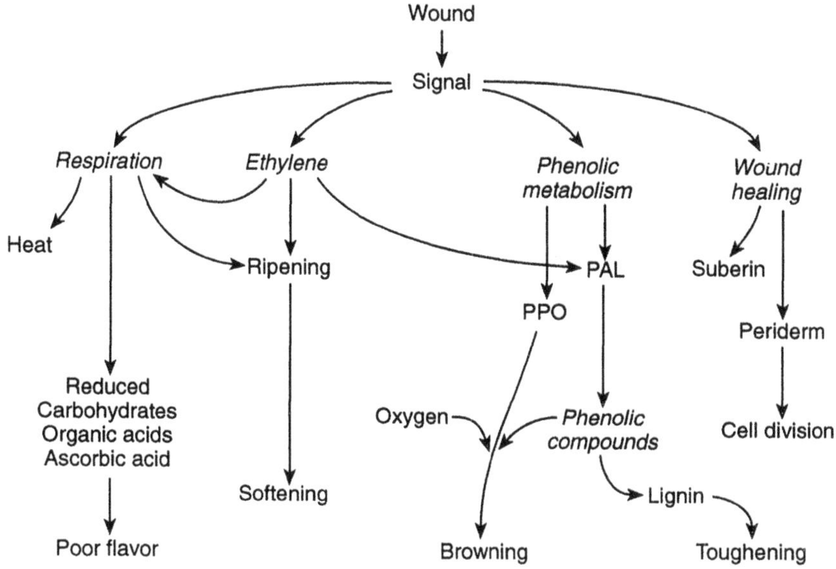

Fig. 10.2. The interrelationship among the many effects of wounding on physiological processes in minimally processed fruits and vegetables.

The nature of the signal is unknown. The wound signal could be purely physical (e.g. a hydraulic or pressure wave), purely physiological (e.g. a wound hormone such as traumatin synthesized specifically in response to wounding), or a combination of physical and physiological processes (e.g. a bioelectrical wave of potential or current). Each possibility has been experimentally identified in one model system or another. Some researchers believe the signal is not purposefully synthesized by wounded tissue, but that it is a byproduct of the initial wound response of injured cells (e.g. pectic fragments of degraded cell walls).

Using multiple displacement transducers ranged along a single seedling wheat leaf, Malone (1992) showed that localized wounding (scorching with a flaming match) caused a rapid and systemic hydraulic signal that could be measured as an increase in leaf thickness. Later, the same technique was used to test for wound-induced hydraulic signals in seedlings of apples, tomatoes, *Chenopodium amaranticolor*, soybeans, sunflowers, faba beans, peas, capsicum, durum wheat, *Phaseolus vulgaris*, *Nicotiana benthamiana*, cucumbers and spinach, and in larger plants of cherries, limes, maize and tomatoes

(Boari and Malone, 1993). Rapid and systemic increases in leaf thickness were induced by localized wounding in every plant tested. Changes in leaf thickness were closely related to leaf water potential and the hydraulic signal traveled at 10–20 cm/s. Increases in thickness preceded changes in the surface electric potential also brought on by wounding. A slower component of the wound-induced hydraulic signal was identified as a xylem-borne mass flow of water and solutes released at the wound site. It was proposed that hydraulic signals could form part of a widespread mechanism for coordination of the plant response to localized stress.

Later, Malone *et al.* (1994) showed that localized wounding caused rapid and systemic induction of proteinase inhibitors in tomato. Although systemic hydraulic signals are transmitted from wound sites in tomato, they did not by themselves induce the proteinase inhibitors. However, the hydraulic signal was nevertheless an essential requirement for the systemic induction by localized treatments. They also showed that mass flows similar to those associated with the wound-induced hydraulic signal can convey a variety of solutes rapidly throughout the shoot. It appears that long-distance wound signalling in tomato occurs by xylem transmission of soluble elicitors in the mass flows induced by wounding.

Fragments of the cell wall are known to elicit a number of wound related responses and are rapidly produced following wounding. Pectic oligomers induced a rapid, transient increase in ethylene biosynthesis when added to pear cells in suspension culture (Campbell and Labavitch, 1991). Both the rapid increase and decrease in biosynthesis appeared to be precisely regulated components of the ethylene response to these pectic oligomers.

The wound response of several plant species involves the activation of proteinase inhibitor genes and the accumulation of those proteins at the local site of injury and systemically throughout the unwounded aerial regions of the plant (Wildon *et al.*, 1992). A possible mobile chemical signal linking the local wound stimulus to the distant systemic response could be systemin, a polypeptide isolated from tomato leaves, and the most powerful inducer of proteinase inhibitor synthesis known (Ryan *et al.*, 1993). It appears to activate the proteinase inhibitor genes in response to wounding via an octadecanoid-based intracellular signaling cascade. However, the speed of transmission is high for the transport of a chemical signal in the phloem (Wildon *et al.*, 1992). Other signals could be involved. The wound response of tomato plants may be inhibited by agents that affect ion transport. Wounds induced by heat or physical injury produces electrical activity that has similarities to animal epithelial conduction system. Translocation in the phloem of tomato seedlings can be completely inhibited by chilling with no effect on the systemic accumulation of transcripts for proteinase inhibitor and their activity in response to injury, and without hindrance to propagated electrical signals (Wildon *et al.*, 1992).

Movement of the wound signal The wound signal can be transmitted over considerable distances at speeds of from 0.5 cm/h in cut lettuce leaves (Fig. 10.3) (Ke and Saltveit, 1989) to 0.2 cm/min in pea epicotyl (Saltveit and Dilley, 1978). Wounding one leaflet of *Mucuna deeringiana* or *Pueraria lobata* leaves caused a significant reduction in isoprene emissions which propagated through the plant at 0.13 cm/s. (Loreto and Sharkey, 1993). This reduction occurred even when the wounded leaf was 52 cm from the measured leaflet. Induction of ethylene synthesis in the etiolated pea segments and PAL synthesis in lettuce occurred several cm from the site of excision. The rate at which the response (e.g. ethylene or PAL synthesis) increases is very large for the first few wounds (Ke and Saltveit, 1989; Elstner and Konze, 1976). As wounding continues, each additional wound induces less of a response, until there is an actual decrease in the response with further wounding (Fig. 10.4). As the signal moves out from the site of injury, an ever increasing area is induced, but the level of induction decreases as the distance from the injury increases. It appears that as it spreads out from the site of injury, the signal dissipates, either by dilution or by metabolic inactivation (Fig. 10.5). The decline in the rate of induction upon progressive injury could be caused by the overlapping of areas induced by individual wound events (Ke and Saltveit, 1989). The velocity of many wound signals is similar to the rate of propagation of electrical signals such as action potentials and variation potentials.

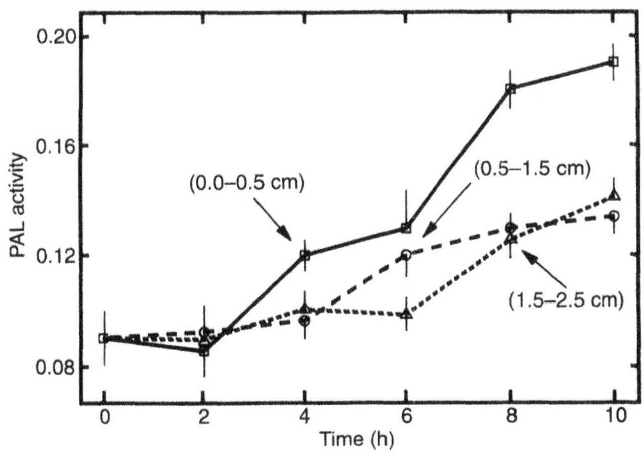

Fig. 10.3. Induction of PAL activity as affected by the distance from the wound and the time after wounding (Ke and Saltveit, 1989). The numbers in parentheses represent the distance from the wounded surface. The arrow represents the time when PAL activity started to increase. The vertical error bars represent the standard error of that mean.

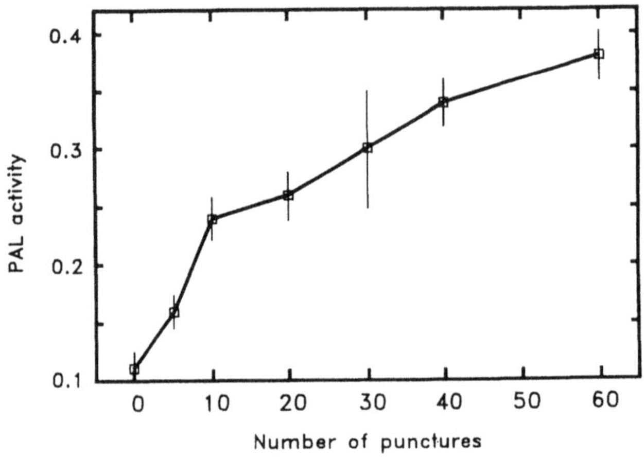

Fig. 10.4. Activity of PAL in wounded lettuce leaf tissue (Ke and Saltveit, 1989). Wounding was done by puncturing an 8 cm square of midrib tissue with a sterile 26-gauge hypodermic needle. All measurements were done on the 2nd day of incubation. The vertical error bars represent the standard error of that mean.

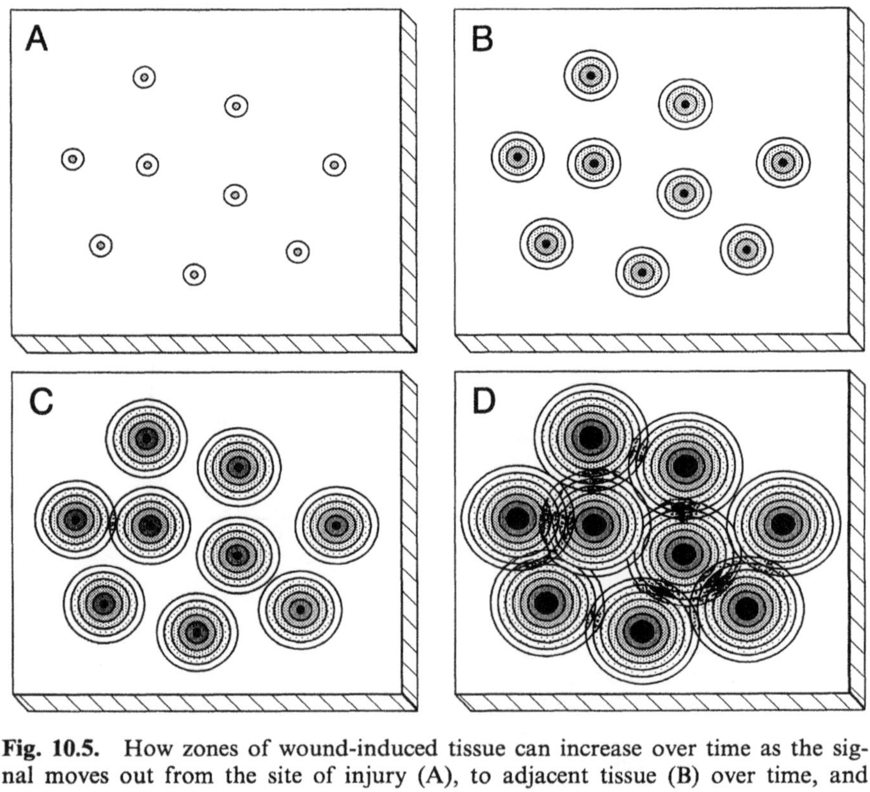

Fig. 10.5. How zones of wound-induced tissue can increase over time as the signal moves out from the site of injury (A), to adjacent tissue (B) over time, and how the zones can overlap (C and D) to form zones of multiple induction.

Membrane depolarization Following injury, the membranes of adjacent cells exhibit increased permeability to a number of cellular components. The breakdown in the normal semipermeable intercellular barriers allows the mixing of cellular and vacuolar compounds with many uncontrolled and unregulated reactions and the formation of many toxic compounds. Wounding causes rapid and profound changes in the permeability and membrane polarization of plant cells (Koopowitz *et al.*, 1975; MacKlon and Higinbotham, 1968). For example, wounding pea epicotyls and cucumber hypocotyls caused a rapid and transient membrane depolarization of the cells extending about 10 mm for pea and 40 mm for cucumber from the cut surface (Stahlberg, and Cosgrove, 1992; 1994; 1995). The wound-induced depolarization reached a maximum within 2 min for pea, while it was more sustained in cucumber. The shape, amplitude and velocity of the depolarization characterized it as a slow-wave potential indicating that the propagating depolarization was caused by a surge in water uptake. Cyanide suppressed the electrical signal at the treated position but did not inhibit its transmission or its development in untreated tissue. It appeared that depolarization was caused by a transient shut-down of the plasma membrane proton pump rather than large ion fluxes. The slow wave of electrical potential cound have been spread in the stem via a hydraulic surge occurring upon relief of the negative xylem pressure after the hydraulic resistance of the root has been removed by excision.

Calcium levels within cells are important because they profoundly affect the stability of cellular membranes (Ferguson, 1984) and many signal transduction pathways (Maruyama *et al.*, 1984). Wounding disrupts calcium distribution within tissues and may alter many physiological processes. Application of calcium during minimal processing can delay wound-induced senescent processes (Rolle and Chism, 1987).

Products of membrane disorganized Immediately following wounding there is a release of lipids from membranes of damaged cells. The deesterification of the lipids to produce free fatty acids has been attributed to the sequential action of a number of enzymes (Shewfelt and Erickson, 1991). The lipids and free fatty acids released and the resultant products of their oxidation are physiologically active. In tomato and alfalfa leaves, wound signals generated by insect or pathogen attacks appear to induce the release of linolenic acid from plant cell membrane lipids into the cytoplasm where it is rapidly converted to jasmonic acid, or perhaps a closely related derivative such as methyl jasmonate, which serves as a signal to regulate the expression of proteinase inhibitor genes (Farmer *et al.*, 1992). Repair of membranes in injured cells requires active lipid synthesis (Picchioni *et al.*, 1994). Preventing membrane dissolution or promoting membrane repair could prolong shelf-life of minimally processed fruits and vegetables.

Later and distant physiological effects

Elevated ethylene production Ethylene is a gaseous plant hormone that is usually produced at very low levels in non-ripening and vegetative tissue. Wounding rapidly induces enzymes in the ethylene biosynthetic pathway with a resultant large burst of ethylene synthesis occurring within minutes or hours of the injury (Abeles *et al.*, 1992). In many tissues, such as lettuce, the burst in ethylene is transitory and lasts only a few hours (Ke and Saltveit, 1989). In other tissues, induced ethylene synthesis persists for many days and can induce profound physiological changes; the most obvious being the initiation of ripening in climacteric fruit (Abeles *et al.*, 1992). Wound-induced ethylene accelerated events that normally take place over an extended period in intact avocado (Starrett and Laties, 1991) and tomato (Li *et al.*, 1992) fruits by stimulating the onset of climacteric ethylene. Wound-induced ethylene from 3 mm thick disks of avocado mesocarp peaked at approximately 18 h, while the climacteric ethylene peaked at approximately 72 h. A commensurate respiratory stimulation accompanied each ethylene peak. An inhibitor of ethylene synthesis (AVG) prevented wound and climacteric ethylene production, and the respiratory peaks and ripening. Suppressing wound-induced ethylene synthesis in sliced mature-green tomatoes allowed them to be stored and subsequently ripened under controlled conditions (Mencarelli and Saltveit, 1988).

Wound-induced ethylene also affects leaf senescence. The senescence of harvested spinach leaves was accelerated by the application of 1-aminocyclopropane-1-carboxylic acid (ACC), the immediate ethylene precursor, and was retarded by application of aminoethoxyvinylglycine (AVG), an inhibitor of ethylene biosynthesis (Philosoph-Hadas *et al.*, 1991). Wounding the leaves caused a burst of ethylene production that promoted senescence of both detached spinach leaves and leaf discs. Wound induced leaf senescence was dependent on the rate of ethylene production, and on the degree of wounding. It appears that wounding is a primary event in the induction of leaf senescence, while ethylene regulates the rate at which senescence proceeds.

The metabolic pathway for wound induced ethylene synthesis is thought to be the same as for normal and ripening related ethylene synthesis (Abeles *et al.*, 1992). However, wound-induced ethylene synthesis is often accompanied by a large increase in ethane production (Elstner and Konze, 1976) which would suggest that another pathway, or a parallel ethane-producing pathway, is also induced by wounding.

Elevated respiration Respiration is also initially increased by wounding in many plant tissues. The production of carbon dioxide increases, as does oxygen consumption and heat production. Enhanced respiration, coupled with decreased gas diffusion due to liquid on the surface and blocking pores, can elevate carbon dioxide and deplete oxygen to levels that induce anaerobic respiration. Although plant tissues are quite tolerant to the products of anaerobic respiration and can survive periods of anaerobic respiration with the concomitant production and accumulation of ethanol and acetaldehyde, these and other products of anaerobic fermentation can impart an off-flavour which, while not phytotoxic, would severely reduce the quality of the packaged produce. Most of these fermentative products are quickly lost from the tissue upon exposure to air for a day or two. However, since one of the most valued aspects of minimally processed fruits and vegetables is their immediate usefulness, such a delay would prove antithetical to their attractiveness.

Wound-induced increases in respiration also cause a rapid decrease in stored reserves. Excising the apical 7 cm from an asparagus spear doubled the rate of loss of soluble carbohydrates, amino acids and proteins (Saltveit and Kasmire, 1985). The increase in oxygen consumption by wounded tissue entails more than just that used by respiration. Of primary importance are those oxidative reactions that lead to tissue browning. Oxidative browning reactions in minimally processed fruits and vegetables have been covered in reviews by Matheis (1983), McEvily *et al.* (1992), and Nicolas *et al.* (1994).

Altered protein synthesis Changes in protein and RNA levels occur in fruits and vegetables upon wounding. Marked differences in mRNA populations for chitinase, sorbitol-6-phosphate dehydrogenase, acyl carrier proteins, the large subunit of rubisco, cellulase, and a cationic peroxidase were detected in a comparison of unwounded and wounded tomato fruit tissue (Mehta *et al.*, 1991). Since some RNAs increased while others decreased in abundance, it appears that wound-induced alterations in gene expression are probably controlled at the level of transcription (Harikrishna *et al.*, 1991).

Wounding carrot (*Daucus carota* L.) root tissue induces the synthesis of several secreted proteins which appear to have structural roles in cell walls (Showalter and Varner, 1987). Analysis of cDNA clones encoding for the induced proteins revealed the presence of a signal sequence (Chen and Varner, 1985). Wounding mesocarp tissue of winter squash fruits induced rapid increases in the activity in ACC synthase, ACC oxidase, PAL (phenylalanine ammonia-lyase) and peroxidase (POD) (Hyodo *et al.*, 1993). 2,5-Norbornadiene, an inhibitor of ethylene action, reduced the increases in PAL and POD activities and lignin formation. However, ethylene did not have an effect on the wound-induced accumulation of a highly anionic peroxidase in tomato fruit tissues (Sherf and Kolattukudy, 1993). It appears that developmental and wound regulated expression of many wound-induced genes occurs by separate or overlapping signal transduction pathways.

The induction or increase in activity of specific wound-associated enzymes (e.g. ACC synthase, PPO and PAL) can be used to ascertain the level of injury experienced by the tissue. In certain cases where the induction of an enzyme significantly affects the shelf-life and quality of minimally processed fruits and vegetables, measurements of its activity and inducibility can be used to predict subsequent rates of quality loss. Browning of processed lettuce is a major cause of quality loss and it is tightly coupled to induction of the enzymes of phenolic metabolism, mainly PAL (Couture *et al.*, 1993). Ethylene-induced PAL activity in uninjured lettuce gave a good indication of the storage life of minimally processed lettuce. Development of more assays to predict the susceptibility and response of tissue to minimal processing are needed so that marketing decisions can be quickly made for the many lots of product that arrive for processing with varying levels of wound susceptibility. Depending on the stresses experienced during growth and harvesting, the same lettuce cultivar can vary greatly in its browning potential and its shelf-life. An assay to predict the shelf-life of minimally processed lettuce would allow processers to market loads of browning-susceptible lettuce to local markets where the minimum shelf-life is needed, while sending browning-resistant loads to distant markets where the maximum shelf-life is needed.

Interaction between wounding and other stresses. Other abiotic stresses, such as gamma irradiation (Chervin *et al.*, 1992) and heat (Brodl and Ho, 1992) diminish wound-induced responses. For example, irradiating grated carrots with 2 kGy gamma radiation reduced subsequent respiration by 50% and ethylene production by 80% (Chervin *et al.*, 1992). Lowering the wound-induced respiration by irradiation halved sugar consumption compared to non-irradiated tissue and significantly improved quality and shelf-life. Heat stress redirected the wound-induced synthesis of several secreted proteins by carrot root tissue. Upon recovery from the 40 °C stress, the carrot tissue resumed the synthesis and secretion of extension and other cell wall proteins. It appears that several abiotic stresses have a higher priority than wounding in directing cellular metabolism. This hierarchical response structure for various stresses could be used to modulate the response of tissue to wounding by imposing another stress that would supersede the wound-induced responses during minimal processing with other more benign responses. For example, if a brief heat-shock during minimal processing of lettuce could divert protein synthesis from enhancing the activity of enzymes of phenolic metabolism to producing innocuous heat shock proteins, then wound-induced physiological browing could be significantly reduced.

References

Abeles, F.B., Morgan P.W. and Saltveit, M.E. (1992). *Ethylene in plant biology*, 2nd edn. Academic Press, New York.

Avena-Bustillos, R.J., Cisneros-Zevallos, L.A., Krochta, J.M. and Saltveit, M.E. (1993). *Trans. Amer. Soc. Agric. Eng.* **36**, 801–5.

Avena-Bustillos, R.J., Cisneros-Zevallos, L.A., Krochta, J.M. and Saltveit, M.E. (1994). *Postharvest Biology and Technology* **4**, 319–29.

Boari, F. and Malone, M. (1993). *J. Expt. Bot.* **44**, 741–6.

Bolin, H.R. and Huxsoll, C.C. (1991). *J. Food Sci.* **56**, 60–7.

Brecht, J.K. (1995). *HortScience* **30**, 18–22.

Brodl, M.R. and Ho, T.D. (1992). *Physiol. Plant.* **86**, 253–62.

Campbell, A.D. and Labavitch, J.M. (1991). *Plant Physiol.* **97**, 699–705.

Chen, J. and Varner, J.E. (1985). *Proc. Natl. Acad. Sci. USA* **82**, 4439–43.

Chervin, C., Triantaphylides, C., Libert, M.F., Siadous, R. and Boisseau, P. (1992). *Postharvest Biol. Technol.* **2**, 7–17.

Cisneros-Zevallos, L., Saltveit, M.E. and Krochta, J.M. (1995). *J. Food Sci.* **60**, 320–3, 333.

Cisneros-Zevallos, L., Saltveit, M.E. and Krochta, J.M. (1996). *J. Food Sci.* **60**, in press.

Couture, R., Ke, D., Cantwell, M.I. and Saltveit, M.E. (1993). *HortSci.* **28**, 723–5.

Elstner, E.F. and Konze, J.R. (1976). *Nature* **263**, 351–2.

Farmer, E.E., Johnson, R.R. and Ryan, C.A. (1992). *Plant Physiol.* **98**, 995–1002.

Ferguson, I.B. (1984). *Plant Cell Environ.* **7**, 477–89.

Harikrishna, K., Paul, E., Darby, R. and Draper, J. (1991). *J. Expt. Bot.* **42**, 791–9.

Howard, L.R. and Griffin, L.E. (1993). *J. Food Sci.* **58**, 1065–7.

Huxsoll, C.C., Bolin, H.R. and King, A.D. (1989). In *Quality factors of fruits and vegetables—Chemistry and technology* (ed. J.J. Jen), pp. 203–15. American Chemical Society, Washington DC.

Hyodo, H., Hashimoto, C., Morozumi, S., Ukai, M. and Yamada, C. (1993). *Acta. Hort.* **343**, 264–9.

Ke, D. and Saltveit, M.E. (1989). *Physiol. Plant.* **76**, 412–18.

Koopowitz, H., Dhyse, R. and Fosket, D.E. (1975). *J. Exp. Bot.* **26**, 131–7.

Krochta, J.M., Baldwin, E.A. and Nisperos-Carriedo, M.O. (1994). *Edible coatings and films to improve food quality*. Technomic Publishing Co., Lancaster, PA.

Li, N., Parsons, B.L., Liu, D. and Mattoo, A.K. (1992). *Plant Mol. Biol.* **18**, 477–87.

Loreto, F. and Sharkey, T.D. (1993). *Plant Cell Environ.* **16**, 563–70.

MacKlon, A.E.S. and Higenbotham, N. (1968). *Plant Physiol.* **43**, 888–92.

Malone, M. (1992). *Planta* **187**, 505–10.

Malone, M., Alarcón, J.J. and Palumbo, L. (1994). *Planta* **193**, 181–5.

Maruyama, K., Mikawa, T. and Ebashi, S. (1984). *J. Biochem.* **95**, 511–19.

Matheis, G. (1983). *Z. Lebensm. Unters. Forsch.* **176**, 454–62.

McEvily, A.J., Iyengar, R. and Otwell, W.S. (1992). *Crit. Rev. Food Sci. Nutr.* **32**, 253–73.

Mehta, R.A., Parsons, B.L., Mehta, A.M., Nakhasi, H.L. and Mattoo, A.K. (1991). *Plant Cell Physiol.* **32**, 1057–65.

Mencarelli, F. and Saltveit, M.E. (1988). *J. Amer. Soc. Hort. Sci.* **113**, 742–5.

Nicolas, J.J, Richard-Forget, F.C., Goupy, P.M., Amiot, M.J. and Aubert, S.Y. (1994). *Crit. Rev. Food Sci. Nutr.* **34**, 109–57.

Philosoph-Hadas, S., Meir, S. and Aharoni, N. (1991). *Physiol. Plant.* **83**, 241–6.

Picchioni, G.A., Watada, A,E., Roy, S., Whitaker, B.D. and Wergin, W.P. (1994). *J. Food Sci.* **59**, 597–605.

Rolle, R.S. and Chism, G.W. III (1987). *J. Food Quality* **10**, 157–77.

Ryan, C.A., Pearce, G., Johnson, S., McGurl, B., Orozco-Cardenas, M., Farmer, E.E. and Fritig, B. (1993). In *Mechanisms of plant defense responses* (ed. M. Legrand.), pp. 196–201. Proceedings of the 2nd International Conference of the European Foundation for Plant Pathology, Strasbourg.

Saltveit, M.E. (1988). *J. Amer. Soc. Hort. Sci.* **113**, 569–72.
Saltveit, M.E. and Dilley, D.R. (1978). *Plant Physiol.* **62**, 710–12.
Saltveit, M.E. and Kasmire, R.F. (1985). *HortSci.* 20, 1114–16.
Sherf, B.A. and Kolattukudy, P.E. (1993). *Plant J.* **3**, 829–33.
Shewfelt, R.L. and Erickson, M.C. (1991). *Trends Food Sci. Technol.* **6**, 152–4.
Showalter, A.M. and Varner, J.E. (1987). In *The biochemistry of plants: A comprehensive treatise*, Vol. 11 (ed. A. Marcus), pp. 485–520. Academic Press, San Diego, CA.
Stahlberg, R. and Cosgrove, D.J. (1992). *Planta* **187**, 523–531.
Stahlberg, R. and Cosgrove, D.J. (1994). *Plant, Cell Environ.* **17**, 1143–51.
Stahlberg, R. and Cosgrove, D.J. (1995). *Plant Cell Environ.* **18**, 33–41.
Starrett, D.A. and Laties, G.G. (1991). *Plant Physiol.* **97**, 720–9.
Watada, A.E., Abe, K. and Yamauchi, N. (1990). *Food Technol.* **44**, 116, 118, 120–2.
Wildon, D.C., Thain, J.F., Minchin, P.E.H., Gubb, I.R., Reilly, A.J., Skipper, Y.D., Doherty, H.M., O'Donnell, P.J. and Bowles, D.J. (1992). *Nature* **360**, 62–5.

11. Role of phytoalexins and other antimicrobial compounds from fruits and vegetables in postharvest disease resistance

JULIEN MERCIER

Department of Plant and Microbial Biology,
111 Koshland Hall, University of California, Berkeley,
CA 94720, USA

Introduction

Postharvest diseases cause important losses in horticultural production and affect all species of fruit and vegetables during storage, transit and retail. Some storage diseases begin as latent or quiescent infections in the field, developing only later during storage (Eckert and Ratnayake, 1983; Verhoeff, 1974). This is typical of anthracnose diseases of fruits caused by various species of *Colletotrichum*. Other pathogens cause wound infections during harvesting and handling and the subsequent storage of the crop. Diseases caused by *Penicillium* spp. in citrus and pome fruits are good examples of wound infection, as these pathogens will rarely attack intact fruits. In addition, some pathogens like *Botrytis cinerea,* the cause of grey mould rot, can cause both latent and wound infections, depending on the host attacked. Once infection has taken place, secondary spread to an adjacent host is possible, causing nesting of infection. Regardless of how these diseases start, there is always a period before or after harvest when fruit and vegetables exhibit considerable resistance to both types of rotting organisms, before eventually becoming fully susceptible as they approach maturity or senescence (Brown, 1989; Eckert and Ratnayake, 1983; Verhoeff, 1974). Such resistance indicates that defence mechanisms from the host could participate in restricting or slowing down subsequent disease development during storage.

In plants, a number of constitutive and inducible defence mechanisms may be implicated in their resistance to fungal and bacterial pathogens (Hahlbrock and Scheel, 1987). Antimicrobial secondary metabolites might play a role either constitutively as preformed inhibitors, sometimes called phytoncides, or as inducible compounds known as phytoalexins. In addition, some secondary metabolites with no antimicrobial activity can play a role indirectly by inhibiting microbial enzymes (Jersch *et al.*,

1989). Both preformed inhibitors and phytoalexins are a diverse group of compounds synthesized through a number of metabolic pathways (Stoessl, 1982).

Preformed inhibitors are present in healthy tissues and their concentration usually changes little during the course of an infection (Schönbeck and Schlösser, 1976). They are thus the first line of defence against invaders, along with the cuticle at the surface of leaves and fruits. I include in this group some 'stress metabolites' with antimicrobial activity which are present in healthy tissues but can increase in response to wounding or other stresses (Haard, 1983). These compounds are on the borderline between true preformed compounds and the inducible phytoalexins, which usually accumulate only in response to microbes or more specific elicitors.

Phytoalexins are synthesized and accumulate locally at the site of infection in response to elicitor molecules released by the pathogen or the host (Darvill and Albersheim, 1984; Dixon, 1986; Grisebach, 1983; Hahlbrock and Scheel, 1987). Some abiotic elicitors, such as heavy metals, can cause cell damage and release endogenous elicitors, triggering phytoalexin synthesis (Dixon, 1986). During the infection process, phytoalexin accumulation is accompanied by a number of other inducible defences such as cell wall modifications, defence enzymes and hypersensitive cell death (Hahlbrock and Scheel, 1987).

There is a considerable amount of information on the role of antimicrobial compounds, both preformed and inducible, in disease resistance of plants (Kuć, 1976; Mansfield, 1982; Schönbeck and Schlösser, 1976). Most of this information, however, has little relevance to postharvest pathology, having been obtained from plant species or organs which are not normally stored; or from cultured cells, which do not behave physiologically like differentiated plant organs. Furthermore, many studies of phytoalexins have focused on interactions of plants with non-pathogens to assess their role in non-host resistance. This review will focus on the importance of antimicrobial compounds found in fruits and vegetables in relation to disease resistance after harvest. For this reason, it will primarily discuss information obtained about storage pathogens of plant organs which are normally stored, although results on root, tuber and fruit slices will be included as they provide useful information on wound resistance in these crops (Table 11.1). Because of the number of defence mechanisms known in plants, it is always difficult to impute resistance to only one factor. Most likely, resistance results from the additive or synergistic effect of various defences operating together (Kuć, 1987). In a few cases, however, there is circumstantial evidence linking constitutive and inducible antimicrobial compounds with disease resistance in stored fruit and vegetables. I will present cases where these compounds are likely to play a role and discuss their significance for postharvest technology.

Table 11.1. Crops in which phytoalexins were associated with postharvest disease resistance

Crop	Phytoalexin	Chemical group	Pathogen
Apple	Benzoic acid	Phenolic acid	*Nectria galligena* *Pezicula malicorticis*
Capsicum (Bell pepper)	Capsicannol	Sesquiterpene	*Botrytis cinerea* *Glomerella cingulata*
Carrot	6-Methoxymellein	Isocoumarin	*Botrytis cinerea* *Mycocentrospora acerina* *Sclerotinia sclerotiorum*
Citrus	Scoparone	Coumarin	*Penicillium digitatum*
Lettuce	Lettucenin A	Terpene	*Botrytis cinerea*
Potato	Rishitin	Sesquiterpene	*Erwinia carotovora* *Phytophthora infestans*
Sweet potato	Ipomeamarone Scopoletin Umbelliferone	Furanoterpene Coumarin Coumarin	*Ceratocystis fimbriata*

Role of phytoalexins in disease resistance

Association of phytoalexin accumulation with infection

The first indication of the possible involvement of a phytoalexin in disease resistance is its association with infection by storage pathogens. Even though the elicitation of a phytoalexin has already been reported in a host after inoculation with pathogens or non-pathogens, one cannot assume that all storage pathogens elicit a given phytoalexin. The evidence for the involvement of a phytoalexin is stronger when a correlation exists between its accumulation and the resistance of inoculated tissues (Fig. 11.1). With a few exceptions, there is no absolute resistance/susceptibility in postharvest pathology. We should talk of degrees of susceptibility, rather than true resistance. A higher accumulation of phytoalexins in lesions with restricted development ('arrested', 'resistant' or 'restricted' lesions) compared to spreading lesions ('susceptible' or 'invasive' lesions) has been shown for a number of crop-pathogen interactions (e.g. Adikaram *et al.*, 1988; Goodliffe and Heale, 1978; Lyon *et al.*, 1975). In such cases, the resistance accompanying phytoalexin accumulation does not prevent disease but prevents or delays extensive rotting of plant organs. Accumulation of a phytoalexin in infected tissues to levels found inhibitory *in vitro* to a pathogen, especially when it occurs early during the infection process, is another indication of its possible role as a resistance factor. *In vitro* testing of the antimicrobial activity of a phytoalexin will, however, only give a rough measure of its inhibitory activity in tissue depending on the conditions and parameters tested (Smith, 1982). Finally, the rate of phytoalexin accumulation is particularly important as it can determine the outcome of an interaction, susceptible hosts usually accumulating phytoalexin at a slower rate (Kuć 1991) (Fig. 11.1). A summary of the information about host-pathogen interactions where phytoalexins are likely to contribute to postharvest disease resistance follows.

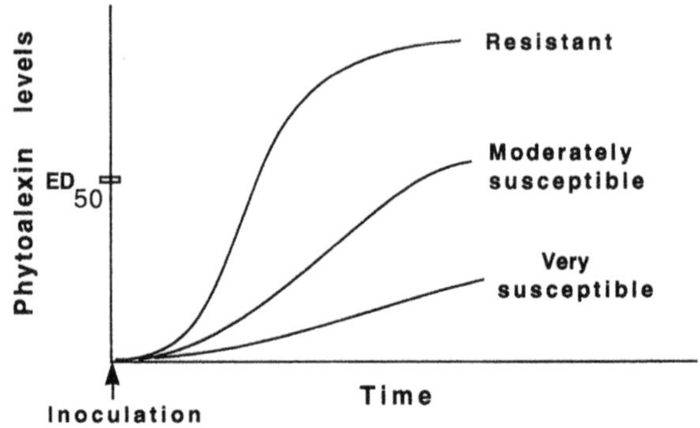

Fig. 11.1. Phytoalexin accumulation in cultivars with various degrees of disease susceptibility. Note short lag phase and rapid rate of accumulation in the resistant cultivar.

Apple

Apple fruits accumulate benzoic acid to inhibitory concentrations in arrested lesions caused by *Nectria galligena* and *Pezicula malicorticis* (Brown and Swinburne, 1971; Noble and Drysdale, 1983; Swinburne, 1973). Fruits of cultivars producing high levels of benzoic acid contain infection by *N. galligena* better than cultivars with lower amounts of the phytoalexin (Saindrenan and Bompeix, 1982). There was a delay of 20 days between inoculation and containment of lesions, however, presumably because of the delay necessary for accumulating benzoic acid (Saindrenan and Bompeix, 1982). Benzoic acid accumulation is relatively slow in comparison with phytoalexins from other plants, being detected 4 days after inoculation with *N. galligena* with a maximum reached in 12 days at 22 °C (Swinburne and Brown, 1975). The development of *N. galligena* during storage could be due to the degradation of benzoic acid in infected tissue or a rise in fruit pH, which reduces its inhibitory activity (Brown and Swinburne, 1971). Benzoic acid, however, was only induced by fungi exhibiting protease activity (Swinburne, 1973). Thus, this phytoalexin is not implicated in resistance against the wound pathogens *Botrytis cinerea* and *Penicillium expansum*, which have not been reported to produce an extracellular protease and do not elicit benzoic acid accumulation.

Bell pepper

Bell pepper (*Capsicum*) fruits accumulate capsicannol (a complex of sesquiterpenoid phytoalexins) and capsidiol in response to wound infection by *B. cinerea* and *Glomerella cingulata* (Adikaram *et al.*, 1982; 1988). Capsicannol was detectable within one day after inoculation and its accumulation continued for 5 days in lesions resistant to *B. cinerea*, while it stopped accumulating after 2 days in susceptible lesions. Capsidiol accumulated more slowly and was not associated with resistance to either pathogen (Adikaram *et al.*, 1982; 1988). Furthermore, capsidiol was degraded within a few hours *in vitro* by *B. cinerea* and *G. cingulata*, making its role in disease resistance more doubtful (Adikaram *et al.*, 1982; Stoessl, *et al.*,1973).

Citrus

Citrus fruits of various species and the related kumquat (*Fortunella margarita*) accumulate scoparone (6,7- dimethoxycoumarin) when held at 36 °C after inoculation with *Penicillium digitatum* (Ben-Yehoshua *et al.*, 1992; Kim *et al.*, 1991). In lemons, only low amounts of scoparone were detected in wound-inoculated fruits at 17 °C, presumably because of the rapid decay caused by *P. digitatum* (Kim *et al.*, 1991). Incubating lemons at 36 °C for a period of 3 days, 24 h after inoculation, enhanced scoparone accumulation and prevented decay. Under these conditions, inhibitory levels of scoparone were present in the inoculated flavedo 24 h after the beginning of the heat treatment and accumulation continued for 6 days after the fruits were transferred back to 17 °C (Kim *et al.*, 1991). This heat treatment, which does not induce scoparone by itself, could help restrict growth of the pathogen until the phytoalexin elicitation is completed, or simply accelerate the defence response.

Carrot

The isocoumarin 6-methoxymellein is the major phytoalexin accumulated in carrot root tissues in response to infection by the storage pathogens *B. cinerea*, *Mycocentrospora acerina* and *Sclerotinia sclerotiorum* (Coxon *et al.*, 1973; Davies and Lewis, 1981; Goodliffe and Heale, 1978; Harding and Heale, 1981; Kurosaki and Nishi, 1983). In carrots held at low temperature and inoculated with *B. cinerea* or *M. acerina*, 6-methoxymellein was accumulated to inhibitory concentrations at the site of resistant lesions (Davies and Lewis, 1981; Goodliffe and Heale, 1978). Roots which had lost 10 or 15% of their fresh weight in water were more susceptible to *B. cinerea* and had a reduced capability of producing 6-methoxymellein (Goodliffe and Heale, 1978).

Beside 6-methoxymellein, a number of other phytoalexins were elicited in response to pathogens but did not reach inhibitory concentrations in infected tissues: *p*-hydroxybenzoic acid, 6-hydroxymellein, eugenin and scopoletin (Coxon *et al.,* 1973; Davies and Lewis, 1981, Harding and Heale, 1981). Harding and Heale (1981) reported the accumulation of the polyacetylene falcarinol in carrot slices inoculated with *B. cinerea*, where it could contribute to disease resistance. Yates (1987), however, did not find changes in falcarinol levels after the exposure of carrots to ultraviolet light, a treatment which elicits phytoalexins in many plant species. Also, carrots contain large amounts of the closely related compound falcarindiol, which occurs as a preformed inhibitor in healthy roots (Garrod *et al.,* 1978). Due to the similarity of both compounds, changes in their levels should be monitored together in order to confirm the status of falcarinol as a phytoalexin of carrot.

Lettuce

Lettuce leaves inoculated with *B. cinerea* accumulate the terpenoid lettucenin A (Bennett *et al.,* 1994). In response to conidial inoculum of this fungus, the phytoalexin accumulated within 3 days at the site of inoculation to levels totally inhibitory to spore germination and germ tube growth. Such inoculation with conidia caused only a low incidence of spreading lesions, in contrast to mycelium of *B. cinerea*, which apparently overwhelmed the phytoalexin response.

Potato

The large number of different phytoalexins and stress metabolites in potato tubers makes the assessment of their individual role in disease resistance more difficult (Kuć, 1976; Lyon, 1989; Price *et al.,* 1976). The presence and amount of the various phytoalexins detected in inoculated tissues varies with the pathogen, the cultivar and conditions of incubation (Ghanekar *et al.,* 1984; Lyon, 1984; Price *et al.,* 1976; Varns *et al.,* 1971). The sesquiterpene rishitin may be associated with the resistance to bacterial soft rot caused by *Erwinia carotovora*, which it inhibits *in vitro* (Lyon and Bayliss, 1975). Amounts of rishitin detected in tissues with restricted rot were twofold higher than in tissues showing extensive rot (Lyon *et al.,* 1975). Higher incubation temperatures, which make tubers more prone to soft rot, also negatively affected rishitin accumulation in favour of phytuberin, which does not inhibit *E. carotovora* (Ghanekar *et al.,* 1984; Lyon, 1984; Lyon and Bayliss, 1975). Low oxygen, which also suppresses rishitin accumulation, favoured the development of soft rot (Ghanekar *et al.*, 1984). Rishitin also appears to be associated with resistance to *Phytophthora infestans*, the late blight fungus which can cause rotting of tubers in storage. Unlike other storage pathogens, *P. infestans* has an elaborate virulence genes system corresponding to resistance genes in potato. Incompatible races of the pathogen cause a rapid accumulation of rishitin in the tuber of resistant cultivars (Sato *et al.*, 1971; Varns *et al.*, 1971). In compatible interactions, virulent races of the pathogen suppress rishitin accumulation (Varns *et al.*, 1971). So far, the role of phytoalexins in the resistance of potato tubers to other fungal pathogens is not clear and requires more study (Corsini and Pavek, 1980; Harris and Dennis, 1976; Price *et al.*, 1976).

Sweet potato

Root slices of sweet potato inoculated with *Ceratocystis fimbriata* accumulate the furanoterpene ipomeamarone and the coumarins umbelliferone and scopoletin (Akazawa and Wada, 1961; Minamikawa *et al.*, 1963). The accumulation of these three compounds was faster in root slices of a resistant cultivar, especially between 24 and 72 h after inoculation. For example, ipomeamarone levels 72 h after inoculation were above 30 mg/g (fresh weight) in the resistant cultivar, compared to only 13 mg/g in the susceptible cultivar, where levels never increased above 20 mg/g by 120 h (Akazawa and Wada,1961). At the same time, umbelliferone levels in the resistant cultivar were more than three times higher than in the susceptible cultivar (Minamikawa *et al.*, 1963). While scopoletin was rapidly degraded *in vitro* by *C. fimbriata*, only 20% of the umbelliferone was degraded by the fungus after 4 days. A number of other sesquiterpenes could also be phytoalexins in sweet potato (Schneider *et al.*, 1984).

Other crops

Phytoalexins have been detected in other fruit and vegetables but their role in postharvest disease resistance has been little studied. Green tomato fruits inoculated with the foliar pathogen *Fulvia fulva* produced the sesquiterpene rishitin and the polyacetylenes falcarindiol, *cis*-tetradeca-6-ene-1,3-diyne-5,8-diol and probably falcarinol (De Witt and Flack, 1979; De Witt and Kodde, 1981). It is not known whether storage pathogens can elicit the accumulation of these compounds and what are their fate during the ripening of tomato. Eggplant (*Solanum melongena*) was found to produce lubimin and other sesquiterpenes in response to *B. cinerea*, *Fusarium oxysporum* and several non-pathogens (Ward *et al.*, 1975). Grape berries exposed to ultraviolet light accumulate resveratrol (3,5,4′-trihydroxystilbene), a phytoalexin considered important in leaf disease resistance but one that has been little studied in relation to postharvest infection (Creasy and Coffey, 1988; Jeandet *et al.*, 1991). A number of unidentified fungitoxic compounds, which could be involved in the latency of anthracnose, were detected in the peel of green banana inoculated with *Colletotrichum musae* (Brown and Swinburne, 1980).

Two phytoalexins, 1,3-dion-5-octyl-cyclopentane and 1,3-dion-5-hexyl-cyclopentane, have been detected in onion inoculated with the non-pathogen *B. cinerea* (Dmitriev *et al.*, 1990). Stored celery accumulates the furocoumarins psoralen, bergapten (5-methoxypsoralen), xanthotoxin (8-methoxypsoralen) and isopimpinellin in response to *Sclerotinia sclerotiorum* and *Erwinia carotovora* pv. *carotovora* (Chaudhary *et al.*, 1985; Surico *et al.*, 1987). These furocoumarins are also accumulated in parsnip infected with *Phoma complanata* (Cerkauskas and Chiba, 1990). Bean (*Phaseolus vulgaris*) pod tissues accumulate the pterocarpan phaseollin rapidly in response to incompatible races of *Colletotrichum lindemuthianum* and the non-pathogen *Monilinia fructicola* (Cruickshank and Smith, 1988).

Change in phytoalexin accumulation potential and disease susceptibility

The phytoalexin accumulation potential, that is, the amount of phytoalexin that can be accumulated in tissues after inoculation or elicitor treatment, usually diminishes in fruit and vegetables during ripening or storage. Such a drop in accumulation potential, if coincident with an increased susceptibility to infection, could be taken as further evidence for a role of phytoalexins in disease resistance (Fig. 11.2). In apple and carrot, the drop in capability to produce phytoalexin after inoculation correlates with an increased susceptibility to storage pathogens (Goodliffe and Heale, 1978; Noble and Drysdale, 1983; Swinburne, 1973). Such a drop in phytoalexin potential can be seen after treatment of maturing grapes and stored carrots with an abiotic elicitor, with a concomitant increase in susceptibility to *B. cinerea* (Creasy and Coffee, 1988, Mercier *et al.*, 1993*a*).

Association of phytoalexin elicitation with induced resistance

Other evidence indicating a role of phytoalexins in disease resistance is the enhanced resistance to pathogens accompanying the induction of phytoalexins by elicitors. A variety of chemical or physical agents can be used as elicitors to induce phytoalexin accumulation before inoculation with a pathogen. Elicitor treatments can, however, induce other defence mechanisms in addition to phytoalexins. Therefore, these results cannot be taken as a definitive proof that phytoalexins are important in disease resistance. On the other hand, the lack of induced resistance after phytoalexin accumulation can be used to dismiss a phytoalexin as a major factor in resistance. Several cases of induced resistance to storage pathogens after phytoalexin elicitation have been reported and are discussed below.

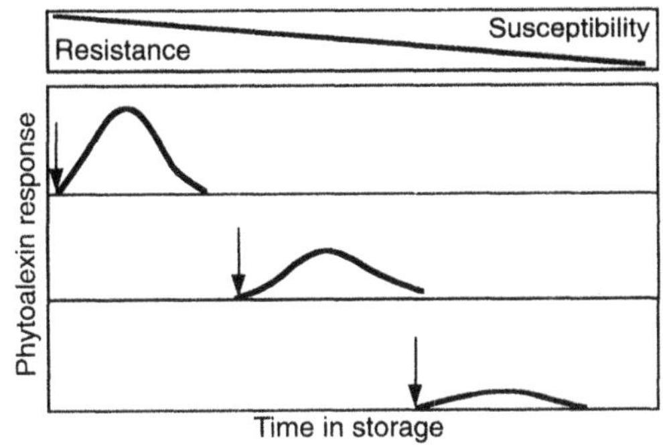

Fig. 11.2. Gradual decline of phytoalexin accumulation potential and disease resistance during storage. Arrows indicate time of inoculation or elicitor treatment.

Capsicannol accumulation in bell pepper fruits can be induced with cell wall hydrolysates of *Glomerella cingulata* (Adikaram *et al.*, 1982). Wounded tissue starts accumulating capsicannol rapidly after elicitor treatment to levels higher than tissues inoculated with *G. cingulata.* No rot developed on wounded fruits treated with cell wall hydrolysates 24 or 48 h before inoculation with *Botrytis cinerea* (Adikaram *et al.*, 1988). This treatment was not as effective in protecting intact (unwounded) fruits, presumably because of lower penetration of the elicitor. Intact fruits treated with elicitor accumulated capsicannol at a faster rate after inoculation than non-treated fruits and showed a higher resistance to *B. cinerea.*

Carrots, kumquats and lemons exposed to low-dose short-wave ultraviolet light (UV-C, peak emission at 254 nm) prior to storage accumulated phytoalexins and became more resistant to storage pathogens (Ben-Yehoshua *et al.*, 1992; Mercier *et al.*, 1993*a*,*b*; Rodov *et al.*, 1992). In kumquat and citrus held at 17 °C, ultraviolet treatment reduced decay when inoculation with *Penicillium digitatum* took place 2 days later, when scoparone levels were rapidly increasing, and 9 days later, when scoparone accumulation had approached peak levels (Ben-Yehoshua *et al.*, 1992; Rodov *et al.*, 1992). The range of ultraviolet doses most effective in inducing scoparone in lemons were also the most effective in preventing decay. Exposure to ultraviolet induced 6-methoxymellein accumulation and resistance to *B. cinerea* and *Sclerotinia sclerotiorum* in carrot slices incubated at 1 and 4 °C (Mercier *et al.*,

1993*b*). Ultraviolet also induced resistance to *B. cinerea* in whole carrots stored at 1 °C (Mercier *et al.*, 1993a). There was a significant correlation between the levels of 6-methoxymellein in the peel of ultraviolet-treated carrots and their resistance to *B. cinerea.* The ultraviolet dose optimal for inducing 6-methoxymellein accumulation was also the most effective for inducing disease resistance (Fig. 11.3). Ultraviolet had only a local effect and non-exposed areas of treated roots did not accumulate 6-methoxymellein, nor developed disease resistance (J. Mercier, unpublished).

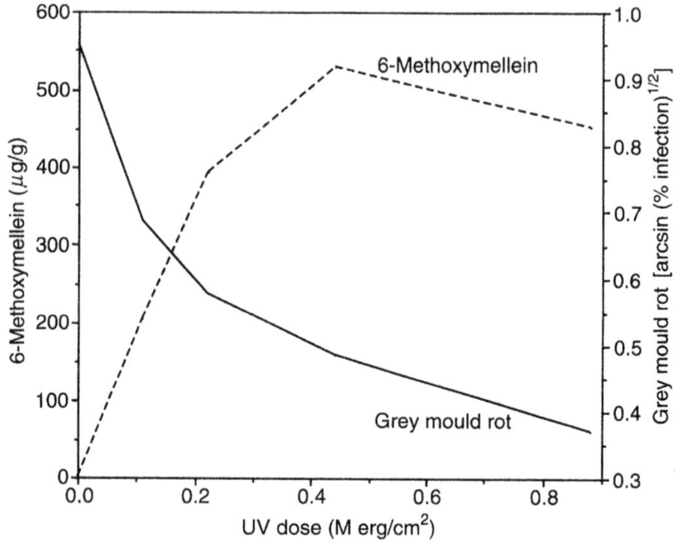

Fig. 11.3. Effect of UV treatment of carrots stored at 1 °C on 6-methoxymellein accumulation (measured 25 days after treatment) and susceptibility to grey mould rot (carrots inoculated with *Botrytis cinerea* 25 days after treatment, infection measured 48 days later). Data from Mercier *et al.* (1993*a*).

A systemic resistance to *B. cinerea* was also induced in wounded carrot roots at 1 °C by prior inoculation of the lower half of the roots with mycelium of *B. cinerea* or *S. sclerotiorum* (Mercier and Arul, 1993). Such pre-inoculation induced a systemic accumulation of 6-methoxymellein in the upper half of unchallenged roots to levels inhibitory to *B. cinerea* (J. Mercier, unpublished). Although the accumulation of 6-methoxymellein in carrot roots was always considered a local phenomenon, it is possible that a combination of inoculation with certain conditions such as severe wounding causes the phytoalexin to accumulate at a site distant from that of infection.

Preformed antimicrobial compounds

Preformed antimicrobial compounds occur at levels inhibitory to storage pathogens in a number of plant foodstuffs. These compounds could be especially important in the resistance of intact fruits, roots and tubers, as high amounts are found in the peel and outer tissues of apple (Ndubizu, 1976), carrot (Garrod *et al.*, 1978), citrus (Ben-Yehoshua *et al.*, 1992), mango (Droby *et al.*, 1987), potato (Allen and Kuć, 1968) and tomato (Verhoeff and Liem, 1975). High concentrations of inhibitory compounds in the peel could explain why the surface is rarely penetrated by most pathogens in the absence of injuries. Another indication supporting the role of preformed compounds in postharvest resistance is the concomitant drop in levels of these compounds in ripening fruits and the decline of resistance to latent infections. The antimicrobial diene and monoene in avocado (Prusky *et al.*, 1991*a*), 5-substituted resorcinols in mango (Droby *et al.*, 1986), tomatine in tomato (Schönbeck and Schlösser, 1976; Verhoeff and Liem, 1975) and 4-coumaryl quinate in apple (Ndubizu, 1976) are all present in high amounts in immature fruits and decline to low levels as they reach full maturity.

The best-studied case of a preformed compound acting in postharvest resistance is probably the diene (1-acetoxy-2-hydroxy-4-oxo-heneicosa-12,15-diene) which is present in high amount in the skin and flesh of firm avocado fruits showing resistance to *Colletotrichum gloeosporioides* (Prusky *et al.*, 1982, 1991*a*). While the skin of freshly harvested fruits contains 1500–2000 μg/g of the diene, spore germination of *C. gloeosporioides* is totally inhibited *in vitro* at 790 μg/ml (Prusky *et al.*, 1982). As avocado fruits ripen, levels of the diene drop to 200 μg/g in the peel and 20 μg/g in the flesh. This drop in diene levels is associated with the appearance of symptoms caused by *C. gloeosporioides* (Prusky *et al.*, 1982; 1991*a*). Cultivars retaining higher levels of the diene during ripening have delayed development of symptoms. Treatment of fruits with tocopherol, which inhibits diene oxidation by lipoxygenases, retarded the appearance of disease symptoms (Prusky *et al.*, 1983). Storing fruits in a modified atmosphere (9% carbon dioxide, 13% oxygen) slowed down diene breakdown and maintained disease resistance (Prusky *et al.*, 1988). Diene breakdown in avocado fruits is affected by their content of epicatechin, an inhibitor of lipoxygenase. Fruits of cultivars with high epicatechin content maintain higher diene levels after harvest and are more resistant to *C. gloeosporioides* (Prusky *et al.*, 1988). The amounts of diene in avocado are not static and can fluctuate widely during storage. The compound in peel was found to drop from 2010 to 120 μg/g in the day following harvest but to increase back to 3000 μg/g one day later (Prusky *et al.*, 1991*b*). Heat treatment (5 or 10 min at 55 °C), which did not affect the initial decline in diene but prevented its recovery, increased the susceptibility of fruits to *C. gloeosporioides*. Conversely, exposure of fruits to 30% carbon dioxide atmosphere for 24 h enhanced diene and epicatechin levels and reduced decay (Prusky *et al.*, 1991*c*).

Mango, which contains a mixture of antifungal 5-substituted resorcinols, is another case where preformed compounds could be important in disease resistance. As for diene in avocado, the drop of resorcinols in ripening mango fruits coincides with the appearance of symptoms caused by *Alternaria alternata* (Droby *et al.*, 1986). The breakdown of resorcinols and decay by *A. alternata* were enhanced by ethylene. Hypobaric storage, which removes ethylene, delayed breakdown of the compounds, as well as disease expression. Kitts, a mango cultivar that retains high levels of resorcinols during ripening, had a slower appearance of disease after inoculation with *A. alternata* than the cultivar Tommy Atkins, in which levels dropped more rapidly. There are a few other cases where the breakdown of preformed inhibitors is related with increased disease susceptibility. In immature apples, the breakdown of p-coumaryl quinate correlated with increased susceptibility to *B. cinerea* (Ndubizu, 1976). In cold-stored celery, the drop in marmesin levels coincided with the development of rot (Afek *et al.*, 1995).

Preformed antimicrobial compounds could also play a role in disease resistance at the wound site. In carrot, falcarindiol is released by broken oil ducts, accumulates at the wound surface for about 2 days and remains at high levels for up to 7 days (Lewis *et al.*, 1983). Such a rapid accumulation of falcarindiol on fresh wounds could be important in protecting them against pathogens until other defence mechanisms start to operate. Tomatine from tomato fruits is most inhibitory to fungi at pH 6.0 or above, while the fruit pH remains between 4 and 4.5 (Schönbeck and Schlösser, 1976). Wounding causes the pH to increase to 5.7, making tomatine more effective against fungal pathogens. *B. cinerea* and other pathogens can, however, lower the pH at the wound site, thus reducing the inhibitory effect of tomatine. Artificially increasing the wound pH of green fruits to 8.0 made them resistant to various pathogens (Schlösser, 1975).

Some preformed compounds can be inducible under certain conditions. Wounding was found to induce the accumulation of diene in avocado (Prusky *et al.*, 1990), resorcinols in mango (Droby *et al.*, 1987) and solanine in potato (Allen and Kuć, 1968). In mango, this accumulation of resorcinols in peeled fruits was accompanied by an increased resistance to *A. alternata*. The inoculation of freshly harvested avocados with *C. gloeosporioides*, but not *Penicillium digitatum* or *Diplodia natalensis*, caused diene levels to increase (Prusky *et al.*, 1990). Marmesin accumulated in celery stored at 20 °C after inoculation with *B. cinerea* (Afek *et al.*, 1995). Phenolic acids present in healthy periderm of potato increased after inoculation with *Erwinia carotovora* (Ghanekar *et al.*, 1984).

Some preformed secondary metabolites with no antimicrobial activity can act as enzyme inhibitors. The inhibition of cell-wall degrading enzymes produced by pathogens is another means of preventing invasion of plant tissues. Tannins from fruits, such as proanthocyanidins, which bind non-specifically to proteins, were found to inhibit fungal enzymes involved in the degradation of fruit tissues (Cole and Wood, 1961; Greene and Morales, 1967; Jersch *et al.*, 1989). The importance of proanthocyanidins in the resistance of immature strawberry fruit to *B. cinerea* was demonstrated by Jersch *et al.* (1989). Proanthocyanidins from strawberry fruit, which are broken down during ripening, inhibit polygalacturonase produced by *B. cinerea*, a key enzyme in pathogenicity. There was a strong negative correlation between the proanthocyanidin content of unripe fruits of different cultivars and their susceptibility to *B. cinerea*. Furthermore, the lower susceptibility of certain cultivars could be due to the retention of proanthocyanidins in the receptacle area because of slower ripening of this area, where infections usually commence. Beside inhibiting pectic enzymes, proanthocyanidins were found also to inhibit enzymes of *B. cinerea* involved in the degradation of phytoalexins of grapes (Pezet *et al.*, 1992). Although simple phenolics are usually not inhibitory to pectic enzymes, Lyon and McGill (1989) found that caffeic acid present in potato tubers inhibited polygalacturonase from *E. carotovora*.

Effect of storage conditions on antimicrobial compounds

There is relatively little information on the effect of storage conditions on the fate of inducible and preformed antimicrobial compounds in fruit and vegetables. In cases where these compounds are associated with postharvest disease resistance, optimizing storage conditions for maintaining these defence factors may have potential for extending the shelf-life of a given crop. One could assume that factors which slow down metabolic processes, such as modified atmospheres or low temperatures, should slow down the breakdown of preformed compounds and the synthesis of phytoalexins. The few studies done on this subject support this assumption. In the case of phytoalexins, however, storage conditions can also affect the kind and quantity of phytoalexins accumulated in a host. We know little of the effect of storage conditions prior to elicitation on the phytoalexin accumulation potential later during storage.

Cheema and Haard (1978) reported that storage temperature prior to elicitation was a determining factor in induced accumulation of rishitin in potato tuber. Tubers incubated for 1 month at 4 °C had higher potential for accumulating rishitin than tubers previously stored at 25 °C. In carrot tissue exposed to ultraviolet, the accumulation of 6-methoxymellein was more rapid at 20 °C but higher concentrations were eventually found at 1 and 4 °C, possibly because of a slower degradation of this phytoalexin at low temperature (Mercier *et al.*, 1993*b*). In potato, cooler temperatures (10–15 °C) were optimal for the accumulation of rishitin and phenolic acids in response to *E. carotovora* (concomitant with a higher resistance to this pathogen), while phytuberin accumulated better at warmer temperatures (20–30 °C) (Ghanekar *et al.*, 1984; Lyon, 1984). Low temperatures (5 and 10 °C), however, were detrimental to the accumulation of 5-substituted resorcinols and the development of resistance to *A. alternata* in peeled mango fruits (Droby *et al.*, 1987). Short exposure to high temperature was detrimental to diene levels in avocado (Prusky *et al.*, 1991b) but beneficial to scoparone accumulation in inoculated lemon (Kim *et al.*, 1991).

A modified atmosphere with high carbon dioxide was beneficial for preserving diene levels in avocado (Prusky *et al.*, 1988). Storage in 10% carbon dioxide or fluctuating carbon dioxide levels (5–10%, 0–10%) adversely affected benzoic acid accumulation in infected apples (Swinburne, 1973). Similarly, in bean, a high carbon dioxide atmosphere was inhibitory to phaseolin production in wounded bean pods during storage (Doherty and Buescher, 1978). Phytoalexin synthesis being accompanied by an increase in metabolism, the presence of oxygen is required for normal elicitation to take place. Low oxygen conditions inhibited the synthesis of antimicrobial compounds in potato (Ghanekar *et al.*, 1984) and mango (Droby *et al.*, 1987). The presence of ethylene in storage induced the accumulation of 6-methoxymellein in carrot (Lafuente *et al.*, 1989) but enhanced the breakdown of 5-substituted resorcinols in mango (Droby *et al.*, 1986).

Inducing phytoalexin for controlling postharvest diseases

If phytoalexins are important in disease resistance, could we perhaps use them for controlling postharvest diseases? Considering that at present there exist few alternatives to fungicides, which are subject to increasing public and regulatory concern, this possibility should not be overlooked. Although spraying phytoalexins is probably not effective or economical, their elicitation in plant tissue could be considered. Inducing phytoalexin accumulation with an elicitor has the advantage of making a susceptible host more disease resistant without altering it genetically. Usually, the lag phase between the beginning of an infection and the accumulation of phytoalexins reduces their effectiveness in stopping invaders. Triggering an early phytoalexin accumulation with elicitors allows the treated tissues to develop resistance early in the infection process (Fig. 11.4). Such induction of phytoalexins in growing plants has been considered not to be promising because of the energetic drain and possible physiological damage it can cause to the plant (Kuć, 1991). In the case of stored fruits and vegetables, however, phytoalexin synthesis does not have to compete with growth. Furthermore, storage at low temperature slows down the process of elicitation and degradation considerably, allowing for an extended effect on host resistance. The are already a few cases where biotic or abiotic elicitors of phytoalexins have induced disease resistance in stored produce (Adikaram *et al.*, 1988; Ben-Yehoshua *et al.*, 1992; Mercier *et al.*, 1993*a*; Rodov *et al.*, 1992). These results show that phytoalexin induction is possible for controlling disease, although such treatment will have to be perfected for practical application. Effect of storage conditions, cultivars and possible physiological damage will require more studies. Any elicitor of phytoalexins that is safe for humans could be considered for inducing disease resistance. Even microbial antagonists used for the biological control of storage diseases could be considered, as phytoalexins are often elicited by non-pathogens.

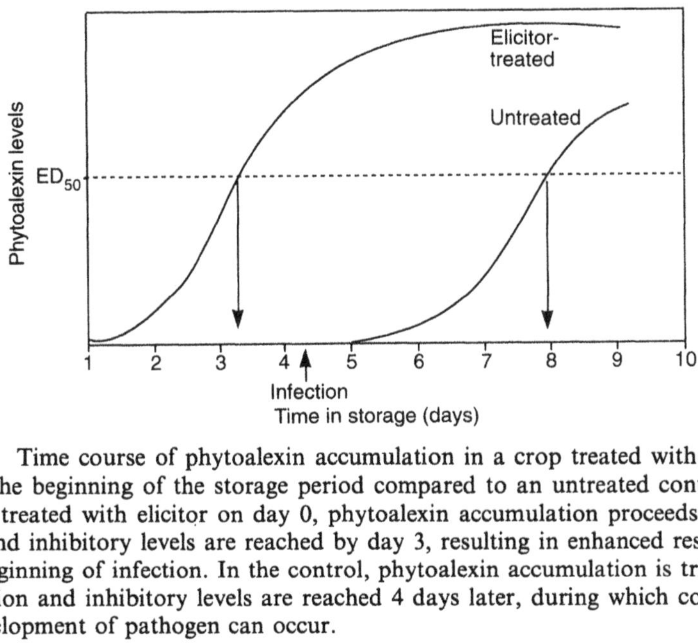

Fig.11.4. Time course of phytoalexin accumulation in a crop treated with an elicitor at the beginning of the storage period compared to an untreated control. In the crop treated with elicitor on day 0, phytoalexin accumulation proceeds immediately and inhibitory levels are reached by day 3, resulting in enhanced resistance at the beginning of infection. In the control, phytoalexin accumulation is triggered by infection and inhibitory levels are reached 4 days later, during which considerable development of pathogen can occur.

Do we want to induce such toxic compounds in produce that we intend to eat? Firstly, phytoalexins are a heterogeneous group of compounds with some being potential carcinogens (e.g. psoralens) while others are closely related to preservatives commonly used in food. Secondly, wounded tissues are usually more responsive to elicitor treatment than intact tissues, meaning that phytoalexin accumulation is likely to be contained in wounded areas, which are most prone to infection. Such association of wounding and phytoalexin accumulation have been shown with bell pepper and carrot respectively treated with cell-wall preparations and ultraviolet (Adikaram *et al.*, 1982; Mercier *et al.*, 1993*a*). Thirdly, if phytoalexin accumulation takes place in the peel and that peel is not normally consumed, or if the phytoalexin is not stable to cooking, then consuming elicitor-treated produce should not pose a problem. In ultraviolet-treated carrot, peeling or boiling removed most of the 6-methoxymellein, with no off-flavour remaining after peeling (Mercier *et al.*, 1994). Finally, phytoalexins are eventually degraded in elicitor-treated tissues, as shown for scoparone in kumquat and capsicannol in bell pepper (Adikaram *et al.*, 1982; Rodov *et al.*, 1992). For each crop, these factors should be considered before the possibility of inducing phytoalexins is dismissed as a potential postharvest treatment. Such induction may already occur in produce subjected to treatments used for extending their shelf-life such as heat, coating, ozone or gamma radiation.

Breeding for higher levels of antimicrobial compounds

For storage diseases, there are rarely 'resistant' cultivars as opposed to 'susceptible' ones (with the notable exception of late blight of potato and anthracnose of bean). Cultivars, when they vary in their degree of susceptibility, can be better described as being more or less susceptible. For example, although no carrot cultivar is resistant to storage rot caused by *Sclerotinia sclerotiorum*, varietal differences in susceptibility exist (Finlayson *et al.*, 1989). Thus, major resistance genes against such diseases rarely exist and breeding programs for fruits and vegetables have rarely considered postharvest diseases resistance as a major selection criteria. Traditionally, the control of such diseases has relied on the pre- and postharvest application of fungicides. Cultivars of fruit and vegetables have been shown to vary in their potential to accumulate phytoalexins or in their levels of preformed inhibitors (Creasy and Coffee, 1988; Droby *et al.*, 1986; Mercier *et al.*, 1993*c*; Pezet and Pont, 1992; Prusky *et al.*, 1991*a*; Saindrenan and Bompeix, 1982). Thus, in cases where these compounds are associated with disease resistance, the potential exists to select cultivars higher in antimicrobial compounds. Such selection could yield cultivars with higher 'horizontal' resistance, that is, a resistance not specific to a certain pathogen.

Several considerations must be taken into account, however, before starting to select for a given compound in a crop. In addition to concerns about human consumption, which have already been addressed in the previous section, a strong correlation between levels of antimicrobial compounds and disease resistance must be established. For example, the content of glycoalkaloids in the peel of potato tubers correlated poorly with resistance to wound inoculation with *Fusarium solani* var. *coeruleum* and *Phoma exiguea* var. *foveata* (Olsson, 1987). Furthermore, it has been suggested that potato cultivars should be bred with low amounts of glycoalkaloids to avoid bitterness and toxicity to humans (Sinden, 1987). In other cases, however, antimicrobial compounds appear to correlate well with disease resistance (Creasy and Coffee, 1988; Pezet and Pont, 1992; Saindrenan and Bompeix, 1982). In comparing cultivars for phytoalexin accumulation, care should be taken that amounts in different cultivars do not just reflect the amount of disease or the growth of the pathogen used for induction. In such case, the most infected cultivars will probably also have the highest phytoalexin production. For this reason, the use of physical or chemical elicitors such as ultraviolet or cupric chloride allows the comparison of a similar tissue area for phytoalexin accumulation and eliminates differences caused by variation in the area of infection or the reaction to the pathogen (Creasy and Coffee, 1988; Mercier *et al.*, 1993*c*; Rouxel *et al.*, 1990). If several unrelated antimicrobial compounds occur in the same plant, it may be difficult to relate disease resistance to only one compound. For example, the potential for 6-methoxymellein accumulation in carrot cultivars varied independently from the amount of preformed falcarindiol (Mercier *et al.*, 1993c). As 6-methoxymellein and falcarindiol are likely to participate in disease resistance, both compounds should be monitored in such breeding program.

The rate of phytoalexin accumulation or the persistence of a preformed inhibitor may be more important than their actual levels found in plant tissues. In sweet potato, a cultivar with a rapid rate of ipomeamarone accumulation was more resistant to *C. fimbriata* (Akazawa and Wada, 1961). In mango and avocado, cultivars with a slower breakdown of preformed inhibitors proved to be the most disease resistant, although the initial amounts were in the same range as the other cultivars (Droby *et al.*, 1986; Prusky *et al.*, 1982, 1988). Thus, the selection for retention of preformed inhibitor levels or rapid phytoalexin synthesis may prove more attractive than the selection for simply higher levels. Little data still exists on varietal differences in antimicrobial compounds of most crops. Data from wild relatives could also be useful, as these plants are often used in breeding programmes as a source of disease resistance.

Concluding remarks

Although we have significant evidence supporting the role of phytoalexins and preformed inhibitors in postharvest disease resistance, we still need definitive proof which would confirm their alleged role. Maher *et al.* (1994) reported that transgenic tobacco with suppressed levels of phenylalanine ammonia lyase, a key enzyme in phenyl propanoid biosynthesis, had an enhanced susceptibility to *Cercospora nicotianae*. Such genetic suppression in the synthesis of an antimicrobial compound, if related to disease susceptibility, would supply the definitive proof of their role against storage pathogens. Most produce, however, especially tree fruits, are not as easily manipulated at the genetic molecular level as is tobacco. Also, our knowledge of antimicrobial compounds and defence mechanisms in general in postharvest systems is still rather fragmentary. For these reasons, such definitive proof that can be obtained from molecular approaches may not be within reach for most of the host–pathogen systems discussed in this review, at least in the near future. Basic studies on the occurrence of antimicrobial compounds in fruits and vegetables should be continued, focusing mainly on their effect on storage pathogens and their interaction with postharvest technology. Such studies could prove useful to improve the postharvest management of these crops.

References

Adikaram, N.K.B., Brown, A.E. and Swinburne, T.R. (1982). *Physiol. Plant Pathol.* **21**, 161–70.

Adikaram, N.K.B., Brown, A.E. and Swinburne, T.R. (1988). *J. Phytopathol.* **122**, 267–73.

Afek, U., Aharoni, N. and Carmili, S. (1995). *Phytopathology* **85**, 711–14.

Akazawa, T. and Wada, K. (1961). *Plant Physiol.* **36**, 139–44.

Allen, E.H. and Kuć, J. (1968). *Phytopathology* **58**, 776–81.

Ben-Yehoshua, S., Rodov, V., Kim, J.J. and Carmeli, S. (1992). *J. Agric. Food Chem.* **40**, 1217–21.

Bennett, M.H., Gallagher, M.D.S., Bestwick, C.S., Rossiter, J.T. and Mansfield, J.W. (1994). *Physiol. Mol. Plant Pathol.* **44**, 321–33.

Brown, A.E. and Swinburne, T.R. (1971). *Physiol. Plant Pathol.* **1**, 469–75.

Brown, A.E. and Swinburne, T.R. (1980). *Phytopath. Z.* **99**, 70–80.

Brown, G.E. (1989). *Phytopathology* **79**, 1381–4.

Cerkauskas, R.F. and Chiba, M. (1990). *Can. J. Plant Pathol.* **12**, 349–57.

Chaudhary, S.K., Ceska, O., Warrington, P.J. and Ashwood-Smith, M.J. (1985). *J. Agric. Food Chem.* **33**, 1153–7.

Cheema, A.S. and Haard, N.F. (1978). *Physiol. Plant Pathol.* **13**, 233–40.

Cole, M. and Wood, R.K.S. (1961). *Ann. Bot.* **25**, 435–52.

Corsini, D.L. and Pavek, J.J. (1980). *Physiol. Plant Pathol.* **16**, 63–72.

Coxon, D.T., Curtis, R.F., Price, K.R. and Levett, G. (1973). *Phytochemistry* **12**, 1881–5.

Creasy, L.L. and Coffee, M. (1988). *J. Amer. Soc. Hort. Sci.* **113**, 230–4.

Cruickshank, I.A.M. and Smith, M.M. (1988). *J. Phytopathol.* **122**, 307–16.

Darvill, A.G. and Albersheim, P. (1984). *Annu. Rev. Plant Physiol.* **35**, 243–5.

Davies, W.P. and Lewis, B.G. (1981). *New Phytologist* **89**, 109–19.

De Witt, P.J.G.M. and Flach, W. (1979). *Physiol. Plant Pathol.* **15**, 257–67.

De Witt, P.J.G.M. and Kodde, E. (1981). *Physiol. Plant Pathol.* **18**, 143–8.

Dixon, R.A. (1986). *Biol. Rev.* **61**, 239–91.

Dmitriev, A.P., Tverskoy, L.A., Kozlovsky, A.G. and Grodzinsky, D.M. (1990). *Physiol. Mol. Plant Pathol.* **37**, 235–44.

Doherty, J. and Buescher, R. (1978). *J. Sci. Food Agric.* **29**, 853–6.

Droby, S., Prusky, D., Jacoby, B. and Goldman, A. (1986). *Physiol. Mol. Plant Pathol.* **29**, 173–83.

Droby, S., Prusky, D., Jacoby, B. and Goldman, A. (1987). *Physiol. Mol. Plant Pathol.* **30**, 285–92.

Eckert, J.W. and Ratnayake, M. (1983). In *Postharvest Physiology and Crop Preservation* (ed. M. Lieberman) pp. 247–64. Plenum Press, New York.

Finlayson, J.E., Pritchard, M.K. and Rimmer, S.R. (1989). *Can. J. Plant Pathol.* **11**, 313–16.

Garrod, B., Lewis, B.G. and Coxon, D.T. (1978). *Physiol. Plant Pathol.* **13**, 241–6.

Ghanekar, A.S., Padwal-Desai, S.R. and Nadkarni, G.B. (1984). *Potato Res.* **27**, 189–99.

Goodliffe, J.P. and Heale, J.B. (1978). *Physiol. Plant Pathol.* **13**, 241–6.

Greene, L. and Morales, C. (1967). *Turrialba* **17**, 447–9.

Grisebach, H. (1983). In *Secondary metabolism and differentiation in fungi* (ed. J.W. Bennett and A. Ciegler), pp. 377–428. Marcel Dekker. New York.

Haard, N.F. (1983). In *Postharvest physiology and crop preservation* (ed. M. Lieberman), pp. 299–314. Plenum Press, New York.

Hahlbrock, K. and Scheel, D. (1987). In *Innovative approaches to plant disease control* (ed. I. Chet), pp. 229–54. Wiley, New York.

Harding, V.K. and Heale, J.B. (1981). *Physiol. Plant Pathol.* **18**, 7–15.

Harris, J.E. and Dennis, C. (1976). *Physiol. Plant Pathol.* **9**, 155–65.

Jeandet, P., Bessis, R. and Gautheron, B. (1991). *Amer. J. Enol. Viticulture* **42**, 41–6.

Jersch, S., Scherer, C., Huth, G. and Schlösser, E. (1989). *Z. PflKrankh. PflSchutz.* **96**, 365–78.

Kim, J.J., Ben-Yehoshua, S., Shapiro, B., Henis, Y. and Carmeli, S. (1991). *Plant Physiol.* **97**, 880–5.

Kuć, J. (1976). In *Physiological plant pathology, encyclopedia of plant physiology* **4** (ed. R. Heitefuss and P.H. Williams), pp. 632–52. Springer-Verlag, New York.

Kuć, J. (1987). In *Innovative approaches to plant disease control* (ed. I. Chet), pp. 255–73. Wiley, New York.

Kuć, J. (1991). In *Mycotoxins and phytoalexins* (ed. R.P. Sharma and D.K. Salunkhe), pp. 595–603. CRC Press, Boca Raton, FL.

Kurosaki, F. and Nishi, A. (1983). *Phytochemistry* **22**, 669–72.

Lafuente, M.T., Cantwell, M., Yang, S.F. and Rubatzky, V. (1989). *Acta Hort.* 258, 523–34.

Lewis, B.C., Garrod, B. and Sullivan, C. (1983). *Trans. Br. Mycol. Soc.* **80**, 183–4.

Lyon, G.D. (1984). *Phytopath. Z.* **111**, 236–43.

Lyon, G.D. (1989). *Plant Pathol.* **38**, 313–39.

Lyon, G.D. and Bayliss, C.E. (1975). *Physiol. Plant Pathol.* **6**, 177–86.

Lyon, G.D. and McGill, F.M. (1989). *Potato Res.* **32**, 267–74.

Lyon, G.D., Lund, B.M., Bayliss, C.E. and Wyatt, G.M. (1975). *Physiol. Plant Pathol.* **6**, 43–50.

Maher, E.A., Bate, N.J., Ni, W., Elkind, Y. and Dixon, R.A. (1994). *Proc. Natl. Acad. Sci. USA* **91**, 7802–6.

Mansfield, J.W. (1982). In *Phytoalexins* (ed. J.A. Bailey and J.W. Mansfield), pp. 253–88. Wiley, New York.

Mercier, J. and Arul, J. (1993). *Can. J. Plant Pathol.* **15**, 281–3.
Mercier, J., Arul, J. and Julien, C. (1993*a*). *J. Phytopathol.* **139**, 17–25.
Mercier, J., Arul, J., Ponnampalam, R. and Boulet, M. (1993*b*). *J. Phytopathol.* **137**, 44–54.
Mercier, J., Ponnampalam, R., Bérard, L.S. and Arul, J. (1993*c*). *J. Sci. Food Agric.* **63**, 313–17.
Mercier, J., Arul, J. and Julien, C. (1994). *Food Res. Intl.* **27**, 401–4.
Minamikawa, T., Akazawa, T. and Uritani, I. (1963). *Plant Physiol.* **38**, 493–97.
Ndubizu, T.O.C. (1976). *J. Hort. Sci.* **51**, 311–19.
Noble, J.P. and Drysdale, R.B. (1983). *Physiol. Plant Pathol.* **23**, 207–16.
Olsson, K. (1987). *J. Phytopath.* **118**, 347–57.
Pezet, R. and Pont, V. (1992). In *Recent advances in Botrytis research.* Proc. 10th International Botrytis Symposium, Heraklion, Greece (ed. K. Verhoeff, N.E. Malathrakis and B. Williamson) pp. 93–8. Pudoc Scientific, Wageningen.
Pezet, R., Pont, V. and Hoang-Van, K. (1992). In *Recent advances in Botrytis research.* Proc. 10th International Botrytis Symp., Heraklion, Greece (ed. K. Verhoeff, N.E. Malathrakis and B. Williamson), pp. 87–92. Pudoc Scientific, Wageningen
Price, K.R., Howard, B. and Coxon, D.T. (1976). *Physiol. Plant Pathol.* **9**, 189–97.
Prusky, D., Karni, L., Kobiler I. and Plumbley, R.A. (1990). *Physiol. Mol. Plant Pathol.* **37**, 425–35.
Prusky, D., Keen, N.T., Sims, J.J. and Midland, S.L. (1982). *Phytopathology* **72**, 1578–82.
Prusky, D., Keen, N.T. and Eaks, I. (1983). *Physiol. Plant Pathol.* **22**, 189–98.
Prusky, D., Kobiler, I. and Jacoby, B. (1988). *J. Phytopathol.* **123**, 140–6.
Prusky, D., Kobiler, I., Fishman, Y., Sims, J.J., Midland, S.L. and Keen, N.T. (1991*a*). *J. Phytopathology* **132**, 319–27.
Prusky, D., Plumbley, R.A. and Kobiler, I. (1991*b*). *Plant Pathol.* **40**, 45–52.
Prusky, D., Plumbley, R.A. and Kobiler, I. (1991*c*). *Physiol. Mol. Plant Pathol.* 39, 325–34.
Rodov, V., Ben-Yehoshua, S., Kim, J.J., Shapiro, B. and Ittah, Y. (1992). *J. Amer. Soc. Hort. Sci.* **117**, 788–92.
Rouxel, T., Renard, M., Kollmann, A. and Bousquet, J.F. (1990). *Euphytica* **46**, 175–81.
Saindrenan, P. and Bompeix, G. (1982). *Fruits* **37**, 249–57.
Sato, N., Kitazawa, K. and Tomiyama, K. (1971). *Physiol. Plant Pathol.* **1**, 289–95.
Schlösser, E. (1975). *Z. PflKrankh. PflSchutz.* **82**, 475–84.
Schneider, J.A., Lee, J., Naya, Y., Nakanishi, K., Oba, K. and Uritani, I. (1984). *Phytochemistry* **23**, 759–64.
Schönbeck, F. and Schlösser, E. (1976). In *Physiological plant pathology, Encyclopedia of plant physiology* **4** (ed. R. Heitefuss and P.H. Williams), pp. 653–78. Springer-Verlag, New York.
Sinden, S.L. (1987). *Acta Hort.* **207**, 41–7.
Smith, D.A. (1982). In *Phytoalexins* (ed. J.A. Bailey and J.W. Mansfield), pp. 218–52. Wiley, New York.
Stoessl, A. (1982). In *Phytoalexins* (ed. J.A. Bailey and J.W. Mansfield), pp. 133-80. Wiley, New York.
Stoessl, A., Unwin, C.H. and Ward, E.W.B. (1973). *Phytopathology* **63**, 1225–31.
Surico, G., Varvaro, L. and Solfrizzo, M. (1987). *J. Agric. Food Chem.* **35**, 406–9.

Swinburne, T.R. (1973). In *Fungal pathogenicity and the plant response* (ed. R.J.W. Byrde and C.V. Cutting), pp. 365–82. Academic Press, London.
Swinburne, T.R. and Brown, A.E. (1975). *Physiol. Plant Pathol.* **6**, 259–64.
Varns, J.L., Kuć, J. and Williams, E.B. (1971). *Phytopathology* **61**, 174–7.
Verhoeff, K. (1974). *Annu. Rev. Phytopathol.* **12**, 99–110.
Verhoeff, K. and Liem, J.I. (1975). *Phytopath. Z.* **82**, 333–38.
Ward, E.W.B., Unwin, C.H., Hill, J. and Stoessl, A. (1975). *Phytopathology* **65**, 859–63.
Yates, S.G. (1987). *ACS Symp. Series* **330**, 295–9.

12. ACC oxidase gene family: characterization and down-regulation by genetic manipulation

M. BOUZAYEN, M. FERRER, P. GUILLEN, R. AYUB, S. BIDONDE, M. BEN AMOR, M. GUIS, S. RAMASSAMY, H. ZEGZOUTI, J.C. PECH and A. LATCHÉ

Ecole Nationale Supérieure Agronomique, UA INRA, 145, Avenue de Muret, 31076 Toulouse Cedex, France

Introduction

The phytohormone ethylene controls many stages of plant development such as ripening of climacteric fruits, leaf and flower senescence, responses to wounding and invasion by pathogens (Abeles *et al.*, 1992). The ethylene biosynthetic pathway includes two main steps:

- the conversion by ACC synthase of *S*-adenosylmethionine into *1*-amino-cyclopropane-*1*-carboxylic acid (ACC)
- the conversion of ACC into ethylene by ACC oxidase (Kende, 1993).

On the other hand, the ethylene precursor ACC can be metabolized by the ACC-*N*-malonyl transferase to a conjugated form so far identified as malonyl-ACC (Yang and Hoffman, 1984). Genes encoding ACC synthase (Sato and Theologis, 1989; Van der Straeten *et al.*, 1990; Nakajima *et al.*, 1990; Olson *et al.*, 1991; Park *et al.*, 1992) were first isolated and it has been shown that, in many plant varieties, ACC synthase is encoded by a multigene family displaying differential expression during plant development. Subsequently, transgenic tomato plants expressing antisense construction of these genes have been obtained (Oeller *et al.*, 1991). Despite the major attention given to ACC oxidase, formely called ethylene-forming enzyme (EFE), the enzyme has long been recalcitrant to extraction and purification. A few years ago, a cDNA clone, pTOM13, isolated from ripe tomato, was shown to strongly inhibit ethylene production when expressed in antisense orientation (Hamilton *et al.*, 1990). The deduced amino acid sequence of this clone showed substantial homology to a soluble flavanone hydroxylase, which gave important clues for successful extraction and purification of ACC oxidase. The ultimate identification of ACC oxidase gene was given by functional expression of pTOM13 cDNA *in Saccharomyces cerevisiae*

(Hamilton *et al.*, 1991) and in *Xenopus* oocytes (Spanu *et al.*, 1991). Screening a genomic library with pTOM13 probe revealed three genomic clones now refered to as *ACO1, ACO2* and *ACO3.* Full sequencing of these clones revealed that pTOM13 cDNA corresponds to *ACO1* (Holdsworth *et al.*, 1988; Köck *et al.*, 1991), and subsequently cDNA clones corresponding to *ACO1* and *ACO2* were isolated (Bouzayen *et al.*, 1993). We show in this study that *ACO2* and *ACO3* are transcriptionally active in tomato and demonstrate that both genes encode a functional ACC oxidase by heterologous expression in *S. cerevisiae.*

In the second part of this work we report on the down regulation of ethylene biosynthesis using antisense RNA strategy. Control of fruit ripening through genetic manipulation of ethylene production has been only achieved with tomato (Hamilton *et al.*, 1990; Oeller *et al.*, 1991; Klee *et al.*, 1991; Good *et al.*, 1994). We describe here the generation of transgenic melons (*Cucumis melo*, var. Reticulatus, Naud) expressing an antisense ACC oxidase gene.

PCR amplification of *ACO2* and *ACO3* cDNA from tomato.

Total RNA from etiolated seedlings, green fruit, breaker fruit, wounded leaves and wounded fruits were used for first strand cDNA synthesis and subsequent PCR amplification. Gene-specific primers encompassing the coding region of *ACO2* and *ACO3* were designed to amplify the corresponding cDNA. Successful amplification was achieved for the *ACO2* clone only from tomato seedlings, whereas *ACO3* cDNA was amplified from breaker and mature green fruit treated 15 h with exogenous ethylene. In both cases, the 1 kb fragment obtained was consistent with the size of the coding regions of *ACO2* and *ACO3*. Any contamination with genomic DNA was ruled out since the size of the *ACO2* and *ACO3* genomic clones is 2.5 kb and 1.6 kb, respectively.

The PCR products were cloned into pCR II vector to generate pEFE2 and pEFE 3, respectively. The analysis of the nucleotide sequences of the cloned fragment revealed their identity to the *ACO2* and *ACO3* genes. However, both cDNAs displayed one base mutation compared to the corresponding genomic clone. In pEFE2 this feature led to the exchange of Arg118 by Ala. In pEFE3 the point mutation resulted in the convertion of Gly 241 into Asp.

Comparative analysis of ACO1, ACO2 and ACO3 peptide sequences

Nucleotide sequence analysis (Table 12.1) showed that ACO1 and ACO3 share higher homology (94%) between them than with ACO2 (85%). Optimal alignment of the open reading frame of the three clones revealed that ACO3 contains a four amino acid deletion spanning amino acids 266–271 and a five amino acid insertion between amino acid 310 and 316. The search for the content in charged groups indicated that ACO2 protein contains more positively charged amino acids resulting in a significantly higher isoelectric point (Table 12.2). Sequence analysis indicated that none of the three proteins have an obvious transit peptide or hydrophobic membrane-spanning domain whereas all three proteins contain a putative glycosylation site.

Table 12.1. Amino acid and nucleotide sequence homology between tomato ACC oxidase genes

	% Homology		
Sequence	ACO1/ACO2	ACO1/ACO3	ACO2/ACO3
Nucleotide	86	94	85
Amino acid	89	94	88

Table 12.2. Isolectric point and charged group content of ACO1, ACO2 and ACO3 encoded proteins

Charged groups	ACO1	ACO2	ACO3
pI	5.0	6.1	4.9
Positively charged	45	50	43
Negatively charged	66	63	65

Cloning and expression in *S. cerevisiae*

The *Xba* I-*Bam* HI fragments of pEFE2 and pEFE3 were ligated into the *Bgl* II site between the phosphoglycerate kinase (PGK) promoter and terminator of the yeast expression vector pBEJ15 (Hadfield *et al.*, 1990). These clones were named pYE2 and pYE3 and used to transform *Saccharomyces cerevisiae* JRY 188. Based on the complementation of the *leu2* mutation in *S. cerevisiae* by a functional gene on pBEJ15 vector, transformed yeasts were first screened on synthetic complete plates without leucine. The presence of the ACO clones in the transformed colonies was thereafter checked by PCR amplification using specific primers for each clone. Selected colonies corresponding to each clone were grown on liquid medium and their capacity to convert ACC to ethylene tested. Figure 12.1 shows that all three lines of transformed yeast were able to convert exogenous ACC to ethylene while neither untransformed cells nor yeast transformed with native pBEJ15 could produce ethylene. However, the three yeast lines displayed different levels of ethylene production. The *ACO1* line showed the highest activity, *ACO3* was slightly lower and *ACO2* reached only 40% of the maximum activity of *ACO1*. These differences in ACC oxidase activity could not be attributed to different transcription levels as northern analysis showed similar accumulation of ACO transcripts in the three yeast lines. Further biochemical characterization of the three isoforms will show whether the difference in activity could account for differences in optimum pH, affinity for substrates (ACC and oxygen) or requirement for cofactors (iron, carbon dioxide and ascorbate).

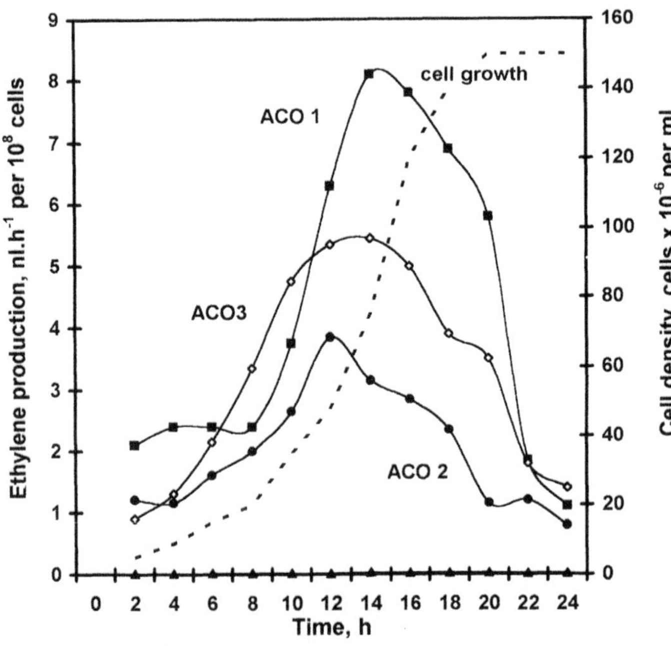

Fig. 12.1. Activity of ACC oxidase isoforms, ACO1, ACO2, and ACO3 in *S. cerevisiae*. The capacity of transformed yeast to convert ACC to ethylene was determined using cell culture grown in liquid medium. The optical density of the yeast culture was measured at A_{600}. A 1 ml sample was incubated for 30 min at room temperature with 0.5 mM ACC in 5 ml glass bottles sealed with Subaseal vaccine caps, and ethylene in the headspace was measured by gas chromatography. (■) ACO1 line , (●) ACO2 line, (□) ACO3 line, (▲) untransformed line, (----) cell density.

Transgenic melons expressing ACC oxidase antisense cDNA

So far genetic manipulation of fruit ripening has been confined exclusively to tomato (Gray *et al.*, 1992). Tomatoes in which ethylene production had been inhibited have been generated using antisense constructs of ACC synthase (Oeller *et al.*, 1991) and ACC oxidase (Hamilton *et al.*, 1990) or sense constructs of ACC deaminase (Klee, 1993) and *S*-adenosylmethionine hydrolase (Good *et al.*, 1994). Cantaloupe melon of the Charentais type represents a good target for inhibiting ethylene production and fruit ripening because of its short shelf-life associated with a sharp climacteric phase and high ethylene production rate.

In order to manipulate the ripening process of a cultivated pure line (cv Védrantais) of Charentais cantaloupe, cotyledons were transformed with disarmed *Agrobaterium tumefaciens* LBA4404 containing the binary vector harboring a NOS/NPTII chimeric gene and an antisense construct of the MEL1 cDNA encoding melon ACC oxidase (Balagué *et al.*, 1993) under the transcriptional control of CaMV35S promoter and the transcript 7 and 5 terminators of the octopine-type Ti plasmid pTiA6. *Agrobacterium*-mediated transformation was performed using 5 day old cotyledons cut into six sections. Explants were first precultured for 3 days in a regeneration medium (Murashige and Skoog, 1962) supplemented with 2.2 μM 6-benzylaminopurine and 0.54 μM α-naphtaleneacetic acid and then soaked 30 min in *Agrobacterium* suspension (10^6–10^8 cells/ml), blotted dry and placed in the same medium for 3 days of co-cultivation. The regeneration procedure was adapted from Roustan *et al.* (1992). It consisted of transferring explants in Petri dishes containing the regeneration medium supplemented with 300 mg/l kanamycin and 500 mg/l carbenicillin. After about one month, young shoots were excised and transferred into medium devoid of hormones and supplemented with 50 mg/l kanamycin and 500 mg/l carbenicillin. After 3–4 weeks, rooted plantlets were transplanted to sterile mixture of soil and vermiculite (4 : 1, v/v) for acclimatation during 2 weeks before transfer to the greenhouse.

Among several transformants, the clone exhibiting the strongest inhibition of wound-induced ethylene production in leaves has been selected. Molecular and genetic studies performed after self-pollination on the R1 progeny had shown that this transformant had incorporated two copies of the T-DNA at the same locus. Northern and Western blot studies showed that expression of endogenous ACC oxidase gene was considerably reduced in transgenic fruit as compared to wild type. To monitor internal ethylene concentration of fruit attached to the vine, an external gas collection apparatus was used as described by Salveit (1993). Gas samples were taken daily, early in the morning, through a serum stopper and quantified on a gas chromatograph equipped with an alumina column and flame ionization detector (Mansour *et al.*, 1986). Figure 12.2 shows that, while untransformed fruit developed a classical sharp peak of ethylene production, no climacteric peak of ethylene production could be observed in antisense fruit both on the vine (Fig. 12.2a) and after detachment (Fig. 12.2b). At the time when control fruit were at the climacteric peak, ethylene production of transgenic fruit was reduced by more than 99%. As a consequence, transgenic fruit exhibited strong inhibition of ripening with a great reduction of chlorophyll degradation and softening and absence of the abscission zone in the peduncle. After more than 2 weeks of storage at room temperature, transgenic fruits were still green and firm while control fruits were yellow, fully senescent and rotten. However, treatment of detached transgenic fruit with 50 ppm ethylene reversed the antisense phenotype leading to the yellowing of the rind, degradation of the chlorophyll and formation of the abscission zone at the same level as in control wild type fruit. One interesting observation was that no significant difference in the colour of the flesh could be detected between transgenic and untransformed fruits, indicating that the synthesis of carotenoid pigments was ethylene-independent. This observation is consistent with previous data showing that the phytoene synthase gene in cantaloupe melon was expressed well before the onset of the climacteric (Karvouni *et al.*, 1995).

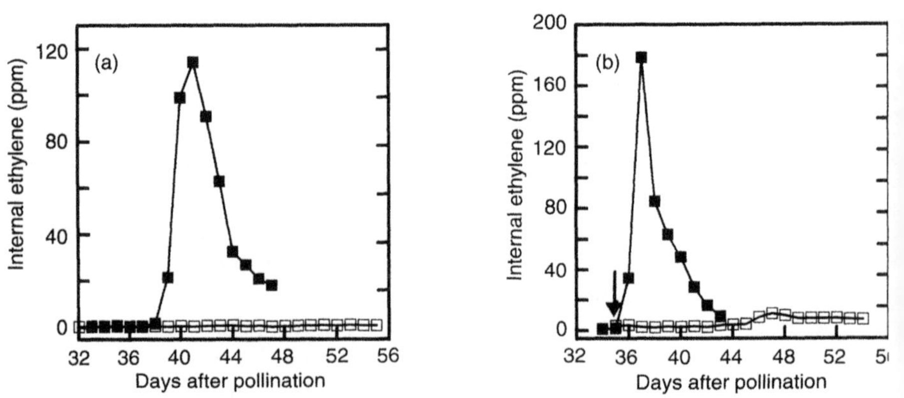

Fig. 12.2. Ethylene production by wild type and antisense ACC oxidase melon fruit, (a) on and (b) off the vine. Internal ethylene concentration of wild type (■) and antisense (□) fruit was measured using an external gas collector apparatus (Salveit, 1993). The arrow indicates the harvest day.

Conclusion

Many plant developmental processes are under control of ethylene. Therefore, understanding the regulation of ethylene biosynthesis at the genomic level is a crucial step towards the control of these processes. In this study we showed that the three members of ACC oxidase multigene family are transcriptionally active in tomato. Although the heterologous expression in yeast demonstrated that all three genes encode functional proteins, the ACC oxidase isoforms displayed different level of activity. In the last part of this work we described the inhibition of ethylene production and fruit ripening in cantaloupe charentais melon through genetic manipulation. Melon proves to be a suitable plant model for the study of the role of ethylene in controlling some ripening related secondary metabolisms such as aroma biosynthesis that can not be adressed with tomato.

Acknowledgements

The authors are grateful to Cathy Sigro for typing the manuscript. This work was supported by INRA (AIP Matural), the EU (AAIR III, 94. 1961) and the NATO (Grant 962/93)

References

Abeles, F.B., Morgan, P.W. and Saltveit, M.E. Jr. (1992). *Ethylene in plant biology*, 2nd edn. Academic Press, San Diego, CA.

Balagué, C., Watson, C.F., Turner, A.J., Rougé, P., Picton, S., Pech, J.C. and Grierson, D. (1993). *Eur. J. Biochem.* **212**, 27–34.

Bouzayen, M., Cooper, W., Barry, C., Zegzouti, H., Hamilton, A.J. and Grierson D. (1993) In *Cellular and molecular aspects of the plant hormone ethylene*. (ed. J.C Pech, A. Latché, and C. Balagué), pp. 76–81. Kluwer, Dordrecht.

Good, X., Kellogg, J.A., Wagoner, W., Langhoff, D., Matsumura, W. and Bestwick, R.K. (1994) *Plant Mol. Biol.* **26**, 781–90.

Gray, J., Pieton, S., Shabbeer, J., Schuch, W. and Grierson, D. (1992). *Plant. Mol. Biol.* **19**, 69–87.

Hadfield, C., Jordan, B., Mount, R.C., Pretorius, G.H.J. and Burak, E. (1990). *Curr. Genet.* **18**, 303–13.

Hamilton, A.J., Lycett, G.W. and Grierson, D., (1990). *Nature* **346**, 284–7.

Hamilton, A.J., Bouzayen, M. and Grierson, D. (1991) *Proc. Nat. Acad. Sci. USA* **88**, 7434–7.

Holdsworth, M.J., Schuch, W. and Grierson, D. (1988). *Plant Mol. Biol.* **11**, 81–8.

Karvouni, Z., John, I., Taylor, J.E., Watson, C.F., Turner, A.J. and Grierson, D. (1995). *Plant Mol. Biol.* **27**, 1153–62.

Kende, H. (1993). *Annu. Rev. Plant Physiol. Plant Mol. Biol.* **44**, 283–307.
Klee, H.J. (1993). *Plant Physiol.* **102**, 911–16.
Klee, H.J., Hayford, M.B., Kretzmer, K.A., Barry, G.F. and Kishore, G.M. (1991). *Plant Cell* **3**, 1187–93.
Köck, M., Hamilton, A.J. and Grierson, D. (1991). *Plant Mol. Biol.* **17**, 141–2.
Mansour, R., Latché, A., Vaillant, V., Pech, J.C. and Reid, M.S. (1986). *Physiol. Plant.* **66**, 495–502.
Murashige, T. and Skoog, F. (1962). *Physiol. Plant.* **15**, 437–97.
Nakajima, N., Mori, H., Yamazaki, K. and Imaseki, H. (1990). *Plant Cell Physiol.* **31**, 1021–9.
Oeller, P.W., Min-Wong, L., Taylor, L.P., Pike, D.A. and Theologis, A., (1991). *Science* **254**, 437–9.
Olson, D.C., White, J.A., Edelman, L., Harkins R.N. and Kende H. (1991). *Proc. Natl. Acad. Sci. USA* **88**, 5340–4.
Park, K.Y., Drory, A. and Woodson, W.R. (1992). *Plant Mol. Biol.* **18**, 377–86.
Roustan, J.P., Latché, A. and Fallot, J. (1992). *J. Plant Physiol.* **140**, 485–8.
Salveit, Jr. M.E. (1993). *Physiol. Plant.* **89**, 204–10.
Sato, T. and Theologis, A. (1989). *Proc. Natl. Acad. Sci. USA* **86**, 6621–5.
Spanu, P., Reinhardt, D. and Boller T. (1991). *EMBO J.* **10**, 2007–13.
Van der Straeten, D., Van Wiemeersch, L., Goodman, H.M. and Van Montagu, M. (1990). *Proc. Natl. Acad. Sci. USA* **87**, 4859–63.
Yang, S.F. and Hoffman, N.E. (1984). *Annu. Rev. Plant Physiol.* **35**, 155–89.

13. Options for the genetic manipulation of astringent and antinutritional metabolites in fruit and vegetables

MARK P. ROBBINS, ADRIAN D. BAVAGE and PHILLIP MORRIS

Cell Manipulation Group, Institute of Grassland and Environmental Research, Aberystwyth Research Centre, Plas Gogerddan, Aberystwyth, Dyfed SY23 3EB, UK

Introduction

Fruit and vegetables produce an array of secondary metabolites that are derived from primary metabolic pathways. Such secondary compounds can have a profound effect on the value of individual fruit and vegetable varieties. For example, anthocyanins, carotenoids and chalcones contribute to fruit colour; whereas other secondary compounds such as glucosinolates and volatile phenolics play an important part in flavour characteristics.

In this chapter we discuss both astringent and antinutritional secondary compounds. Initially we consider some examples relating to forage and feed seeds and subsequently examples that concern fruit and vegetables. In further sections we focus on condensed tannins, which are important secondary product polymers and are of interest in that they are both astringent and antinutritional. We also describe *Lotus corniculatus,* which is a convenient model system for the genetic modification of these metabolic end products. Finally, we describe some current and future strategies for modifying levels and structures of condensed tannins and opportunities for technology transfer to fruit and vegetable varieties.

Astringent and antinutritional metabolites

Astringent compounds can be characterized by their ability to complex with (and cross-link) protein molecules. It is this property which helps to define the taste characteristics of astringent metabolites (see Clifford, Chapter 5 this volume). There are two major classes of secondary compound, the hydrolysable tannins and the condensed tannins, which have well-described astringency characteristics. Such properties are at least partly dependent upon the relatively high number of hydroxyl groups found in hydrolysable and condensed tannin molecules (Haslam, 1989). Whilst moderate astringency can be regarded as a pleasurable sensation, extreme astringency can be considered as a deterrent to consumption of foodstuffs (Pierpoint, 1985).

By contrast, antinutritional end products decrease the nutritive value of foodstuffs. In this context it can be difficult to distinguish between antinutritional metabolites and compounds which are in themselves toxic. For the purposes of this chapter we use the term antinutritional in its widest sense, to include antinutritional compounds that have also been described as having toxic effects.

Animal feedstuffs

Table 13.1 outlines some metabolites from animal feedstuffs that have been described as antinutritional and/or astringent. Of course, there are cases where an individual crop contains significant levels of several secondary metabolites but we have tried to select examples where an individual metabolite is of major importance. Glucosinolates and derived metabolites have deleterious effects which have been well described. Such compounds are a particular problem when rapeseed meals are fed to poultry and pigs. The active components are believed to be the hydrolysis products of glucosinolates; isothiocyanates, nitriles and thiocyanates which are released on ingestion (Duncan and Milne, 1989). Saponins are another troublesome class of secondary compounds and are frequently found at significant levels in leguminous feed seeds. Although difficult to quantify accurately, saponins can now be regarded as toxic metabolites (Heaney and Fenwick, 1989). Condensed tannins are of particular interest and are important antinutritional secondary compounds both in feed seeds and forages (reviewed by Mueller-Harvey and McAllan, 1992; Reed, 1995).

When considering condensed tannins, it is interesting to note that monogastric livestock animals, such as pigs and chickens, are particularly sensitive to polyphenolics in their diet. Tannin-protein binding decreases protein availability and this phenomenon limits the effective use of tannin-containing cultivars of faba bean and sorghum as feedstuffs for monogastric animals (Griffiths, 1989). With ruminant livestock, the position is more complex as moderate levels of condensed tannins are beneficial; tannin-protein binding is believed to decrease rates of microbial degradation and therefore improves retention of dietary nitrogen (Tanner *et al.*, 1994). High levels of polyphenolics in forage crops such as African browse species and *Desmodium*, however, bind dietary protein and reduce the nutritive value of foliage of these legume crops in third world agricultural systems (van Hoven and Furstenburg, 1992).

Table 13.1. Examples of antinutritional and astringent metabolites in feed, forage and browse

Secondary metabolite	Crop species
Feed seed	
Glucosinolates	Brassicas, e.g. rapeseed (*Brassica napus*)
Saponins	Legume seeds
Condensed tannins	Sorghum (*Sorghum vulgare*)
	Faba bean (*Vicia faba*)
	Amaranthus (*Amaranthus* sp.)
	Rapeseed (*Brassica napus*)
	Mesquite (*Prosopis juliflora*)
Free and esterified phenolic acids	Rapeseed (*Brassica napus*)
	Oat (*Avena sativa*)
Forage	
Polymeric condensed tannins	Birdsfoot trefoil (*Lotus corniculatus*)
	Big trefoil (*Lotus uliginosus*)
	Sainfoin (*Onobrychis viciifolia*)
	Crown vetch (*Coronilla varia*)
	Lespedeza (*Lespedeza cuneata*)
Saponins	Alfalfa (*Medicago sativa*)
Cyanogenic glycosides	White clover (*Trifolium repens*)
	Birdsfoot-trefoil (*Lotus corniculatus*)
Oestrogenic isoflavonoids	White clover (*Trifolium repens*)
	Red clover (*Trifolium pratense*)
	Alfalfa (*Medicago sativa*)
Browse	
Polymeric condensed tannins	Desmodium (*Desmodium ovalifolium*)
	Acacia (*Acacia cyanophylla*)
	Sesbania (*Sesbania seban*)
	Flemengia (*Flemengia macrophylla*)
	Dioclea (*Dioclea guianiensis*)
	Leucaena (*Leucaena leucocephala*)
	Khejri (Prosopis cineraria)

Human food

Some examples of astringent and antinutritional metabolites found in fruit and vegetable crops intended for human consumption, are given in Table 13.2. Once again, it is difficult to distinguish between toxic and antinutritional compounds. Cyanogenic glycosides, glycoalkaloids and non-protein

Table 13.2. Examples of antinutritional and astringent metabolites in fruit and vegetable crops

Secondary metabolite	Crop species
Cyanogenic glycosides	Cassava (*Manihot esculenta*)
Glycoalkaloids	Potato (*Solanum tuberosum*)
Non-protein amino acids	Chickpea (*Lathyrus sativus*)
Polymeric condensed tannins	Banana (*Musa sapientum*)
	Cocoa (*Theobroma cacao*)
	Plum (*Prunus domesticum*)
	Grape (*Vitis vinifera*)
	Gooseberry (*Ribes grossularia*)
	Blueberry (*Vaccinium corymbosum*)
	Blackcurrant (*Ribes nigrum*)
	Strawberry (*Fragaria ananassa*)
	Apple (*Malus pumila*)
	Coconut (*Cocos nucifera*)
	Kiwifruit (*Actinidia chinensis*)
Flavans and oligomeric condensed tannins	Grape (*Vitis vinifera*)
	Tea (*Camellia sinensis*)
	Blackcurrant (*Ribes sanguinium*)
	Blackberry (*Rubus fruticosus*)
	Raspberry (*Rubus idaeus*)
	Hops (*Humulus lupulus*)
	Rhubarb (*Rheum rhaponticum*)
	Cocoa (*Theobroma cacao*)
	Avocado (*Persia gratissima*)
	Apple (*Malus sylvestris*)
	Peach (*Prunus persica*)
Hydrolysable tannins	Strawberry (*Fragaria ananassa*)
	Blackberry (*Rubus fruticosus*)
	Raspberry (*Rubus idaeus*)
	Tea (*Camellia sinensis*)
	Pomegranate (*Punica granatum*)
Caffeic acid derivatives	Apple (*Malus pumila*)
	Coffee (*Coffea arabica*)
	Tomato (*Lycopersicon esculentum*)
	Cucumber (*Cucumis sativus*)
Furanocoumarins	Celery (*Apium graveolens*)
	Parsley (*Petroselinum crispum*)

amino acids can all be considered as undesirable from the point of view of human nutrition, and options for lowering the dietary intake of such compounds are of interest. Focusing again on condensed tannins, there may well be cases where these products have negative nutritional effects with reference to the non-availability of dietary protein. An additional antinutritional effect of condensed tannins may be a decrease in the availability of non-haem iron in high-tannin diets (reviewed by Pierpoint, 1985). Strategies for decreasing the levels of deleterious end products could therefore be of use in improving the nutritional value of fruit and vegetable species in developing countries.

In the context of products consumed in developed countries it is clear that the taste characteristics of a range of fruit and vegetable species depend at least in part on condensed tannins. Foo and Porter (1981) have described the structure of tannins in a range of edible fruits. Fruits contain both hydrolysable and condensed tannins, and the latter class of tannins are important components in fruits as varied as apple, kiwifruit, gooseberry, redcurrant and banana. Products derived from fruits and vegetables also have a value that depends upon condensed tannins and their consequent astringency characteristics. Fruit juices, tea, coffee, wine and spices are obvious examples where the value of these products depends upon condensed tannins and the detailed fine structure of tannin polymer molecules. Therefore, strategies for changing both the level and the structure of tannins may well be of value in fruit, vegetables and derived products.

A system for manipulating condensed tannins

In view of the wide range of crops which biosynthesize tannins and because of our interest in herbage, we have selected a model forage species to test strategies for the genetic modification of polyphenolics and related pathways. *Lotus corniculatus* (birdsfoot trefoil) has a number of useful characteristics as a model plant and these have been reviewed by Robbins *et al.* (1992).

To summarize briefly some of the advantages of this system, *Lotus corniculatus* is a readily transformable and regenerable tannin-positive crop species. Therefore gene constructs can easily be introduced into *Lotus* and resultant transgenic plants analysed for changes in tannin biosynthesis. Another advantage is that, when using *Agrobacterium rhizogenes*-mediated cotransformation, independently transformed lines can be analysed both as root cultures and as transgenic plants. Finally, the development of clonal recipient genotypes now allows the analysis of the effect of a construct in lines with differential levels of flux into polyphenolic biosynthesis.

A flowchart indicating how this laboratory generates and analyses transgenic *Lotus* lines harbouring sense and antisense gene constructs is shown in Fig. 13.1. As can be seen, cotransformed lines may be analysed at the root-culture level or at the whole plant level. As root cultures exist predominantly as a single tissue type, the analysis of tannin structure is relatively straightforward. In this paper we report results with sense and antisense gene constructs studied at the root-culture level.

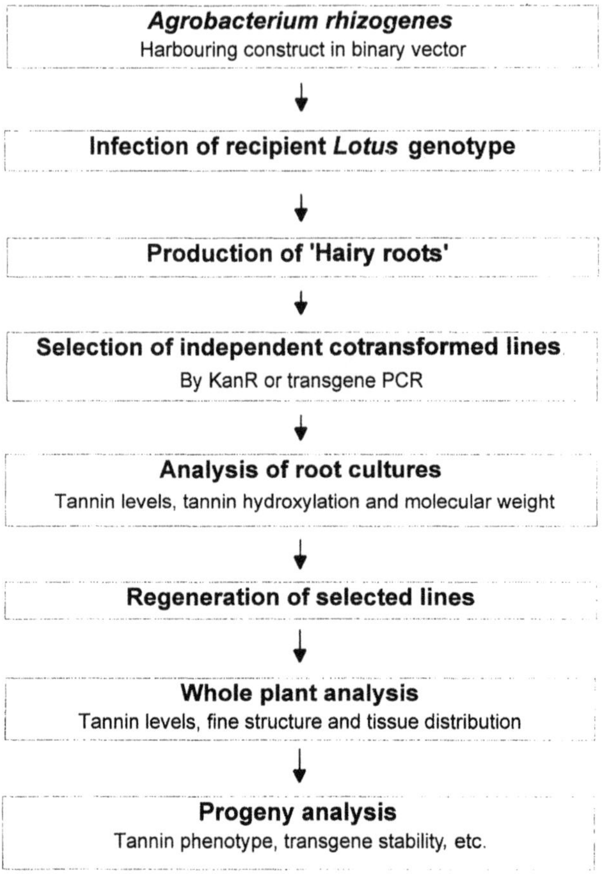

Fig. 13.1. Scheme for the production and analysis of *Lotus* lines with modified tannin phenotypes.

Condensed tannin biosynthesis

Elucidating secondary pathways in plants is a far from straightforward process. In the case of condensed tannin biosynthesis original work by Stafford has clarified issues relating to the late enzymology of the pathway (Stafford and Lester, 1984, 1985) and she has indicated the importance of NADPH reductases in this biosynthetic pathway. Additionally, studies on proanthocyanidin mutants in barley (Jende-Strid, 1991) have given information on the number of enzymatic steps in the pathway. Figure 13.2 gives a representation of condensed tannin biosynthesis in *Lotus* and other plant species. Enzymes in bold have been cloned from at least one species of higher plant. A significant point to note about this metabolic pathway is that condensed tannin polymers are produced by the condensation of flavan-3,4-diols and catechin-type units. Tannin precursors are metabolically derived from dihydroflavonols and the hydroxylation pattern of polymers depends upon the degree of hydroxylation of the available precursors. In most higher plants, condensed tannins appear to be made up primarily from cyanidin-type and delphinidin-type units (Porter, 1986).

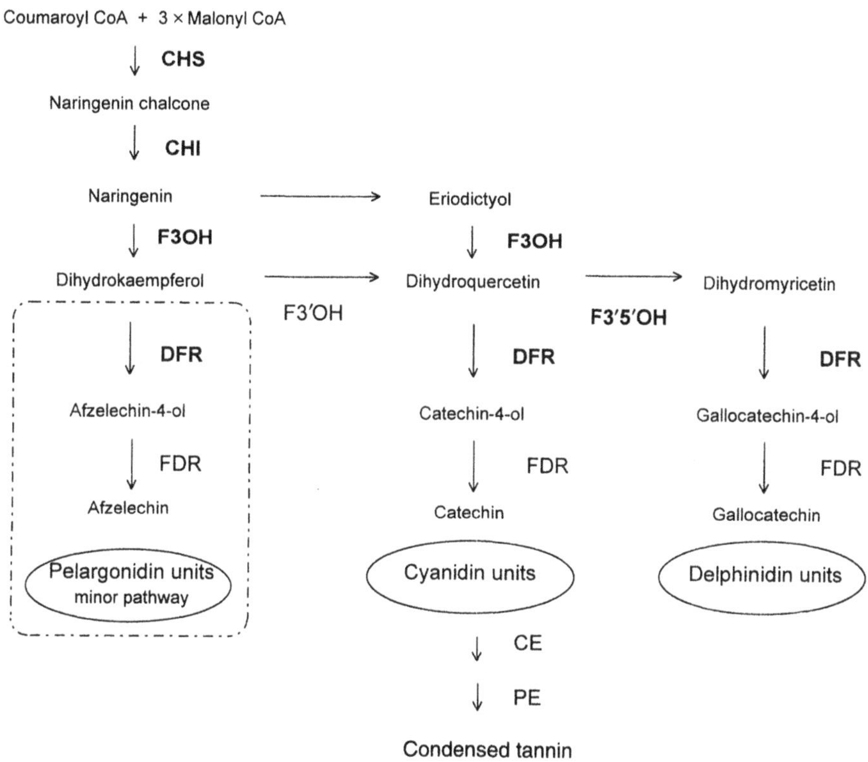

Fig. 13.2. Condensed tannin pathway and key enzymatic steps. CHS (chalcone synthase), CHI (chalcone isomerase), F3OH (flavanone-3 hydroxylase), F3′OH (flavonoid-3′ hydroxylase), F3′5′OH (flavonoid-3′5′ hydroxylase), DFR (dihydroflavonol reductase), FDR (flavan-3,4-diol reductase), CE (condensing enzyme), PE (polymerizing enzyme).

In view of the current information regarding the biosynthesis of condensed tannins, there are a number of possible options for trying to modify tannin levels and structures. We have focused on gene targets that should modify levels of individual dihydroflavonols and consequently would be predicted to alter tannin levels and tannin structure. Strategies that we have considered include: modification of flavanone 3 hydroxylase (F3OH), changes in flavonoid 3′5′hydroxylase (F3′5′OH) and alterations at the level of dihydroflavonol reductase (DFR). Although some data indicate that F3OH may be important in the regulation of the pathway, there is currently no detailed understanding of the substrate specificity of F3OH (Forkmann, personal communication). We have therefore concentrated our work on the enzymatic steps catalysed by DFR and F3′5′OH.

Modifying tannin structures using DFR

Antisense expression of DFR

In initial work, we examined the consequences of antisensing DFR, which would be predicted to decrease tannin levels in effective antisense events. Carron *et al.* (1994) introduced a heterologous antisense DFR construct into three recipient genotypes of *Lotus corniculatus*. In two of the genotypes, S33 and S50, lines were produced with significantly decreased levels of tannins when analysed as root cultures. Decreased tannin levels have also been noted in some lines when analysed as whole plants (Robbins *et al.*, 1994; Robbins, unpublished).

These observations demonstrate that there are genuine options for down-regulating tannin biosynthesis in this model system. A subsequent point to note is that these results also indicate that, in successful antisense lines, DFR becomes rate limiting with reference to the quantity of condensed tannin in *Lotus* root cultures. Carron *et al.* (1994) also noted that polymer structure was modified in antisense down-regulated lines and that tannins from down-regulated lines showed decreased levels of hydroxylation. A possible biochemical explanation for this observation may be that the *Lotus* DFR accepts both dihydroquercetin and dihydromyricetin as substrates but that dihydroquercetin is the preferred substrate at lower enzyme concentrations.

Sense expression of DFR

Following on from work using DFR in antisense, we have analysed effects resulting from the expression of sense DFR constructs.

In this context there are two particular points of interest. Firstly, successful sense expression would be expected to cause an increase in total tannin levels if DFR levels are rate limiting in *Lotus*. Secondly, DFR is an enzyme which has been reported to have differing substrate specificities when extracted and analysed from different higher plant species (Heller and Forkmann, 1993). Some DFR genes encode enzymes which preferentially accept dihydrokaempferol and dihydroquercetin (*A. majus* and *Z. mays*) while others display a preference for dihydroquercetin and dihydromyricetin (*P. hybrida* and presumably *L. corniculatus*). Therefore, introducing DFR genes with differing substrate specificities should cause predictable changes in tannin hydroxylation.

We have introduced sense *A. majus* and sense *P. hybrida* gene constructs into *Lotus*, and preliminary results are indicated in Fig. 13.3. Up-regulated lines were noted both with the *A. majus* gene and also the *P. hybrida* gene, and hydroxylation patterns were as predicted. In fact, one sense *A. majus* DFR line (ADFR10) was both up-regulated and also yielded significant quantities of pelargonidin (about 10%) when condensed tannin was hydrolysed. With reference to down-regulated lines, two down-regulated and possibly cosuppressed lines (ADFR02 and ADFR07) were found when expressing the full length sense *A. majus* DFR gene. These down-regulated lines showed no significant alteration in hydroxylation when compared with controls.

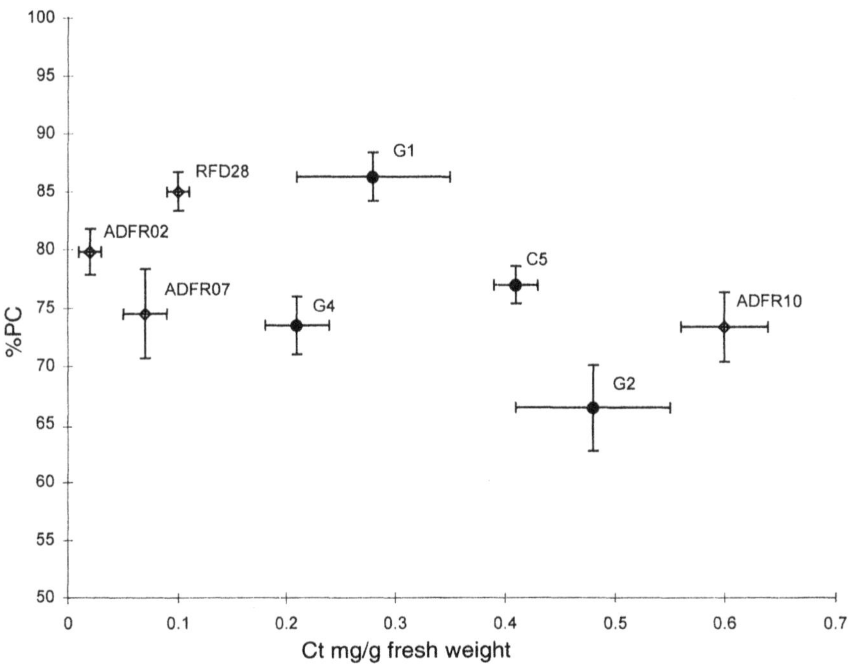

Fig. 13.3. Tannin levels and structures in *Lotus* root cultures harbouring sense and antisense *A. majus* DFR constructs in genotype S50. G1, G2, G4, C5 (control lines). RFD28 (antisense DFR), ADFR02, ADFR07 (cosuppressed sense DFR), ADFR10 (overexpressing sense DFR).

Future options for the genetic modification of condensed tannins

The work reported here indicates that sense and antisense manipulations of DFR can be used to produce a useful range of tannin levels and tannin structures. We hope to analyse potential changes in tannin molecular weights in the near future. There are a number of other strategies and approaches relating to condensed tannins that have not yet been tested. One novel option includes the use of other structural genes to modify tannin biosynthesis. It is interesting to note that antisense expression of an inducible CHS gene from *P. vulgaris* results in an up-regulation of tannin biosynthesis (Colliver *et al.*, 1994). Other genes have not been tested however and genes encoding terminal enzymatic steps (flavan 3,4 diol reductase, condensing enzyme and polymerising enzyme) may be attractive targets for an antisense approach.

Another comparatively untested strategy is the use of regulatory genes to cause global changes in the regulation of tannin biosynthesis. Expression of *Sn*, a maize anthocyanin regulatory gene, in *Lotus* has been reported to have dramatic effects upon pathway regulation (Crea *et al.*, 1994). An additional approach could be to develop methodologies which induce tannin turnover in mature and senescent tissues. This could be especially important with regard to fruit and vegetable species. Condensed tannins can be regarded as antimicrobial and anti-insecticidal end products and therefore may be important components relating to plant defence during the growth of harvestable fruit and vegetable tissues (see Harborne, Chapter 18 this volume). A strategy to induce tannin turnover at the time of harvest would take advantage of the defence-related properties of condensed tannins and yet could decrease the levels of antinutritional polyphenolics before consumption by the consumer.

Technology transfer to fruit and vegetable species

The possibility of designing tannin structures in higher plants may have considerable applications in horticultural and food-related industries. Complex tannin structures in wines can be regarded as being of value while the taste of fruit juices depends upon ratios of 'hard' and 'soft' tannins. A number of other beverages, such as tea, have qualities strictly dependent on tannin structure.

Regarding alterations in tannin levels, there may be future fruit and vegetable varieties which become acceptable after down-regulation of excessive levels of condensed tannins in appropriate tissues. We therefore hope that plant breeders with interests in fruit and vegetable species will consider and evaluate these new technologies.

Returning to other astringent and antinutritional metabolites, there are real opportunities for the down-regulation of a number of other metabolic pathways. Of particular importance may be opportunities to decrease the levels of antinutritional and toxic metabolites. Once such metabolites are definitively identified it should be possible to devise suitable molecular strategies for their down-regulation. This technology could be of value for plant breeders, farmers and consumers in both developed and developing countries.

Acknowledgements

The authors would like to thank Dr Eunice Carter for helpful advice relating to this manuscript. Thanks also to Teri Evans and Ian Davies for excellent technical assistance.

IGER is core funded by the BBSRC. Adrian Bavage was supported by grant PG 203/536 from the BBSRC Plant Molecular Biology Initiative.

References

Carron, T.R., Robbins, M.P. and Morris, P (1994). *Theor. Appl. Genet.* **87**, 1006–15.

Colliver, S.P., Robbins, M.P. and Morris, P. (1994). *Acta Hort.* **381**, 148–51.

Crea, F., Paolocci, F., Consonni, G., Tonelli, C., Damiani, F. and Arcioni, S. (1994). *Abstracts of the VIII Int. Cong. of Plant Tiss. and Cell Cult.* S8-1, p. 159.

Duncan A.J. and Milne, J.A. (1989). *Aspects Appl. Biol.* **19**, 75–92.

Foo, L.Y. and Porter, L.J. (1981). *J.Sci.Food Agric.* **32**, 711–16.

Griffiths, D.W. (1989). *Aspects Appl. Biol.* **19**, 93–103.

Haslam, E. (1989). *Plant polyphenols—vegetable tannins revisited*, pp. 1–13. Oxford University Press, Oxford.

Heaney, R.K. and Fenwick, G.R. (1989). *Aspects Appl. Biol.* **19**, 1–20.

Heller, W. and Forkmann, G. (1993). In *The Flavonoids. Advances in research since 1986* (ed. J.B.Harborne), pp. 499–536. Chapman & Hall, London.

van Hoven, W. and Furstenburg, D. (1992). *Comp. Biochem. Physiol.* **101A**, 381–5.

Jende-Strid, B. (1991). *Theor. Appl. Genet.* **81**, 668–74.

Mueller-Harvey, I. and McAllan, A.B. (1992). *Adv. Plant Cell Biochem. Biotech.* **1**, 151–217.

Pierpoint, W.S. (1985). In *Biochemistry of plant phenolics* (ed. C. van Sumere and P.J. Lea), pp. 427–51.

Porter, L.J. (1986). In *The Flavonoids. Advances in research since 1980* (ed. J.B.Harborne), pp. 21–62. Chapman & Hall, London.

Reed, J.D. (1995). *J.Anim. Sci.* **73**, 1516–28.

Robbins, M.P., Carron, T.R. and Morris, P. (1992). In *Plant polyphenols* (ed. R.W.Hemingway and P.E.Laks), pp.111–31. Plenum Press, New York.

Robbins, M.P., Morris, P. and Carron, T.R. (1994). *Acta Hort.* **381**, 141–7.

Stafford, H.A. and Lester, H.H. (1984). *Plant Physiol.* **76**, 184–6.

Stafford, H.A. and Lester, H.H. (1985). *Plant Physiol.* **78**, 791–4.

Tanner, G.J., Moore, A.E. and Larkin, P.J. (1994). *Br. J. Nut.* **71**, 947–58.

14. Gibberellins and fruit development

JOSÉ L. GARCÍA-MARTÍNEZ[1] and PETER HEDDEN[2]

[1]Instituto de Biología Molecular y Celular de Plantas, Universidad Politécnica de Valencia–CSIC, Camino de Vera 14, 46022 Valencia, Spain and [2]IACR–Long Ashton Research Station, Department of Agricultural Sciences, University of Bristol, Long Ashton, Bristol BS18 9AF, UK

Introduction

Fruit development is dependent on the action of each of the five major classes of plant hormones. Gibberellins (GAs) play a major role in this process, inducing fruit-set and the expansion of the pericarp, and, as a consequence, they have important practical applications in fruit production. For example, GA_3 is applied to seedless grapes in order to obtain the required size and shape of the berry (Martín, 1983) and to seedless cultivars of mandarin to enhance fruit set (Monselise, 1979). However, despite the importance of GAs in fruit development and the large body of physiological, analytical and biochemical data on this subject, fundamental questions about the source of the active GAs and their mode of action in fruit have yet to be answered unequivocally. The significance of the available information is discussed in this review, which is restricted to fleshy fruit; monocotyledoneous plants are not considered.

The GAs form a large group of natural compounds with over 100 different structures characterized from plants and fungi to date. Only a small number of these compounds are known to have intrinsic biological activity, although biosynthetic precursors often exhibit activity due to metabolism to active forms. Gibberellins are involved in many aspects of plant development. Apart from fruit growth, processes in which GAs have been implicated include stem extension, leaf expansion, flower development and seed germination (Davies, 1995). Dwarf, GA-deficient mutants have been used to demonstrate that GA_1 is the major active GA controlling stem extension in many species (Phinney, 1984). The biologically active forms, as typified by GA_1 (Fig. 14.1), have a C_{19} skeleton, having lost carbon 20 (C-20) by oxidation, with a 19→10 lactone and a 7-carboxylic group. In addition, for stem extension activity (Serebryakov *et al.*, 1984) and probably also for stimulating fruit growth (Santes *et al.*, 1993), the presence of a 3β-hydroxyl group appears to be imporant. The function of the 13-hydroxyl group is

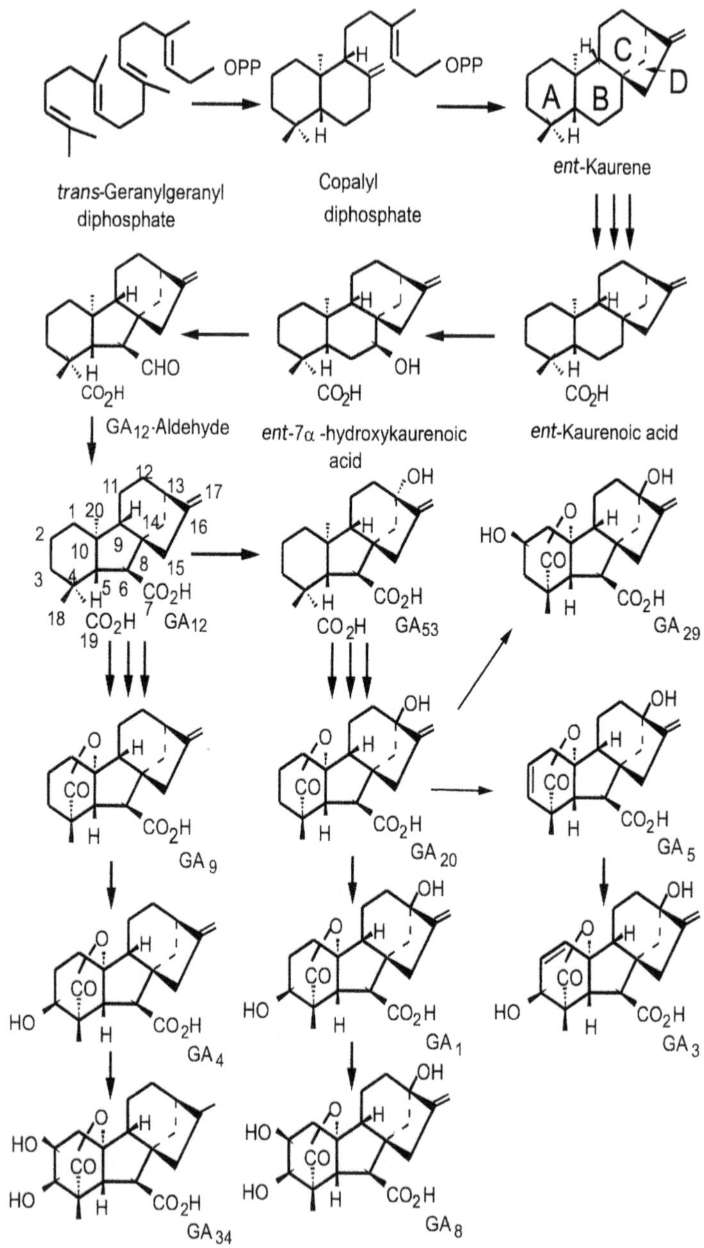

Fig. 14.1. Biosynthetic pathways to the biologically active GAs, GA_1, GA_3 and GA_4, and to the inactive 2β-hydroxylated metabolites. The ring designations are shown for *ent*-kaurene and the numbering of the carbon atoms for GA_{12}.

unclear. Indeed, in some species, particularly within the Cucurbitaceae, this functional group is known to be unnecessary, or even detrimental for activity (Serebryakov *et al.*, 1984).

Gibberellins are present in most actively growing tissues, although their concentrations may vary by several orders of magnitude, depending on the tissue. Highest concentrations are present in developing seeds, which may contain numerous diverse GA structures, the significance of which are unclear. In the following section, we review the identification of GAs in fruit, including seeds, with particular emphasis on those with a presumed function.

Distribution of gibberellins in fruit and seeds

A great deal has been published on the presence of GA-like active compounds in fruit and their changes during development (reviewed by Goodwin, 1978; Pharis and King, 1985). However, since the active substances were not identified conclusively, such results are of limited value and will not be included in this review. We will, therefore, restrict consideration to work where GAs have been characterized by GC–MS, the only unequivocal method of identification currently available for these compounds (Hedden, 1993). Gibberellins that have been identified in fruit tissues by GC–MS are listed in Table 14.1. Most of these are present in developing seeds. Seeds of the Rosaceae contain a particularly large variety of structures, with often multiple hydroxylation in rings A, C and D (see Fig. 14.1). For example, apple seeds contain more than 30 GAs, including compounds with hydroxyl groups at one or more of the 1, 3, 11, 13 and 15 positions (Hedden *et al.*, 1993). 15β-Hydroxylated GAs are also found in pear (Martín *et al.*, 1977), whereas *Prunus* species produce GAs that are hydroxylated at the 12α- and 15β positions, as well as at the more usual 3β- and 13 positions (Yamaguchi and Takahashi, 1976; Bottini *et al.*, 1985; Blake *et al.*, 1993).

There is evidence for two phases of GA production during seed development. In apple, high concentrations of biologically active GAs (GA_1, GA_3, GA_4 and GA_7) are present in seeds at about 20 days after anthesis, when the growth-rate of the fruit is at a maximum (Hedden and Hoad, 1985). A second phase of GA production corresponds to the development of the endosperm and the embryo. It is only at this later stage that the large variety of GAs is apparent. The two phases of GA biosynthesis have also been noted in developing pea seeds. Gibberellin A_1, GA_3 and GA_8 (3-hydroxylated GAs) were identified in both pea seeds and pods harvested 4 and 5 days after anthesis (García-Martínez *et al.*, 1987, 1991*a*). The concentrations of GA_1 and GA_3 were maximal at the time of highest pod growth-rate (4–7 days after anthesis), but present in only trace amounts or undetectable in

Table 14.1. Gibberellins identified in fruits by GC–MS (conjugates have not been included)

Species	Tissue	Gibberellin	Reference
Brassica napus cv Westar	Young fruits	A_1, A_3, A_4, A_8, A_9, A_{17}, A_{19}, A_{20}, A_{24}, A_{29}, A_{34}, A_{51}, A_{53}	Zanewich and Rood, 1993
Citrus reticulata cv Oroval	Ovaries (anthesis)	A_3, A_8, A_{29}	Talón *et al.*, 1990*b*
	Young fruits	A_1, 3-*epi*-A_1, A_4, A_8, A_9, A_{17}, A_{19}, A_{20}, A_{24}, A_{29}, A_{44}, A_{53}, 16,17-H_2, 17-OH A_{53}	Talón *et al.*, 1992
Citrus sinensis cv Valencia	Young fruits	A_3, *iso*-A_3, A_8	Turnbull 1989
	Seeds	A_1, *epi*-A_1, A_{19}, A_{20}, A_{29}, *epi*-A_{29}	
cvs Blanca Comuna and Salustiana	Young fruits	A_1, A_4 (put.), A_{20}, A_{29}	Talón *et al.*, 1990*a*
Citrus unshiu	Ovaries (anthesis)	A_1, A_3, *iso*-A_3, A_8, A_{19}, A_{20}, A_{29}	Talón *et al.*, 1990*b*
cv Owari	Young fruits	A_1, 3-*epi*-A_1, A_4, A_8, A_9, A_{17}, A_{19}, A_{20}, A_{24}, A_{29}, A_{44}, A_{53}, 16,17-H_2, 17-OH A_{53}	Talón *et al.*, 1992
Corylus avellana	Mature seeds (embryo)	A_1, A_9	Williams *et al.*, 1974
Cucurbita maxima cv Gelbe Riesenmelone mit Vernetzter Haut	Immature seeds	A_4, 12-OH A_4, A_{12}, 12-OH A_{12}, A_{13}, 12-OH A_{14}, A_{25}, 12-OH A_{37}, A_{39}, A_{43}, A_{49}	Blechschmidt *et al.*, 1984
	Endosperm	A_1, A_8, A_{12}, A_{13}, A_{20}, A_{24}, A_{25}, A_{28}, A_{36}, 12-OH A_{37}, A_{43}, 12-OH A_{43}, iso-12-OH A_{43}, A_{48}, A_{49}, A_{58}, A_{74}	Lange *et al.*, 1993*a*

Species	Tissue	Gibberellin	Reference
	Embryo	A_1, A_4, A_{13}, A_{17}, A_{23}, A_{25}, 12-OH A_{25}, A_{28}, 2β-OH A_{28} (put.), A_{37}, *epi*-A_{37}, *endo*-A_{37}, 12-OH A_{37}, A_{38}, A_{39}, A_{39} 3-isovalerate, A_{43}, 12-OH A_{43}, A_{48}, A_{49}, A_{58}, A_{74}, A_{85}	Lange *et al.*, 1993*b*
Cytisus laburnum	Embryo	A_1	Picciarelli *et al.*, 1991
	Suspensor A_1		
	Endosperm	A_1, 16,17-H_2, 17-OH A_4, A_8, A_{23}	
Eriobotrya japonica cv Mogi	Immature seeds	A_9, A_{15}, A_{24}, A_{35}, A_{50}, A_{80}, A_{84}	Yuda *et al.*, 1992
Helianthus annuus	Seeds	A_1, A_4, A_{19}, A_{20}, A_{45}, A_{64}, A_{65}, A_{66}, A_{67}, A_{72}, 3-epi-A_{72}, A_{75}, A_{76}, A_{100}, A_{101}, A_{102}	Castellaro *et al.*, 1990; Hutchison *et al.*, 1988; Owen *et al.*, 1996
Lycopersicon esculentum	Seeds	A_1, A_8, A_{15}, A_{17}, A_{19}, A_{20}, A_{24}, A_{25}, A_{29}, A_{44}	Bohner *et al.*, 1988
	Pericarp	A_1, A_8, A_{17}, A_{19}, A_{20}, A_{29}	
	Young fruits	A_1, A_8, A_9, A_{15}, A_{17}, A_{19}, A_{20}, A_{24}, A_{25}, A_{29}, A_{44}, 2ß-OH A_{53}	Koshioka *et al.*, 1994
Malus domestica cvs Cox's Orange Pippin, Dabinett, Sunset and Tremlett's Bitter	Immature seeds	A_1, A_3, A_4, 16,17-$H_2(OH)_2$ A_4, A_7, 16,17-$H_2(OH)_2$ A_7, A_9, 16,17-$H_2(OH)_2$ A_9, A_{12}, 16,17-$H_2(OH)_2$ A_{12}, 16β,17-$H_2(OH)_2$ A_{12}, A_{15}, A_{17}, A_{19}, A_{20}, A_{25}, A_{34}, A_{35}, A_{44}, A_{45}, A_{53}, A_{54}, 3-*epi*-A_{54}, A_{61}, A_{62}, A_{63}, A_{68}, A_{80}, A_{84}, A_{88}, A_{103}, A_{104}, A_{105}, A_{106}, A_{107}, A_{108}	Hedden *et al.*, 1993; Luckwill *et al.*, 1969; Oyama *et al.*, 1996
Marah macrocarpus	Endosperm	A_3, A_4, A_4 diacid, A_7, iso-A_7, A_7 diacid, A_{24}, A_{25}, A_{37}, A_{43}, A_{46}, A_{62}	Beeley *et al.*, 1975; MacMillan and Gaskin 1996
	Embryo	A_3, A_4, A_4 diacid, A_7, iso-A_7, A_{13}, A_{25}, A_{43}, A_{46}	

Table 14.1. *contd*

Species	Tissue	Gibberellin	Reference
Phaseolus coccineus	Embryo	A_1, A_4, A_5, A_6, A_{17}, A_{19}, A_{20}, A_{38}, A_{44}, A_{60}	Albone *et al.*, 1984; Piaggesi *et al.*, 1989
	Testa	A_1, A_5, A_6, A_8, A_8-cat., A_{17}, A_{19}, A_{20}, A_{29}, A_{38}, A_{44}	Albone *et al.*, 1984
	Suspensor	A_1, A_4, A_5, A_6, A_{44}	Piaggesi *et al.*, 1989
Phaseolus vulgaris cv Kentucky Wonder	Embryo	A_1, A_4, A_5, A_6, A_8, A_9, A_{12}, A_{15}, A_{17}, A_{19}, A_{20}, A_{23}, A_{24}, A_{29}, A_{37}, A_{38}, A_{44}, A_{53}	Takahashi *et al.*, 1986
	Testa	A_1, A_4, A_6, A_8, A_9 (put.), A_{15}, A_{17}, A_{19}, A_{20}, A_{23}, A_{24}, A_{29}, A_{37}, A_{38}, A_{44}, A_{53}	
cv Canadian Wonder	Immature seeds	A_1, 16,17-H_2, 17-OH A_1, A_4, A_5, A_6, A_8, A_8-cat., 16,17-H_2, 17-OH A_8, A_{17}, A_{19}, A_{20}, A_{20}-open lact., A_{29}, A_{34}, A_{38}	Albone *et al.*, 1989
	Cotyledon	A_1, A_3, iso-A_3, A_4, A_5, A_6, A_8, A_{17}, A_{20}, A_{20}-open lact., A_{38}	
Pisum sativum cv Progress no. 9	Immature seeds	A_9, A_{17}, A_{20}, A_{29}, A_{29}-cat., 9-OH A_{29}-cat., A_{38}, A_{44}, 2β-OH A_{44}, A_{51}	Frydman *et al.*, 1974; Gaskin *et al.*, 1985
cv Alaska	Immature seeds	A_{17}, A_{19}, A_{20}, A_{29}, A_{29}-cat., 9-OH A_{29}-cat., A_{44}, 2β-OH A_{44}	Gaskin *et al.*, 1985
	Young fruits:		García-Martínez *et al.*, 1991*a*
	Seeds	A_1, A_3, A_8, A_{19}, A_{20}, A_{29}	
	Pods	A_1, A_3, A_8, A_{17}, A_{19}, A_{20}, A_{29}, A_{29}-cat., A_{81}	

Species	Tissue	Gibberellin	Reference
	Young fruits	A_1, A_1 1,10-ene diacid, A_3, *iso*-A_3, 16, 17-$H_2(OH)_2$ A_4, *iso*-A_7 open lact., 16, 17-$H_2(OH)_2$ A_7 (put.), A_8, A_8-cat., 16,17-$H_2(OH)_2$ A_9, 16, 17-$H_2(OH)_2$ A_{12}, A_{13}, 11-OH A_{14} (put.), A_{17}, A_{19}, A_{20}, A_{29}, A_{29}-cat., 16, 17-$H_2(OH)_2$ A_{34}, A_{43}, 12-OH A_{53} (put.), A_{60}, A_{81}, A_{91}	García-Martínez *et al.*, 1987; Santes *et al.*, 1995
Prunus armeniaca	Immature seeds	A_1, A_5, A_{29}, A_{32}	Bottini *et al.*, 1985
Prunus avium cv Stella	Young fruits	A_1, A_3, A_5, A_{19}, A_{20}, A_{29}, A_{32}, A_{85}, A_{86}, A_{87}	Blake and Browning 1994
Prunus cerasus	Immature seeds	A_{32}	Bukovac *et al.*, 1979
Prunus domestica	Young fruits	A_{29}	Reed and Martin 1976
Prunus persica	Immature seeds	A_5, A_{32}, A_{86}	Yamaguchi *et al.*, 1975; Bhaskar *et al.*, 1991
Pyrus communis cvs Barlett and Winter Nelis	Ovules and immature seeds	A_{17}, A_{25}, A_{45}, 3β-OH A_{45} (put.)	Martín *et al.*, 1977
Sechium edule	Embryo	A_1, A_3, A_4, A_7, A_8, A_8-cat., 16,17-$H_2(OH)_2$ A_{12}, A_{17}, A_{25}, A_{29}, A_{38}, A_{44}	Albone *et al.*, 1984
	Endosperm	A_1, A_3, A_4, A_7, 16,17-$H_2(OH)_2$ A_{12}-ald., 16,17-$H_2(OH)_2$ A_{12}, A_{13}, A_{17}, A_{25}, A_{38}, A_{44}	
	Testa	A_1, A_3, A_7, A_8, A_8-cat., A_{12}, 16,17-$H_2(OH)_2$ A_{12}, A_{13}, A_{15}, A_{19}, A_{25}, A_{27}, A_{29}, A_{44}, A_{53}	
Tropaeolum majus	Embryo-suspensor	A_{63}	Picciarelli and Alpi, 1987

ald, aldehyde; cat., catabolite; lact., lactone; put., putative; 16,17-H_2, 17-OH, 16,17-dihydro-17-hydroxy; 16,17-$H_2(OH)_2$, 16,17-dihydro-16,17-dihydroxy.

unpollinated ovaries and in parthenocarpic fruits growing in the absence of pollination (García-Martínez *et al.*, 1991*a*). There is a massive accumulation of GA_{20} and GA_{29} in seeds at about 22 and 27 days after anthesis, respectively (Frydman *et al.*, 1974), but no 3-hydroxylated GAs are detectable in the seeds at this stage. Gibberellin A_{20} and GA_{29} are synthesized mainly in the cotyledons; the latter GA is oxidized to GA_{29}-catabolite in the testa, from where it is transported to the cotyledons prior to seed desiccation (Sponsel, 1983). Since related metabolites (GA_8 and GA_8-catabolite) accumulate in the testa of immature seeds of *Phaseolus coccineus* and *Sechium edule* (Albone *et al.*, 1984) successive oxidation on C-2 could be a general characteristic of seed development.

The early phase of GA production in seeds that results in the production of biologically active compounds is associated with the growth of the fruit and, as will be discussed in more detail later, may provide an important stimulus for fruit development. Indeed, in developing pea fruit, there are much higher concentrations of GAs in the seeds than in the pod (in the case of GA_3, about 80% of the total content of the fruit was in the seeds) (García-Martínez *et al.*, 1991*a*). However, in the *le* mutant, a dwarf plant in which there is a partial block in the 3β-hydroxylation of GA_{20} to GA_1 in the shoot (Ingram *et al.*, 1984), the growth of the fruit is not reduced even though the mutation is also apparently expressed in pods and seeds. It is possible that there is sufficient GA_1 or GA_3, the concentration of which is not altered significantly by the mutation, in the mutant seeds to induce normal pod growth (Santes *et al.*, 1993). It appears likely that GA_1 and/or GA_3 are the active GAs that induce fruit-set and pericarp growth in pea, and that the developing young seeds are an important source of these GAs. In order to obtain more information about the nature of the compounds controlling fruit development in pea, extracts of 4–6 day old pollinated pea fruits were investigated for activity in a fruit-set bioassay (Santes *et al.*, 1995). HPLC fractions containing GA_1 and GA_3 showed, as expected, the highest fruit-set activity. There were, however, other active fractions containing as yet unidentified factors. A total of 28 GAs and related compounds were detected in the extracts, including a series of non-13-hydroxylated GAs and the dihydrodiols of GA_4, GA_7, GA_9 and GA_{12}. The large amounts of GA_7 16, 17-dihydrodiol present suggest that GA_7 synthesis is a major pathway in young pea fruit and that GA_3 may be formed from GA_7 by 13-hydroxylation (Santes *et al.*, 1995).

The suspensor, which is a relatively large organ in legume seeds, has been shown to be extremely rich in GAs (Piaggesi *et al.*, 1989; Picciarelli *et al.*, 1991). In *Phaseolus coccineus* suspensors, the high concentrations of GAs at the heart stage support the idea that this organ supplies GAs to the embryo, which would acquire progressively, upon development, the capacity for GA biosynthesis (Cioni *et al.*, 1976). However, in *Cytisus laburnum*, the concentration of GA_1 in the suspensor, embryo and endosperm was not significantly different at the stage investigated (Picciarelli *et al.*, 1991).

Gibberellins have been identified in pericarp of developing fruit of several other species in addition to pea. Bohner *et al.* (1988) and Koshioka *et al.* (1994) identified a series of 13-hydroxylated and 13-deoxy GAs in both seeds and pericarp of young developing tomato fruit. The GAs could not be detected in ovaries from emasculated flowers, whereas the content of GA_{19} increased rapidly in developing ovaries following the day equivalent to anthesis, both in pollinated and auxin (4-chlorophenoxyacetic acid, CPA)-induced parthenocarpic fruits (Koshioka *et al.*, 1994). A positive correlation between GA_1 content and fruit size of three mutants varying in their cellular size was observed in *Lycopersicon pimpinellifolium*, but not between tomato (*L. esculentum*) fruit of different sizes within a truss (Bohner *et al.*, 1988). The GA content of tomato pericarp declines rapidly following on the onset of ripening (Fraser *et al.*, 1995).

Immature orange fruits of cvs Blanca Comuna (seeded) and Salustiana (seedless) contain GA_1, GA_8, GA_{20} and GA_{29} (Talón *et al.*, 1990*a*). In both cultivars, GA_1 content is maximal at anthesis, when GA_{20} content is relatively low. 13-Hydroxylated GAs (e.g. GA_1) and 13-deoxy GAs (e.g. GA_4 and GA_9) were also identified in fruit of the seedless mandarin cvs Satsuma and Clementine Oroval (Talón *et al.*, 1990*b*, 1992). Amounts of all 13-hydroxylated GAs, except that of GA_8, were higher in Satsuma than in Clementine Oroval, associated with a greater capacity of the former to set parthenocarpically (Talón *et al.*, 1992). In these cases also, the maximum GA content was at anthesis. Since no quantitative differences between cultivars in GA_4 and GA_9 contents were detected, it was suggested that GA_1 may control fruit set and development in Citrus. It is tempting to suggest further that natural parthenocarpy depends on the capacity of fruit to produce sufficient amounts of GAs and other hormones in the absence of fertilized seeds.

Gibberellin metabolism

Much of our knowledge of GA biosynthesis in higher plants comes from work with immature seeds, in which GA concentrations are much higher than in other plant tissues. Cell-free systems from endosperm/nucellus or developing embryos have been essential tools for characterizing the enzymes involved. In particular, endosperm of *Cucurbita maxima* (pumpkin) contains sufficiently high activity to have allowed the purification and cDNA cloning of two of the biosynthetic enzymes (Kamiya *et al.*, 1995; Lange 1994; Lange *et al.*, 1994).

Gibberellin biosynthesis has been reviewed extensively (Graebe, 1987; Sponsel, 1995) and, with the exception of the most recent advances, will be covered only in broad outline here. Pathways leading to the biologically active compounds, GA_1, GA_3 and GA_4, from the common diterpene precursor, geranylgeranyl diphosphate (GGDP), are depicted in Fig. 14.1, which shows also the 2β-hydroxylation steps, from GA_1, GA_4 and GA_{20}, that lead to the production of biologically inactive metabolites. The pathway can be divided into three stages. The first leads to the formation of *ent*-kaurene from mevalonic acid via GGDP, the reactions being catalysed by soluble enzymes, which are localized within plastids (Aach *et al.*, 1995). GGDP is cyclized to *ent*-kaurene in two steps by the A and B activities of *ent*-kaurene synthase. The A activity, which produces the bicyclic intermediate, copalyl diphosphate (CDP), was cloned from *Arabidopsis thaliana* by subtractive hybridization (Sun *et al.*, 1992; Sun and Kamiya, 1994) and from *Zea mays* by transposon tagging (Bensen *et al.*, 1995). The B activity, which converts CDP to *ent*-kaurene, has been cloned recently from *C. maxima* endosperm by PCR, following purification and partial sequencing of the enzyme (Kamiya *et al.*, 1995). cDNA clones for both enzymes encode proteins with an N-terminal sequence that targets them to plastids. The targeting sequence is cleaved on transport of the enzyme into the organelle (Sun and Kamiya, 1994).

GGDP is the common precursor of all diterpenoids. As well as being converted to GAs, it is also an intermediate in the formation of carotenoids and chlorophyll, which are major metabolites in plastids at certain stages of development. For example, during ripening of many fruit, chlorophyll-rich chloroplasts are converted to chromoplasts with an associated massive increase in carotenoid production. It is probable, although not yet shown unequivocally, that each of the various pathways utilize the same plastidic pool of GGDP and would, therefore, be competing directly for this precursor. Carotenoid biosynthesis is initiated by the conversion of GGDP to phytoene, catalysed by phytoene synthase. In ripening fruit of transgenic tomato producing antisense mRNA for a ripening-specific phytoene synthase gene (*Psy1*), carotenoid production is severely reduced (Bramley *et al.*, 1992). This results in delayed ripening and also in increased concentrations of GAs in the fruit, presumably because there is more GGDP available for GA biosynthesis (Fraser *et al.*, 1995). Constitutive overexpression of *Psy1* can cause dwarfism associated with a low GA_1 content in tomato shoots, providing further evidence that the carotenoid and GA pathways compete for GGDP (Fray *et al.*, 1995).

In the second stage of GA biosynthesis, *ent*-kaurene is converted to GA_{12} by membrane-associated mono-oxygenases, probably all belonging to the cytochrome P450 group of enzymes. In developing seeds, 13-hydroxylation of GA_{12} to GA_{53} is also catalysed by a mono-oxygenase, although in spinach leaves this activity is due to a soluble dioxygenase (Gilmour *et al.*, 1986). The mono-oxygenases are present in the endoplasmic reticulum in pumpkin endosperm and developing pea cotyledons (Graebe 1982) and it is, therefore, necessary for the highly hydrophobic *ent*-kaurene to be transported from the plastid for further metabolism. Whether or not this transport is mediated by a specifc protein carrier, as was suggested by Moore *et al.* (1972), remains to be investigated. The *D3* gene of *Zea mays*, which is associated with dwarfism and probably encodes a GA-biosynthetic enzyme (Phinney, 1984), has been cloned by transposon tagging and found to encode a cytochrome P450-dependent mono-oxygenase (Winkler and Helentjaris, 1995). However, the precise step catalysed by the *D3* product is unknown.

The third and final stage of the pathway is catalysed by soluble dioxygenases for which 2-oxoglutarate is a cosubstrate (Hedden and Graebe, 1982). These enzymes, which require ferrous iron and ascorbate for maximum activity, belong to a rapidly-expanding class, members of which are involved in numerous biosynthetic pathways in plants and fungi (Prescott, 1993). Production of GA_1 from GA_{53} requires oxidation at C-20, resulting in its loss as carbon dioxide (Kamiya *et al.*, 1986) with the formation of the C_{19} skeleton, followed by 3β-hydroxylation (see Fig. 14.1). The 2β-hydroxylases also belong to this group of enzymes (Smith and MacMillan, 1984, 1986; Griggs *et al.*, 1991). Oxidation of C-20, although a multistep process, appears to be catalysed by a single enzyme. A GA 20-oxidase that produces predominantly inactive tricarboxylic acids, was cloned following its purification to homogeneity from pumpkin endosperm (Lange, 1994; Lange *et al.*, 1994). Expression of the cDNA clone as a fusion protein in *E. coli* produced a functionally active enzyme, which converted GA_{12} to GA_{25}, with the C_{19} GA, GA_9, also formed, but in very low yield.

Gibberellin 20-oxidases have now been cloned from several other species, including *Arabidopsis thaliana* (Phillips *et al.*, 1995; Xu *et al.*, 1995), *Pisum sativum* and *Phaseolus vulgaris* (García-Martínez *et al.*, 1995), by the use of PCR. Unlike the pumpkin enzyme, these other 20-oxidases produce C_{19} GAs as major products. Three different GA 20-oxidase cDNA clones were isolated from *Arabidopsis*. They were shown to encode functionally identical enzymes, but correspond to genes that are expressed in different tissues: stems, flowers and developing siliquae (seed pods). The level of expression of the last gene was relatively high and it is probably expressed in the immature seeds. Tissue specificity of expression was noted also for GA 20-oxidase genes in *P. vulgaris*; one is expressed in leaves and a second in developing embryos (García-Martínez *et al.*, 1995). There is genetic evidence for other GA-biosynthetic enzymes to suggest that different genes are expressed in developing seeds and vegetative tissues. The *Na* gene of pea is thought to encode an enzyme that catalyses an intermediate reaction in the pathway, possibly GA_{12}-aldehyde synthase (Ingram and Reid 1987). Mutation of this gene produces severe dwarfism, but fruit growth is unaffected and the seeds have normal levels of GA-like substances (Potts and Reid 1983). On the other hand, the *GA4* gene of *Arabidopsis,* which was cloned recently by T-DNA tagging (Chiang *et al.,* 1995) and has been shown to encode a GA 3β-hydroxylase (J. Williams, A.L. Phillips and P. Hedden, unpublished), is expressed in both stems and siliquae. Although mutation of *GA4* leads to dwarfism, highest levels of expression are in the siliquae (Chiang *et al.,* 1995).

Gibberellin biosynthesis in developing seeds results, in many species, in the production of numerous different structures by oxidative reactions not included in the main pathway. Examples of oxidation at C-1, C-9, C-11, C-12 and C-15 are known, often producing polyhydroxylated compounds (see Table 14.1). Little is known about the enzymes that introduce these hydroxyl groups, except in pumpkin endosperm, which has been shown to contain both microsomal (mono-oxygenase) and soluble (dioxygenase) 12α-hydroxylases, which act, respectively, at intermediate and late stages of the pathway (Hedden *et al.,* 1984; Lange *et al.,* 1993*a*). Further structural diversity may arise in seeds from 2β- and 3β-hydroxylation of C_{20} GAs, which are not normally substrates for these hydroxylases in other tissues. Whether these reactions are due to the presence, in seeds, of enzymes with unusually broad substrate specificities or are the result of the very high concentrations of C_{20} GAs, remains to be determined. The function of these seed-derived GAs is unclear. However, deactivation of GAs by 2β-hydroxylation as seeds approach maturity (Albone *et al.,* 1989; Sponsel, 1983) has an important function, as is demonstrated by the *sln* overgrowth mutant of pea (Ross *et al.,* 1995). Although the phenotype of *sln* is characterized by excessive growth of the seedling, particularly in the first few weeks after germination, the gene appears to be expressed mainly in the developing seed. Pea embryos accumulate large amounts of GA_{20} during development and, shortly before maturity, this GA is normally converted by 2β-hydroxylation to GA_{29}, which is further oxidized in the testa to the so-called GA_{29}-catabolite (Sponsel, 1983). The *sln* mutation results in the loss of both reactions such that GA_{20} remains at high levels in the mature seed. On germination, GA_{20} is 3β-hydroxylated to the biologically active GA_1 resulting in an abnormally high growth rate (Ross *et al.,* 1993).

Within seeds, several tissues are capable of at least some steps of GA - biosynthesis. As discussed above, there are often very high rates of GA production in endosperm/nucellus and in the expanding cotyledons. The testa is metabolically active and capable of GA catabolism as seeds approach maturity. The testa may also contribute to GA biosynthesis in young pea ovules (M. J. Rodrigo, J. L. García-Martínez and P. Hedden, unpublished results). In young legume seeds, the suspensor is extremely active for GA-biosynthesis; a cell-free system from *Phaseolus coccineus* suspensors has been shown to contain all the enzyme activities necessary for the conversion of mevalonic acid to GA_1 (Ceccarelli *et al.,* 1979, 1981*a*,*b*). Fruit pericarp may be an important source of GA for developing fruit. The capacity to convert

GA_{12} aldehyde to C_{20} GAs, including GA_1, was shown to be present in young pea pods (Maki and Brenner, 1991). There is evidence that conversion of GA_{19} to GA_{20} in pea pericarp is dependent on the presence of seeds (Ozga *et al.*, 1992). In which case, the seeds, rather than providing GA to promote fruit expansion, may stimulate the pericarp to synthesize biologically active GAs for its own growth. The question of the source of the GAs involved in fruit development, which is discussed further in the following section, is clearly unresolved.

Function of gibberellins in fruit set and development

For simplicity, the early development of the ovary can be divided into three stages (Gillaspy *et al.*, 1993). The first phase is associated with flower development and precedes fruit-set, which is induced by signals arising as a result of fertilization. During this phase, growth is by cell division. In the second phase, cell division in the ovary may resume following fruit-set. This is followed, in the third stage, by cell expansion, which normally accounts for most of the increase in fruit size. There is considerable variation in this developmental process among species, particularly in the relative duration of each stage. For example, in fruit of some species, cell division ceases at anthesis, while in avocado, cell division continues almost until ripening and accounts for almost all of fruit growth (Schroeder, 1953). Gibberellins are known to induce fruit-set and to stimulate cell expansion in fruit. They are probably also involved in the development of the seed.

There is considerable evidence to suggest that seed-derived GAs are necessary for normal fruit development in many species. In peas, applied GAs can reverse completely the absence of growth in pollinated ovaries when the seeds are destroyed by pricking shortly after fertilization (Eeuwens and Schwabe, 1975; Sponsel, 1982; García-Martínez *et al.*, 1991*b*). Ovaries from emasculated flowers develop parthenocarpically following a single application of GAs (GA_1, GA_3, GA_{20}) (García-Martínez and Carbonell, 1980; García-Martínez *et al.*, 1987). Gibberellins are effective until about 3 days after the day equivalent to anthesis, then the unpollinated ovary becomes insensitive and degenerates. Though auxins and cytokinins can also induce parthenocarpic development in pea, GAs are the only hormones to produce fruits morphologically similar to those obtained by pollination (García-Martínez and Carbonell, 1980; Carbonell and García-Martínez, 1985). Different inhibitors of GA-biosynthesis (AMO-1618, ancymidol and triazol and acylcyclohexanedione derivatives) reduce the growth of pollinated pea ovaries, the inhibition being counteracted by applied GA_3 (García-Martínez *et al.*, 1987; Santes and García-Martínez, 1995). Furthermore, the GA-response mutant of pea, *la cry*^{c}, which has a slender phenotype and behaves as if saturated with GA, produces many male-sterile flowers that develop parthenocarpically (Potts *et al.*, 1985), in agreement with the proposal that GAs control fruit set and development in pea.

Genetic evidence from *Arabidopsis* and tomato also indicate an important role for seed-derived GAs in fruit development. Results from crossing wild-type *Arabidopsis* plants with the *ga1* GA-deficient mutant, which lacks *ent*-kaurene synthase activity (Sun and Kamiya, 1994), indicate that the growth of the siliqua depends on GAs from both maternal and embryonic tissues (Barendse *et al.*, 1986). Flowers from the *ga-1* mutant of tomato, which is also blocked at *ent*-kaurene synthase and has a low GA content, produce very few fruit, some of them parthenocarpically, with final size proportional to the number of seeds (*ga-1*/*ga-1*) from a minimum of about 35 seeds per fruit (Groot *et al.*, 1987). Fertilization with pollen from *Ga-1* plants produces seeds (*Ga-1*/*ga-1*) that are much more active in stimulating ovary growth, such that fruit size is proportional to seed number from 10 rather than 35. It would be of interest to know whether the seed GA content is actually higher in *Ga-1*/*ga-1* than in *ga-1*/*ga-1*. Decapitation of pea plants induces parthenocarpic development of unpollinated ovaries (Carbonell and García-Martínez, 1980). There is evidence that this effect is due, at least partially, to accumulation in the ovary of GAs, transported normally to the growing shoot, from mature leaves, which become an alternative source of GAs in the absence of fertilization (Peretó *et al.*, 1988; García-Martínez *et al.*, 1991*a*).

The pea has proved a valuable experimental system for examining the mechanism of GA-induced fruit growth. The pea ovary wall is composed of the three layers exocarp, mesocarp and endocarp; the pericarp of the fruit is produced mainly by enlargement of mesocarp cells following pollination or GA_3 treatment (Vercher *et al.*, 1984). Unpollinated ovaries respond to GAs until about 3 days after anthesis, entering a fruit development pathway as a result of hormone application. Alternatively, in the absence of a stimulus the ovary enters a pathway of senescence, leading to the loss of sensitivity to applied GAs (García-Martínez and Carbonell, 1980), the appearence of new proteolytic activity which degrades ribulosediphosphate carboxylase/oxygenase (rubisco) (Carbonell and García-Martínez, 1985) and early degeneration of the endocarp (Vercher *et al.*, 1984, 1987) and mesocarp cells (Vercher and Carbonell, 1991). Whereas the induction of fruit-set and development does not alter appreciably the patterns of newly synthesized proteins or translatable RNAs, indicating that fruit development is apparently a natural continuation of ovary formation, ovary senescence involves significant changes in gene expression (Sánchez-Beltrán *et al.*, 1992). The thiol-protease that degrades rubisco has been purified and immunolocalized in the tonoplast and dense materials of vacuoles from exocarp and external mesocarp layers of senescent ovaries (Cercós *et al.*, 1993). This activity increases rapidly from two days after anthesis, but its formation is inhibited by GA_3 treatment

(Cercós and Carbonell, 1993). Another thiol-protease, with unknown substrate, has been cloned from senescent ovaries (Granell *et al.*, 1992). It was shown by *in situ* localization to be expressed in the endocarp of unpollinated ovaries before appearence of the proteolytic activity, and in the vascular bundles and ovules. The protease is down-regulated by GA_3 and may be responsible for the proteolytic activity detected using gelatin as a substrate, and which is also localized, histochemically, in the endocarp of senescent ovaries (Vercher *et al.*, 1989). A cDNA clone encoding a lipoxygenase of unknown function in fruit development has also been obtained from unpollinated pea ovaries (Rodríguez-Concepción and Beltrán, 1995). The expression of this gene is localized in the endocarp and mesocarp of the pod, and repressed by fertilization and GA_3 treatment. It would appear that an important function of GA in preventing fruit senescence is to repress the induction of degradative enzymes.

During the rapid, cell-expansion phase of growth, fruit are strong sinks for nutrients. Application of GA_3 to fruit of grape (Weaver *et al.*, 1969) or calamondin (Powell and Krezdorn, 1977) stimulated fruit growth and also the accumulation in the fruit of leaf-applied [^{14}C] carbon dioxide metabolites. In the case of seedless Clementine mandarin, the total contents of nitrogen, phosphorus and potassium in the leaves from leafy inflorescences treated with GA_3 decreased, while those in the fruits increased, relative to untreated inflorescences, within 5–16 days after treatment (García-Martínez and García-Papí, 1979*a*). In strawberry, the application of a mixture of GA_3, IAA and kinetin to one side of the fruit attracted more [^{14}C]sucrose and ^{32}P-labelled metabolites from the leaves to the side of hormone application (Lis and Antoszewski, 1982). In the absence of appropriate time-course studies, however, it is difficult to decide whether increased movement of nutrients to the fruit is a primary effect of GA-action or is rather a consequence of GA-induced fruit growth.

In experiments with pea plants, Jahnke *et al.* (1989) showed by monitoring simultaneously the transport of [^{11}C] carbon dioxide metabolites from the leaf and the growth of the ovary with voltage transducers, that application of GA_3 or pollination increased the sink capacity of the ovary and decreased that of the shoot, within 2 h, before any increase in the growth rate of the ovary could be detected . When [^{14}C]sucrose was applied to the leaves for several hours before application of GA_3 to unpollinated ovaries, it was possible to detect an increase in transport of sucrose into the ovary within 30 min after GA-application (Peretó and Beltrán, 1987). Since discs of GA_3-treated ovaries did not take up more [^{14}C]sucrose *in vitro* than discs from non-treated ovaries, it was concluded that the hormone affects the unloading of sucrose in the ovary. In fact, the degradation of sucrose in the pea ovary is controlled by invertases, rather than by sucrose synthase or sucrose phosphate synthase, and GA_3 increases the activity of soluble neutral invertase within 2 h after treatment; at the same time there is a decrease in soluble acid invertase and an increase in apoplast acid invertase activities (Estruch and Beltrán 1991*a*). Interestingly, protoplasts from pea mesocarp cells release acid invertase into the incubation medium, a process enhanced by GA_3 an hour after its application to the medium (Estruch and Beltrán, 1991*b*). It was found that GA_3 stimulates the transport of acid invertase from the endoplasmic reticulum to the Golgi apparatus (Estruch and Beltrán, 1991*b*). The results suggest that GA_3 stimulates the growth of pea pericarp by enhancing the amount of cellular neutral invertase, leading to increased sucrose metabolism, and of apoplastic acid invertase, which results in increased sucrose unloading into the apoplast (Beltrán *et al.*, 1991).

Evidence that GAs are involved in seed development has been difficult to obtain. Application of inhibitors of GA biosynthesis reduces the GA content of seeds without affecting seed growth or germination capability (Baldev *et al.*, 1965; Zeevaart, 1966; García-Martínez *et al.*, 1987; Santes *et al.*, 1995). Furthermore, the weight of *Arabidopsis* seeds is not affected in the *ga-1* mutant (Barendse *et al.*, 1986), although the effect of the mutation on GA-content was not measured. However, the lh^i mutation in pea, in which *ent*-kaurene oxidase activity is reduced, produces dwarf plants and a reduction in the number and size of seeds (Swain *et al.*, 1993). This mutation, contrary to other mutations in pea that affect GA-biosynthesis, such as *le*, *na* and *ls* is expressed also in the seeds (Swain *et al.*, 1995). The reduced growth and increased abortion of lh^ilh^i seeds, compared with Lh^iLh^i and Lh^ilh^i seeds, was associated with their lower content of active GAs (GA_1 and GA_3) during the days following fertilization (Swain *et al.*, 1995). This indicates that GAs may be important for the early development of the seed and that the relatively high amounts of GAs accumulating later, such as GA_{20} and GA_{29} in pea (Frydman *et al.*, 1974; García-Martínez *et al.*, 1987), are irrelevant both for subsequent seed growth and fruit development. The inability of GA-biosynthesis inhibitors to block seed development in pea and the lack of an effect of applied GAs on seed growth in the lh^i mutant might be due to poor uptake of these chemicals into the seeds at the appropriate time.

Practical application of gibberellins in fruit production

Stimulation of fruit growth, particularly of unseeded fruit, may be of considerable benefit in horticulture. Many fruit respond to GAs, some of which are also sensitive to auxins, but others, particularly those containing only one or few ovules, do not respond to auxin treatments (reviewed by Crane, 1964, 1969; Goodwin, 1978). Thus, GAs, but not auxins induce fruit development in almond (*Prunus amygdalus*), peach (*P. persica*), apricot (*P. armeniaca*), cherry (*P. avium*), plum (*P. domestica*), blueberry (*Vaccinium angustifolium*), cranberry (*V. macrocarpon*), mandarin (*Citrus reticulata*), Washington Navel orange (*C. sinensis*), lime (*C. aurantifolia*), and lemon Eureka (*C. lemon*). Gibberellins are also capable of stimulating fruit development, at least as efficiently as auxins in the following cases: pear (*Pyrus communis*), apple (*Malus domestica*), tomato (*Lycopersicon esculentum*), fig (*Ficus carica*), currant (*Ribes* species), cucumber (*Cucumis sativus*), eggplant (*Solanum melongena*), pepper (*Capsicum* species), and strawberry (*Fragaria* species). The largest commercial use of GA worldwide is in the production of seedless grapes (*Vitis vinifera*), particularly of Thomson's Seedless. The effect of GAs on grape is complicated by the fact that they decrease pollen viability while also stimulating berry growth in the absence of seeds. In general, if applied at early bloom, GAs elongate the rachis of the berries, decrease fruit set, and increase the size of the remaining fruit, creating looser clusters less susceptible to disease during the growing season (Gianfagna, 1995).

Webster and Goldwin (1981) found that the effect of GA_3 on both yield and fruit size in plum cv Victoria is enhanced significantly if the GA is applied together with an auxin [2,4,5-trichlorophenoxypropionic acid (2,4,5-TP) better than 2-naphthoxyacetic acid (NOXA)]. In pear cv Winter Nelis, GA_3 stimulates development of unpollinated ovaries very effectively (Martín *et al.*, 1982). Since it had been suggested that GA_{45} (15β-OH GA_9) is the active GA in pear (Martín *et al.*, 1977), Yuda *et al.*, (1992) compared the relative efficiency of the mixture GA_4 + GA_7 with different 15β-hydroxy GAs in three species of *Pyrus*; the order of activity was: $GA_{4+7} > GA_{45} >$ 15β-OH $GA_{15} >$ 15β-OH GA_{24}. An interesting effect of GA_3 in pear is the delay in embryo sac degeneration (Herrero and Gascón, 1987), resulting in an extended period of ovule receptivity and an increased probability of successful fertilization under suboptimal pollination conditions. In the case of apple cv Cox's Orange Pippin, the mixture of GA_3, 1-naphthylacetic acid (NAA) and 1,3-diphenyl urea (DPU), which induces parthenocarpic development when applied to isolated emasculated flowers (Kotob and Schwabe 1971), has also been shown to increase the production of seeded fruit when applied to entire trees (Williams and Flook, 1980). Indeed the effect of GA_3 in apple is clearly enhanced when mixed with *N*-(2-chloro-4-pyridyl)-*N*-phenylurea (CPU) in the cvs Golden Delicious and Jonagold (Bangerth and Schröder, 1994). Different GAs and mixtures (e.g. GA_4, GA_4 + GA_7, and GA_4 + *iso*GA_7) have been applied successfully to reduce the severity of russeting in certain apple cultivars (Golden Delicious, Cox's Orange Pippin) (Looney *et al.*, 1992). The treatments produced also some decrease in fruit-set and number of seeds per fruit and a reduction in return bloom, although there was less effect with the mixture of GA_4 and *iso*GA_7.

Application of GA_3 to individual leafy inflorescences of seedless Clementine mandarin cv Fino, a self-incompatible cultivar that sets very poorly, promoted a rapid increase of ovary size associated with a remarkable enhancement of fruit-set (García-Martínez and García-Papí, 1979*a*). In seeded Clementine mandarin (cv Monreal), GA_3 also enhanced fruit-set in both pollinated and unpollinated ovaries (García-Martínez and García-Papí, 1979*b*). Optimal results were obtained with GA_3 applications between anthesis and 14 days afterwards. Although application of GA_3 to individual flowers always increased the size of the fruit, when it was applied to entire trees the average weight of the fruits decreased. This was due probably to increased competion for nutrients between the much larger number of fruits present in the GA_3-treated trees (García-Martínez and García-Papí, 1979*b*).

Gibberellins retard peel senescence in fruit. In Citrus, GA_3 is applied during development to avoid disorders of the rind, such as creasing, blemishes, pittings, etc (Monselise, 1979). An undesirable side effect of this kind of treatment is the delay in attaining a satisfactory colour, if GA_3 is applied to the fruits too late in the season (Monselise, 1979). There is high increase in chlorophyllase activity associated with the transformation of chloroplasts into chromoplasts (degreening) which is enhanced by ethylene and counteracted by GA_3 (Trebitsh *et al.*, 1993). Using tomato fruit discs it has also been shown recently that both GA_3 and calcium ions retard colour development and tissue softening (Ben-Arie *et al.*, 1995).

Conclusions

Fruit-set is the changeover from a static condition of the flower ovary to the rapidly growing young fruit (Leopold and Scott, 1952). However, the literature reveals that relatively little work (e.g. Bohner *et al.*, 1988; García-Martínez *et al.*, 1991*a*; Talón *et al.*, 1992; Koshioka *et al.*, 1994) has been carried out on GAs at the time when fruit-set actually happens, that is, at anthesis and shortly after. Nevertheless, work done mainly with pea supports the idea that, at least in this species, fruit-set and development depend on the action of GAs (probably GA_1 and/or GA_3). The fact that one single application of GA induces not only fruit-set, but also the complete development of the fruit suggests that GA acts as a trigger of the entire developmental process. Results obtained with pea indicate that the flower ovary is an organ programmed to senesce unless a positive stimulus, such as that provided by GAs synthesized as a result of fertilization of the ovules, down-regulates the expressión of proteolytic enzymes. At the same time, GAs activate primary metabolism within the ovary by a still unknown mechanism, converting it into an actively growing organ.

Acknowledgements

We thank the DGICYT of Spain for support (grant PB93–0133 to J.L. G.-M.). IACR receives grant-aided support from the Biotechnology and Biological Sciences Research Council of the United Kingdom.

References

Aach, H., Böse, G. and Graebe, J. E. (1995). *Planta* **197**, 333–42.

Albone, K. S., Gaskin, P. MacMillan, J., and Sponsel, V. M. (1984). *Planta* **162**, 560–5.

Albone, K. S., Gaskin, P., MacMillan, J., Smith, V. A. and Weir, J. (1989). *Planta* **177**, 108–15.

Baldev, B., Lang, A. and Agatep, A. O. (1965). *Science* **147**, 155–6.

Bangerth, F. and Schröder, M. (1994). *J. Plant Growth Regul.* **15**, 293–302.

Barendse, G. W. M., Kepczynski, J., Karssen, C. M. and Koornneef, M. (1986). *Physiol. Plant.* **67**, 315–19.

Beeley, L. J., Gaskin, P. and MacMillan, J. (1975). *Phytochemistry* **14**, 779–83.

Beltrán, J. P., Estruch, J. J., Cañas, L. and Parets-Soler, A. (1991). In *Recent Advances in Phloem Transport and Assimilate Compartmentation* (ed. J. L. Bonnemain, S. Delrot, W. J. Lucas and J. Dainty), pp. 275–82. Ouest Ed., Nantes.

Ben-Arie, R., Mignani, H., Greve, L. C., Huysamer, M. and Labavitch, J. M. (1995). *Physiol. Plant.* **93**, 99–107.

Bensen, R. J., Johal, G. S., Crane, V. C., Tossberg, J. T., Schnable, P. S., Meeley, R. B. and Briggs, S. P. (1995). *Plant Cell* **7**, 75–84.

Bhaskar, K. V., Chu, W.-L. A., Gaskin, P., Mander, L. N., Murofushi, N., Pearce, D. W., Pharis, R. P., Takahashi, N. and Yamaguchi, I. (1991). *Tetrahedron Lett.* **32**, 6203–6.

Blake, P. S. and Browning, G. (1994). *Phytochemistry* **35**, 1383–6.

Blake, P. S., Browning, G., Chu, A. W. L. and Mander, L. N. (1993). *Phytochemistry* **32**, 781–4.

Blechschmidt, S., Castel, U., Gaskin, P., Hedden, P., Graebe, J. E. and MacMillan, J. (1984). *Phytochemistry* **23**, 553–8.

Bohner, J., Hedden, P., Bora-Haber, E. and Bangerth, F. (1988). *Physiol. Plant.* **73**, 348–53.

Bottini, R.,Bottini, G., Koshioka, M., Pharis, R. P. and Coombe, B. (1985). *Plant Physiol.* **78**, 417–9.

Bramley, P. M., Teulieres, C., Blain, I., Bird, C. and Schuch, W. (1992). *Plant J.* **2**, 343–9.

Bukovac, M. J., Yuda, E., Murofushi, N. and Takahashi, N. (1979). *Plant Physiol.* **63**, 129–32.

Carbonell, J. and García-Martínez, J. L. (1980). *Planta* **147**, 444–50.

Carbonell, J. and García-Martínez, J. L. (1985). *Planta* **164**, 534–9.

Castellaro, S. J., MacMillan, J., Singh, A. K. and Willis, C. L. (1990). *J. Chem. Soc. Perkin Trans.* **1**, 145–52.

Ceccarelli, N., Lorenzi, R. and Alpi, A. (1979). *Phytochemistry* **18**, 1657–8.

Ceccarelli, N., Lorenzi, R. and Alpi, A. (1981*a*). *Plant Sci. Lett.* **21**, 325–32.

Ceccarelli, N., Lorenzi, R. and Alpi, A. (1981*b*). *Z. Pflanzenphysiol.* **102**, 37–44.

Cercós, M. and Carbonell, J. (1993). *Physiol. Plant.* **88**, 267–74.

Cercós, M., Harris, N. and Carbonell, J. (1993). *Physiol. Plant.* **88**, 275–80.

Chiang, H.-H., Hwang, I. and Goodman, H. M. (1995). *Plant Cell* **7**, 195–201.

Cioni, P. G., Bonnici, A., Alpi, A. and D'Amato, F. (1976). *Planta* **131**, 115–17.

Crane, J. C. (1964). *Annu. Rev. Plant Physiol.* **15**, 303–26.

Crane, J. C. (1969). *HortScience* **4**, 108–11.

Davies, P. (ed.) (1995). *Plant hormones, physiology, biochemistry and molecular biology*. Kluwer, Dordrecht.

Eeuwens, C. J. and Schwabe, W. W. (1975). *J. Exp. Bot.* **26**, 1–14.

Estruch, J. J. and Beltrán, J. P. (1991*a*). *Physiol. Plant.* **81**, 319–26.

Estruch, J. J. and Beltrán, J. P. (1991*b*). *FEBS Lett.* **279**, 303–6.

Fraser, P. D., Hedden, P., Cooke, D. T., Bird, C. R., Schuch, W. and Bramley, P. M. (1995). *Planta* **196**, 321–6.

Fray, R., Wallace A., Fraser, P. D., Valero, D., Hedden, P., Bramley, P. M. and Grierson, D. (1995). *Plant J.* **8**, 693–701.

Frydman, V. M., Gaskin, P. and MacMillan, J. (1974). *Planta* **118**, 123–32.

García-Martínez, J. L. and Carbonell, J. (1980). *Planta* **147**, 451–6.

García-Martínez, J. L. and García-Papí, M. A. (1979a). *Sci. Hort.* **11**, 337–47.

García-Martínez, J. L. and García-Papí, M. A. (1979b). *Sci. Hort.* **10**, 285–93.

García-Martínez, J. L., Sponsel, V. M. and Gaskin, P. (1987). *Planta* **170**, 130–7.

García-Martínez, J. L., Santes, C., Croker, S. J. and Hedden, P. (1991*a*). *Planta* **184**, 53–60.

García-Martínez, J. L., Martí, M., Sabater, T., Maldonado, A. and Vercher, Y. (1991*b*). *Physiol. Plant.* **83**, 411–16.

García-Martínez, J.L., López-Díaz, Y., Ward, D.A., Phillips, A.L., Gaskin, P. and Hedden, P. (1995). *15th Int. Conf. Plant Growth Subst.*, Abstract no. 99. International Plant Growth Substance Association, Minneapolis, USA.

Gaskin, P., Gilmour, S. J., MacMillan, J. and Sponsel, V. M. (1985). *Planta* **163**, 283–9.

Gianfagna, T. J. (1995). In *Plant hormones. Physiology, biochemistry and molecular biology* (ed. P. Davies), pp. 751–73. Kluwer, Dordrecht.

Gillaspy, G., Ben-David, H. and Gruissem, W. (1993). *Plant Cell* **5**, 1439–51.

Gilmour, S. J., Zeevaart, J. A. D., Schwenen, L. and Graebe, J. E. (1986). *Plant Physiol.* **82**, 190–5.

Goodwin, P. B. (1978). In *Phytohormones and related compounds—A comprehensive treatise,* Vol. II (ed. D. S. Letham, P. B. Goodwin, and T. J. V. Higgins) pp. 175–214. North-Holland-Elsevier, Amsterdam.

Graebe, J. E. (1982.) In *Plant growth substances 1982* (ed. P. F. Wareing), pp. 71–80. Academic Press, London.

Graebe, J. E. (1987). *Annu. Rev. Plant. Physiol.* **38**, 419–65.

Granell, A., Harris, N., Pisabarro, N. and Carbonell, J. (1992). *Plant J.* **2**, 907–15.

Griggs, D. L., Hedden, P. and Lazarus, C. M. (1991). *Phytochemistry* **30**, 2507–12.

Groot, S. P. C., Bruinsma, J. and Karssen, C. M. (1987). *Physiol. Plant.* **71**, 184–90.

Hedden, P. (1993). *Annu. Rev. Plant Physiol. Plant Mol. Biol.* **44**, 107–29.

Hedden, P. and Graebe, J. E. (1982). *J. Plant Growth Regul.* **1**, 105–16.

Hedden, P. and Hoad, G. V. (1985). In *Regulation of sources and sinks of crop plants,* Monograph 12 (ed. B. Jeffcoat, A. F. Hawhins and A. D.Stead) pp. 211–24. British Plant Growth Regulator Group, Wantage.

Hedden, P., Graebe, J. E., Beale, M. H., Gaskin, P. and MacMillan, J. (1984). *Phytochemistry* **23**, 569–74.

Hedden, P., Hoad, G. V. H., Gaskin, P., Lewis, M. J., Green, J. R., Furber, M. and Mander, L. N. (1993). *Phytochemistry* **32**, 231–7.

Herrero, M. and Gascon, M. (1987). *Ann. Bot.* **60**, 287–93.

Hutchison, M., Gaskin, P., MacMillan, J. and Phinney, B.O. (1988). *Phytochemistry* **27**, 2695–701.

Ingram, T. J. and Reid, J. B. (1987). *Plant Physiol.* **83**, 1048–53.

Ingram, T. J., Reid, J. B., Murfet, I. C., Gaskin, P., Willis, C. L. and MacMillan J. (1984). *Planta* **160**, 455–63.

Jahnke, S., Bier, D., Estruch, J. J. and Beltrán, J. P. (1989). *Planta* **180**, 53–60.

Kamiya, Y., Takahashi, N. and Graebe, J. E. (1986). *Planta* **169**, 524–9.

Kamiya, Y., Yamaguchi, S., Saito, T., Ait-Ali, T., Swain, S., Abe, H. and Sun, T. (1995). In *15th Int. Conf. Plant Growth Subst.*, Abstract no 40. International Plant Growth Substance Association, Minneapolis, USA.

Koshioka, M., Nishijima, T., Yamazaki, H., Liu, Y., Nonaka, M. and Mander, L. N. (1994). *J. Hort. Sci.* **69**, 171–9.

Kotob, M. A. and Schwabe, W. W. (1971). *J. Hort. Sci.* **46**, 89–93.

Lange, T. (1994). *Planta* **195**, 108–15.

Lange, T., Hedden, P. and Graebe, J. E. (1993*a*). *Planta* **189**, 340–9.

Lange, T., Hedden, P. and Graebe, J. E. (1993*b*). *Planta* **189**, 350–8.

Lange, T., Hedden, P. and Graebe, J.E. (1994). *Proc. Natl. Acad. Sci. USA* **91**, 8552–6.

Leopold, A.C. and Scott, F.I. (1952). *Am. J. Bot.* **39**, 310–17.

Lis, E. K. and Antoszewski, R. (1982). *Planta* **156**, 492–5.

Looney, N. E., Granger, R. L., Chu, C. L., McArtney, S. J., Mander, L. N. and Pharis, R. P. (1992). *J. Hort. Sci.* **67**, 613–18.

Luckwill, L. C., Weaver, P. and MacMillan, J. (1969). *J. Hort. Sci.* **44**, 413–24.

MacMillan, J. and Gaskin, P. (1996). submitted to *Phytochemistry.*

Maki, S. J. and Brenner, M. L. (1991). *Plant Physiol.* **97**, 1359–66.

Martín, G.C. (1983). In *The biochemistry and physiology of gibberellins,* Vol. 2 (ed. A. Crozier), pp. 395–444. Praeger, New York.

Martín, G. C., Dennis, F. G., MacMillan, J. and Gaskin, P. (1977). *J. Amer. Soc. Hort. Sci.* **102**, 16–9.

Martín, G. C., Horgan, R. and Nishijima, C. (1982). *J. Amer. Soc. Hort. Sci.* **107**, 479–82.

Monselise, S. P. (1979). *Sci. Hort.* **11**, 151–62.

Moore, T. C., Barlow, S. A. and Coolbaugh, R. C. (1972). *Phytochemistry* **11**, 3225–33.

Owen, D.J., Mander, L. N., Gaskin, P. and MacMillan, J. (1996). submitted to *Phytochemistry.*

Oyama, N., Yamaguchi, T., Yamane, H., Murofushi, N., Agatsumi, M. P. and Mander, L. N. (1996). submitted to *Biosci. Biotech. Biochem.*

Ozga, J. A., Brenner, M. L. and Reinecke, D. M. (1992). *Plant Physiol.* **100**, 88–94.

Peretó, J. G. and Beltrán, J. P. (1987). *Physiol. Plant.* **69**, 356–60.

Peretó, J. G., Beltrán, J. P. and García-Martínez, J. L. (1988). *Planta* **175**, 493–9.

Pharis, R. P. and King, R. W. (1985). *Annu. Rev. Plant Physiol.* **36**, 517–68.
Phillips, A. J., Ward, D. A., Uknes, S., Appleford, N. E. J., Lange, T., Huttly, A. K., Gaskin, P., Graebe, J. E. and Hedden, P. (1995). *Plant Physiol.* **108**, 1049–57.
Phinney, B. O. (1984). In *The biosynthesis and metabolism of plant hormones*. Society for Experimental Biology Seminar Series **23** (ed. A. Crozier and J. R. Hillman), pp. 17–41. Cambridge University Press, Cambridge.
Piaggesi, A., Picciarelli, P., Lorenzi, R. and Alpi, A. (1989). *Plant Physiol.* **91**, 362–6.
Picciarelli, P. and Alpi, A. (1987). *Phytochemistry* **26**, 329–30.
Picciarelli, P., Piaggesi, A. and Alpi, A. (1991). *Phytochemistry* **30**, 1789–92.
Potts, W. C. and Reid, J. B. (1983). *Physiol. Plant.* **57**, 449–54.
Potts, W.C., Reid, J.B. and Murfet, I.C. (1985). *Physiol. Plant.* **63**, 357–64.
Powell, A. A. and Krezdorn, A. H. (1977). *J. Amer. Soc. Hort. Sci.* **102**, 709–14.
Prescott, A. G. (1993). *J. Exp. Bot.* **44**, 849–61.
Reed, W. and Martin, G.C. (1976). *J. Amer. Soc. Hort. Sci.* **101**, 527–31.
Rodríguez-Concepción, M. and Beltrán, J.P. (1995). *Plant Mol. Biol.* **27**, 887–99.
Ross, J. J., Reid, J. B. and Swain, S. M. (1993). *Aust. J. Plant Physiol.* **20**, 585–99.
Ross, J. J., Reid, J. B., Swain, S. M., Hasan, O., Poole, A. T., Hedden, P. and Willis, C. L. (1995). *Plant J.* **7**, 513–23.
Sánchez-Beltrán, M. J., Carbonell, J., García-Martínez, J. L. and López-Díaz, I. (1992). *Physiol. Plant.* **85**, 69–76.
Santes, C. M. and García-Martínez, J. L. (1995). *Plant Physiol.* **108**, 517–23.
Santes, C. M., Hedden, P., Sponsel, V. M., Reid, L. B. and García-Martínez, J. L. (1993). *Plant Physiol.* **101**, 759–64.
Santes, C. M., Hedden, P., Gaskin, P. and García-Martínez, J. L. (1995). *Phytochemistry* **40**, 1347–55.
Schroeder, C. (1953). *Proc. Am. Soc. Hort. Sci.* **61**, 103–6.
Serebryakov, E. P., Epstein, N. A., Yasinskaya, N. P. and Kaplun, A. B. (1984). *Phytochemistry* **23**, 1855–63.
Smith, V. A. and MacMillan, J. (1984). *J. Plant Growth Regul.* **2**, 251–64.
Smith, V. A. and MacMillan, J. (1986). *Planta* **167**, 9–18.
Sponsel, V. M. (1982). *J. Plant Growth Regul.* 1, 147–52.
Sponsel, V. M. (1983). *Planta* **159**, 454–68.
Sponsel, V. M. (1995). In *Plant hormones, physiology, biochemistry and molecular biology* (ed. P. Davies), pp. 66–97. Kluwer, Dordrecht.
Sun, T. -P. and Kamiya, Y. (1994). *Plant Cell* **6**, 1509–18.
Sun, T. -P., Goodman, H. M. and Ausubel, F. M. (1992). *Plant Cell* **4**, 119–28.
Swain, S. M., Reid, J. B. and Ross, J. J. (1993). *Planta* **191**, 482–8.
Swain, S. M., Ross, J. J., Reid, J. B. and Kamiya, Y. (1995). *Planta* **195**, 426–33.
Takahashi, M., Kamiya, Y., Takahashi, N. and Graebe, J. (1986). *Planta* **168**, 190–9.
Talón, M., Hedden, P. and Primo-Millo, E. (1990*a*). *J. Plant Growth Regul.* **9**, 201–6.
Talón, M., Zacarías, L. and Primo-Millo, E. (1990*b*). *Physiol. Plant.* **79**, 400–6.
Talón, M., Zacarías, L. and Primo-Millo, E. (1992). *Plant Physiol.* **99**, 1575–81.
Trebitsh, T., Goldschmidt, E. E. and Riov, J. (1993). *Proc. Natl. Acad. Sci. USA* **90**, 9441–5.
Turnbull, C. G. N. (1989). *J. Plant Growth Regul.* **8**, 273–82.
Vercher, Y. and Carbonell, J. (1991). *Physiol. Plant.* **81**, 518–26.
Vercher, Y., Molowny, A., López, C., García-Martínez, J. L. and Carbonell, J. (1984). *Plant Sci. Lett.* **36**, 87–91.
Vercher, Y., Molowny, A. and Carbonell, J. (1987). *Physiol. Plant.* **71**, 302–8.

Vercher, Y., Carrasco, P. and Carbonell, J. (1989). *Physiol. Plant.* **76**, 405–11.
Weaver, R. J., Shindy, W. and Kliewer, W. M. (1969). *Plant Physiol.* **44**, 183–8.
Webster, A. D. and Goldwin, G. K. (1981). *J. Hort. Sci.* **56**, 27–40.
Williams, P. M., Bradbeer, J. W., Gaskin, P. and MacMillan, J. (1974). *Planta* **117**, 101–3.
Williams, R. R. and Flook, V. A. (1980). *J. Hort. Sci.* **55**, 275–7.
Winkler, R. G. and Helentjaris, T. (1995). *Plant Cell* **7**, 1307–17.
Xu, Y.-L., Li. L., Wu, K., Peeters, A. J. M., Gage, D. A. and Zeevaart, J. A. D. (1995). *Proc. Natl. Acad. Sci. USA* **92**, 6640–4.
Yamaguchi, I. and Takahashi, N. (1976). *Plant Cell Physiol.* **17**, 611–14.
Yamaguchi, Y., Yokota, T., Murofuchi, N., Takahashi, N. and Ogawa, Y. (1975). *Agr. Biol. Chem.* **39**, 2399–403.
Yuda, E., Nakagawa, S., Murofushi, N., Yokota, T., Takahashi, N., Koshioka, M., Murakami, Y., Pearce, D., Pharis, R. P., Patrick, G. L., Mander, L. N. and Kraft-Klaunzer, P. (1992). *Biosci. Biotech. Biochem.* **56**, 17–20.
Zanewich, K. P. and Rood, S. B. (1993). *J. Plant Growth Regul.* **12**, 41–6.
Zeevaart, J. A. D. (1966). *Plant Physiol.* **41**, 856–62.

15. The possible role of antioxidants in fruit and vegetables in protecting against coronary heart disease

DAVID S. LEAKE

School of Animal and Microbial Sciences, The University of Reading, Whiteknights, PO Box 228, Reading RG6 6AJ, UK

Introduction

There has been a great increase in recent years in interest in the possible effects of antioxidants in the diet in protecting against coronary heart disease (Steinberg *et al.*, 1992). Atherosclerosis is the underlying cause of coronary heart disease, strokes and peripheral arterial disease (atherosclerosis of the arteries of the legs). It is therefore responsible for about half the deaths in the developed world and its incidence in developing countries, although presently low, is increasing.

Atherosclerosis is a disease of large and medium-sized arteries in the body. The arterial wall has three layers and it is the innermost layer, the tunica intima, which is mainly affected by the disease. The intima is normally a thin layer but becomes greatly thickened when an atherosclerotic lesion forms in it.

Atherosclerotic lesions have five main characteristics.

- Monocytes and T-lymphocytes enter the arterial wall from the blood in large numbers. The monocytes then differentiate into macrophages.
- The macrophages accumulate vast amounts of cholesteryl ester in droplets inside them. This gives them a foamy appearance under the light microscope and they are therefore known as 'foam cells'. The resident smooth muscle cells in the arterial wall also accumulate cholesteryl ester droplets and become foam cells, but the extent of cholesteryl ester accumulation in these cells is less than in the macrophages. In later lesions, there is considerable extracellular deposition of cholesteryl esters, which forms a large pool of lipid at the base of the lesions.
- There is a proliferation of smooth muscle cells in the intima.
- The smooth muscle cells secrete large amounts of collagen, which adds bulk to the lesions.
- The macrophages die in the deeper parts of the lesions, mainly at the edge of the lipid pool in advanced lesions, and this may possibly be involved in the progression of the lesions.

Atherosclerotic lesions grow over a period of years or decades and their bulk can impede the flow of blood through the arteries. If this occurs in the coronary arteries, it can cause angina. Most heart attacks (myocardial infarctions) are due to the sudden fissuring of an atherosclerotic lesion (Davies and Thomas, 1984). These fissures cause blood to come into contact with substances inside the lesions that cause a thrombus (a blood clot) to form. This can sometimes spread into the inside of the artery (the lumen) and block the flow of blood, leading to a myocardial infarction if the artery supplies the muscle of the heart, a thrombotic stroke if the artery supplies the brain or peripheral arterial disease if it supplies the legs. Sometimes the artery supplying the brain contains an atherosclerotic lesion that so weakens the wall of the artery that it begins to leak blood into the brain tissue thus causing another type of stroke called a haemorrhagic stroke. Thrombotic strokes outnumber haemorrhagic strokes about three to one.

What causes atherosclerotic lesions, which can cause these tragic consequences, to form in the first place? This is far from clear at present, despite extensive research for many years throughout the world, but the most popular theory at present is that it may be due to low density lipoprotein (LDL) that becomes oxidized locally within the arterial wall. This is the reason for the current intense interest in the possible role of dietary antioxidants in the prevention of cardiovascular disease. Before the possible role of oxidized LDL in atherosclerosis is discussed, it is necessary to describe briefly what LDL is and what its normal function is.

Low density lipoprotein

Lipids are of course insoluble in water but have to be transported around the body in the blood. This problem is solved by packaging them into water-soluble particles of lipid and protein called the plasma lipoproteins. The amphipathic lipids (phospholipids and some non-esterified cholesterol) and the protein are on the outside of the particles and the hydrophobic lipids (cholesteryl esters, triacylglycerols (triglycerides) and some non-esterified cholesterol) are tucked away from water in the interior of the particles.

There are four main types of lipoproteins: chylomicrons, very low density lipoprotein (VLDL), LDL and high density lipoprotein (HDL). Chylomicrons transport mainly triacylglycerols from the small intestine to the adipose tissue and skeletal muscle, whereas VLDL transports mainly triacylglycerols from the liver to adipose tissue and skeletal muscle. VLDL is converted in the circulation to LDL, which contains far less triacylglycerol than VLDL but far more cholesteryl ester as a percentage of its mass. Each LDL particle is about 22 nm in diameter on average and contains about 1600 molecules of cholesteryl esters and 170 molecules of triacylglycerols in its core, about 700 molecules of phospholipid (mainly phosphatidylcholine but also some sphingomyelin) at its surface as a monolayer and about 600 molecules of non-esterfied cholesterol distributed between its core and its surface (Esterbauer *et al.*, 1992). The major fatty acids esterified in lipids of LDL are linoleic acid (*cis*, *cis*-Δ^9, Δ^{12}-octadecadienoate), with about 1100 molecules per LDL particle on average, palmitic acid (hexadecanoate; about 700 molecules), oleic acid (*cis* -Δ^9-octadecanoate; about 450 molecules), arachidonic acid (all *cis* -Δ^5, Δ^8, Δ^{11}, Δ^{14}-eicosatetraenoate; about 150 molecules) and stearic acid (octadecanoate; about 140 molecules). Each LDL particle therefore contains about 1300 molecules of polyunsaturated fatty acids in its lipids, which are highly oxidizable. LDL also contains one molecule of one of the largest polypeptides known, called apolipoprotein B-100 (apo B-100). This protein sits on the surface of the LDL, has an M_r of 513 000 and contains 4536 amino acids.

The function of LDL is to transport cholesterol to cells. To do this, the apo B-100 on the LDL particles binds to LDL receptors on cells and the LDL is then internalized by the cells by receptor-mediated endocytosis and delivered to the lysosomes. The LDL is then broken down by the lysosomal enzymes and its cholesterol is made available to the cells for the synthesis of membranes or steroid hormones or, in the liver, of bile acids.

HDL is thought to carry cholesterol from cells throughout the body, including those in the arterial wall, back to the liver, although the details of how exactly this is done are unclear. HDL is therefore believed to protect against atherosclerosis, whereas LDL is believed to exacerbate it.

Oxidized low density lipoprotein

The oxidative theory of atherosclerosis proposes that it is not LDL itself that is atherogenic (i.e. a factor which promotes atherosclerosis) but LDL that has been oxidized locally in the arterial wall (Steinberg *et al.*, 1989; Witztum and Steinberg, 1991; Leake, 1993; Rice-Evans and Bruckdorfer, 1995) (Fig 15.1). LDL enters the arterial wall from the plasma and is believed to be oxidized locally within the wall by oxidizing agents derived from the cells present in atherosclerotic lesions, namely macrophages, smooth muscle cells, endothelial cells and lymphocytes. The most active type of cell in oxidizing LDL *in vitro* is the macrophage (Morgan and Leake, 1993; Lamb *et al.*, 1992). The oxidized LDL is then believed to bind to scavenger receptors on macrophages and to be internalized rapidly by receptor-mediated endocytosis. The amount of cholesterol in LDL entering the cells is thought to overwhelm the capacity of the macrophages to release it and the cholesterol may therefore accumulate inside the cells, converting them into foam cells.

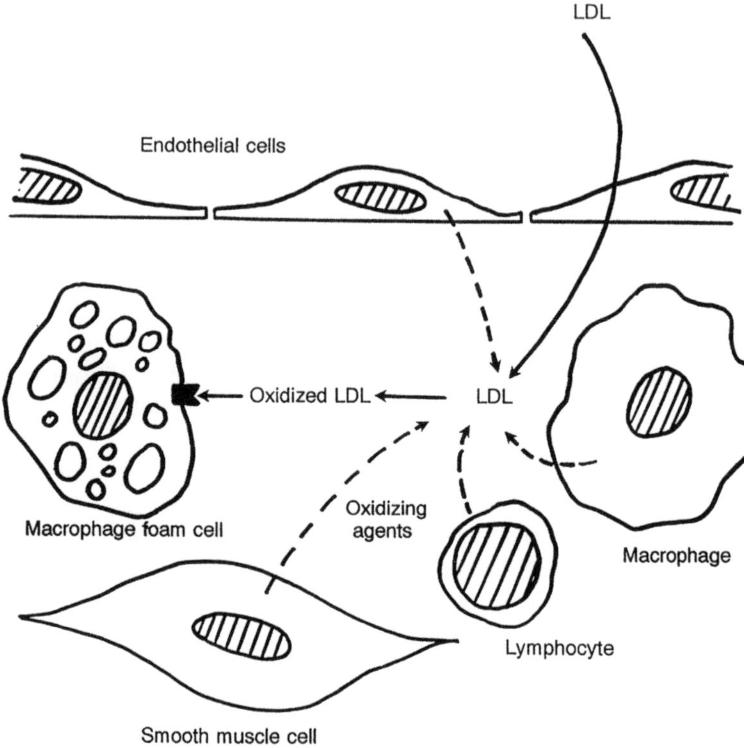

Fig. 15.1. LDL oxidation in an atherosclerotic lesion. LDL is believed to enter the wall of an artery from the plasma and then to be oxidized locally in the wall by oxidizing agents from macrophages, smooth muscle cells, endothelial cells and lymphocytes. The oxidized LDL may then become recognized by various scavenger receptors on the surfaces of macrophages and be internalized by endocytosis and deliver large amounts of cholesterol to these cells. The macrophages may then become cholesterol-laden foam cells and the atherosclerotic lesion may progress.

It used to be thought that a receptor cloned and characterized by Kodama *et al.* (1990) was responsible for internalizing oxidized LDL (but not normal LDL) by macrophages, but since then a number of other scavenger receptors that recognize oxidized LDL have been identified (Stanton *et al.*, 1992; Endemann *et al.*, 1993; Ottnad *et al.*, 1995). It is therefore unclear which ones are involved in taking up oxidized LDL *in vivo*. Scavenger receptors are mainly found on macrophages but a few other cell types (e.g. endothelial cells) express modest levels of them.

As well as converting macrophages into foam cells (at least *in vitro*), oxidized LDL has many other effects that may be of importance in atherosclerosis. For instance, it is directly chemotactic for macrophages and lymphocytes, it activates the genes for cellular adhesion molecules and chemotactic proteins that may attract monocytes and lymphocytes into the arterial wall and it induces apoptosis (programmed cell death) in macrophages (see Rice-Evans and Bruckdorfer, 1995).

What is the evidence that oxidized LDL is actually involved in atherosclerosis? There are several lines of evidence for this.

- Antibodies that recognize oxidized LDL, but not native LDL, stain atherosclerotic lesions in animals and humans (Palinski *et al.*, 1989).
- Some studies (but not all) show that LDL can be extracted from animal or human atherosclerotic lesions that shows chemical signs of oxidation and is taken up faster by macrophages by means of their scavenger receptors (Shaikh *et al.*, 1988; Ylä-Herttuala *et al.*, 1989).
- Animals and humans possess plasma autoantibodies to oxidized LDL. For these autoantibodies to have been generated, there must have been an antigen present and this may have been oxidized LDL or something similar.
- Antioxidants, such as probucol, butylated hydroxytoluene or α-tocopherol (vitamin E) can inhibit atherosclerosis in animals (Carew *et al.*, 1987; Björkhem *et al.*, 1991; Verlangieri and Bush, 1992).
- There is a positive association between the oxidizability of LDL by copper ions and coronary atherosclerosis measured by angiography in survivors of myocardial infarction (Regnström *et al.*, 1992) and the oxidizability of LDL by copper is greater in patients with coronary heart disease than in control subjects (Cominacini *et al.*, 1993). There is also a positive correlation between the oxidizability of LDL by metmyoglobin and the rate of progression of carotid or femoral atherosclerosis as assessed by duplex scanning (Andrews *et al.*, 1995).

How do cells oxidize LDL?

The mechanisms by which cells oxidize LDL are highly controversial. The involvement of superoxide has been proposed (Heinecke *et al.*, 1986; Steinbrecher, 1988; Hiramatsu *et al.*, 1987; Cathcart *et al.*, 1989) but others argue against it (Jessup *et al.*, 1993; Wilkins and Leake, 1994*a*; Wilkins *et al.*, 1994). The activity of lipoxygenases has also been proposed to be involved in LDL oxidation (Parthasarathy *et al.*, 1989; Rankin *et al.*, 1991; Benz *et al.*, 1995) but this has been questioned (Sparrow and Olszewski, 1992; Jessup *et al.*, 1991; Wilkins and Leake, 1994*b*). It was proposed in 1987 by Heinecke *et al.* that cells may oxidize LDL by releasing the thiol-containing amino acid cysteine (Heinecke *et al.*, 1987). Cysteine oxidizes to its disulfide form cystine in the extracellular space and in so doing may produce sulfur- or oxygen-centred free radicals that may attack LDL. Also, when cysteine oxidizes to cystine, iron(III)may be reduced to iron(II) and copper(II) to copper(I). These transition metal ions in their lower valency states may oxidize LDL better than they do in their higher valency states. The original observations of Heinicke *et al.* (1987) have been confirmed (Sparrow and Olszewski, 1993; Graham *et al.*, 1994).

Whatever the mechanism, the oxidation of LDL will involve the abstraction by an unidentified free radical of a hydrogen atom from a bisallylic carbon atom of an esterified polyunsaturated fatty acid in LDL (Esterbauer *et al.*, 1992) (Fig 15.2). The resulting lipid alkyl radical will then very rapidly undergo a rearrangement to form a conjugated diene and molecular oxygen will bind to form a lipid peroxyl radical. This will attack a neighbouring polyunsaturated fatty acid and will itself become a lipid hydroperoxide, with the other polyunsaturated fatty acid becoming a lipid alkyl radical. This alkyl radical will then follow the same pathway as the first one, thereby setting up a chain reaction of lipid peroxidation. This chain reaction is normally kept in check in LDL by the antioxidants it contains. If iron(II) or copper(I) is present, the lipid hydroperoxides will accept an electron from them, forming iron(III) or copper(II) and a lipid alkoxyl radical. This radical is unstable and can either abstract a hydrogen atom from a polyunsaturated fatty acid forming a lipid alcohol or it can cyclase to form an epoxide or, as shown in Fig 15.2, a process known as β-scission can occur and the carbon atom bearing the oxygen radical can have either of its two bonds to its adjacent carbon atoms broken to form an aldehyde and a carbon-centred free radical. These aldehydes can be either on the parent lipid molecule or on the fragment of the fatty acid chain that leaves the parent molecule.

A wide variety of aldehydes are produced by oxidized LDL (Esterbauer *et al.*, 1992), the two major ones in terms of mass being malondialdehyde ($CHO\text{-}CH_2\text{-}CHO$) and hexanal ($CH_3(CH_2)_4CHO$). The aldehydes can combine with the epsilon-amino groups of the lysyl residues of apo B-100 to form Schiff bases, which may later rearrange to other products. When a Schiff base forms the normal positive charge of the lysyl residue is abolished and the LDL particles acquire a greater net negative charge (see Fig. 15.2). The oxidized LDL then becomes recognized by scavenger receptors on macrophages and is internalized rapidly and degraded intracellularly. The uptake of oxidized LDL by macrophages involves more than simply a charge of net charge, however, because the uptake continues to increase rapidly even after the increase in net negative charge (assessed by electrophoretic mobility) has begun to plateau (Carpenter *et al.*, 1994).

LDL oxidation by cells usually requires a source of iron or copper ions, with copper being more potent than iron in catalysing LDL oxidation by cells (Heinecke *et al.*, 1984; Leake and Rankin, 1990; Lamb and Leake, 1994*a*). It has been shown that some advanced atherosclerotic lesions in humans contain catalytically active iron and copper ions (Smith *et al.*, 1992). The

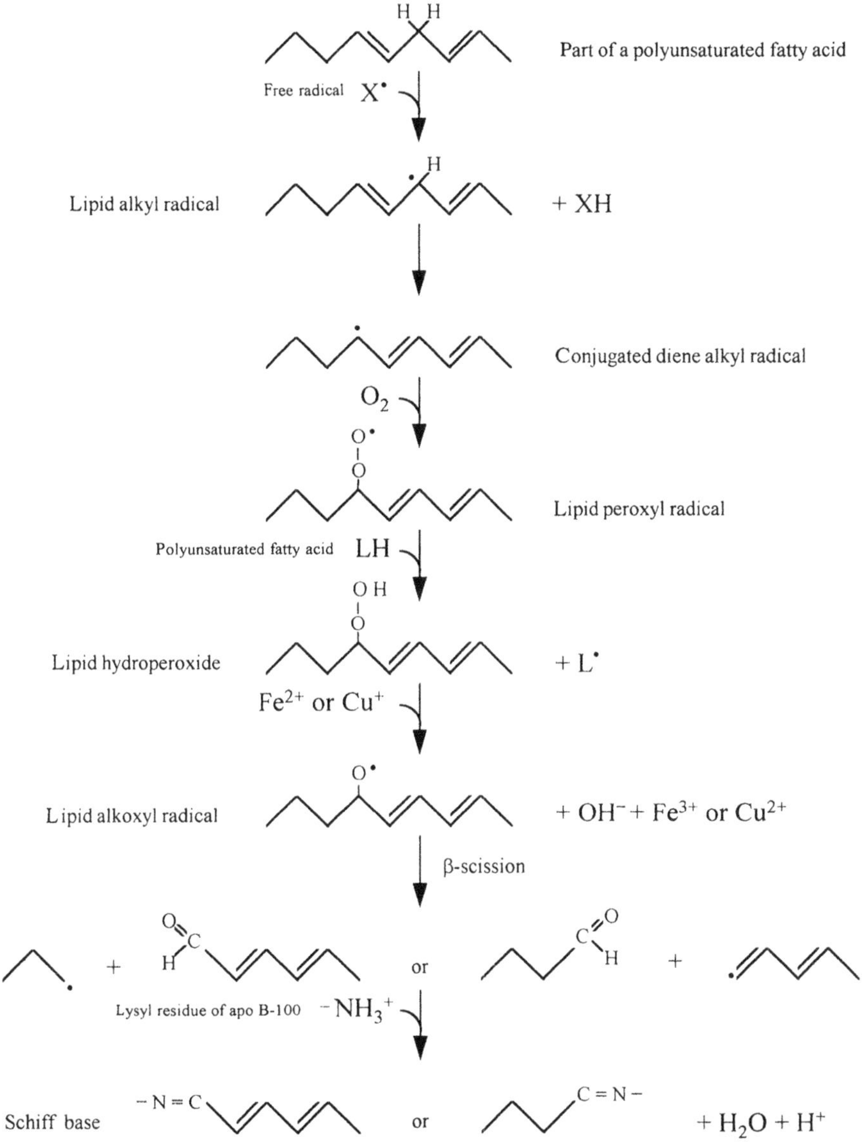

Fig. 15.2. The chemistry of LDL oxidation. A free radical abstracts a hydrogen atom from a polyunsaturated fatty acid in LDL. A double bond in the resulting lipid alkyl radical rearranges to form a conjugated diene and molecular oxygen then binds to form a lipid peroxyl radical. This attacks another polyunsaturated fatty acid, itself becoming a lipid hydroperoxide. In the presence of iron(II) or copper(I), this lipid hydroperoxide breaks down to form a lipid alkoxyl radical, which can fragment to form an aldehyde. This aldehyde may react with a positively charged lysyl residue in apo B-100 of LDL, causing the neutralization of its charge. The LDL then becomes recognized by the scavenger receptors of macrophages. It should be noted that the actual chemistry of LDL oxidation is far more complex than is shown here.

source of this catalytically active iron or copper is unknown, as they are normally tightly sequestered in the body so that they cannot catalyse free radical reactions. It has, however, been shown that transferrin, the protein that carries iron in plasma, is capable of catalysing LDL oxidation at acidic pH but not at pH 7.4 (Lamb and Leake, 1994*b*). Caeruloplasmin, the protein that carries copper in plasma, catalyses LDL oxidation faster at mildly acidic pH than at pH 7.4 (Lamb and Leake, 1994*a*). It has also been shown by other groups that caeruloplasmin can catalyse LDL oxidation at pH 7.4 (Ehrenwald *et al.*, 1994) and that caeruloplasmin can oxidize lipids better when it is exposed to peroxynitrite ($ONOO^-$) (Swain *et al.*, 1994), a molecule that may possibly be formed in atherosclerotic lesions.

The antioxidants in LDL

The antioxidants in LDL are consumed when LDL is oxidized and large-scale lipid oxidation only starts when they have been depleted (Esterbauer *et al.*, 1987; Jessup *et al.*, 1990). Ubiquinol-10 is consumed first, followed by α-tocopherol then γ-tocopherol and then by the carotenoids (Esterbauer *et al.*, 1987; Stocker *et al.*, 1991).

The main antioxidant in LDL appears to be α-tocopherol, with each LDL particle on average containing about 6 molecules of this antioxidant, although this varies widely between individuals and countries (Jessup *et al.*, 1990; Esterbauer *et al.*, 1992). Each LDL particle on average also contains about 0.5 molecules of γ-tocopherol, 0.3 molecules of β-carotene, 0.1 molecules of α-carotene, 0.15 molecules of lycopene, 0.15 molecules of crytoxanthin, 0.02 molecules of cantaxanthin and 0.04 molecules of lutein + zeaxanthin, 0.05 molecules of phytofluene and 0.1 molecules of ubiquinol –10, although these may also vary widely (Esterbauer *et al.*, 1992). Each LDL particle therefore contains on average only about 8 known antioxidant molecules to protect about 1300 molecules of polyunsaturated fatty acids from oxidation.

Effects of dietary antioxidants on LDL oxidation and cardiovascular disease

As mentioned above, there is now a great deal of interest in the possible role of dietary antioxidants in protecting against cardiovascular disease. Most of these antioxidants are derived from the plant components of our diet.

α-Tocopherol

Vegetable oils are the main source of α-tocopherol in the human diet, although smaller amounts are found in many foods, for instance, cereal grains, lettuce and nuts. Smith *et al.* (1993) have used a method employing luminol oxidation to measure the total antioxidant capacity of LDL and have concluded that α-tocopherol accounts for about 80% of this. It has been shown that increasing the amount of α-tocopherol in LDL by oral supplementation with vitamin E decreases the oxidizability of human LDL by macrophages or copper (Jessup *et al.*, 1990; Dieber-Rotheneder *et al.*, 1991; Princen *et al.*, 1992; Jialal and Grundy, 1992; Reaven *et al.*, 1993). The levels of α-tocopherol in LDL can be increased by about 2–4 fold by oral supplementation, but not by any more. The amounts required to raise the levels of α-tocopherol in LDL considerably are in the pharmacological range, that is about 400–800 iu/day. (One international unit of vitamin E is equal to 1 mg of DL-α-tocopherol acetate). For comparison, the average amount in the human diet is about 10 iu/day (although this varies considerable between different regions of the world). It has been shown, however, that even relatively small supplements, such as 25 iu/day will increase the levels of α-tocopherol in LDL a little and give it a little more protection against oxidation (Princen *et al.*, 1995).

α-Tocopherol supplementation delays the start of lipid peroxidation in LDL and thereby increases the lag period before increased macrophage uptake is seen (Jessup *et al.*, 1990). It acts as a chain-breaking antioxidant by scavenging lipid peroxyl radicals and converting them into lipid hydroperoxides before they can attack another polyunsaturated fatty acid and propagate the chain reaction of lipid peroxidation:

$$\text{LOO}\cdot + \alpha\text{toc OH} \rightarrow \text{LOOH} + \alpha\text{toc O}\cdot$$

The α-tocopheroxyl radical is relatively stable because the impaired electron, which makes it a free radical, is delocalized around its aromatic molecular structure.

The efficiency of each molecule of α-tocopherol within LDL in inhibiting LDL oxidation varies quite widely between individuals, for some unknown reason (Esterbauer *et al.*, 1992). This may be of some importance as it may mean that supplementing one person's LDL with a certain amount of α-tocopherol may have more effect than supplementing another person's LDL with the same amount of α-tocopherol.

There is some evidence that the α-tocopheroxyl radical can sometimes attack a polyunsaturated fatty acid to convert it into a lipid alkyl radical and that α-tocopherol, under certain circumstances, can even promote LDL oxidation (Bowry *et al.*, 1992; Bowry and Stocker, 1993). Most workers, however, find that α-tocopherol inhibits rather than promotes LDL oxidation. α-Tocopherol has been shown to inhibit atherosclerosis in primates (Verlangieri and Bush, 1992) and hamsters (Parker *et al.*, 1995).

α-Tocopherol may possibly decrease atherosclerosis in other ways than by inhibiting LDL oxidation, for instance, it can protect cells against the toxic effects of oxidized LDL (Reid and Mitchinson, 1993; Mabile *et al.*, 1995).

Epidemiological studies suggest that α-tocopherol may protect against cardiovascular disease (but, of course, like all epidemiological studies, they do not prove it). Gey *et al.* (1991) have studied various European populations and have found that the mortality from coronary heart disease is lower in regions where the plasma α-tocopherol level is high than in regions where the plasma levels are low. It was found in the USA that the risk of coronary heart disease in nurses (Stampfer *et al.*, 1993) and male health professionals (Rimm *et al.*, 1993) is lower if they were taking vitamin E supplements. There was no association between their dietary vitamin E intake and the incidence of coronary heart disease, emphasizing that pharmacological rather than dietary amounts of α-tocopherol in the USA may be required to protect against atherosclerosis. It should be borne in mind, however, that a lifetime's exposure to a higher than average intake of α-tocopherol in the natural diet may possibly confer some protection against cardiovascular disease, whereas a 'snapshot' of α-tocopherol intake at a certain point in life may not reflect this.

Hodis *et al.* (1995) studied a group of men with coronary heart disease on lipid-lowering drugs and measured the rate of progression of atherosclerotic lesions in their coronary arteries over 2 years by angiography (a radiographic procedure). If they consumed over 100 iu/day of vitamin E (mainly in the form of supplements), the rate of progression of their lesions was slower than if they consumed less vitamin E. It should be noted, however, that this trial was part of a larger one and was not a randomized trial of the effects of taking vitamin E supplements. There was no apparent effect of dietary vitamin E or of total vitamin C intake. The intake of vitamin E in the Scottish Health Study was associated inversely with coronary heart disease in men, but not in women, both before and after adjustment for the other risk factors (Todd *et al.*, 1995). The level of α-tocopherol in adipose tissue in nine European populations, however, was not associated with acute myocardial infarction (Kardinaal *et al.*, 1995).

To prove or disprove that α-tocopherol protects against cardiovascular disease, it will be necessary to carry out clinical trials in which large numbers of subjects who are given α-tocopherol or a placebo and the number of cardiovascular events in both groups are compared over a number of years. Trials of this type are now in progress in a number of countries and we should know the answer to whether or not α-tocopherol protects against cardiovascular disease in the next few years. I say 'should know' because it is always possible that a lifetime's exposure to high α-tocopherol intake may protect against cardiovascular disease whereas exposure for only a few years may not, as atherosclerosis is a disease that takes decades to develop.

γ-Tocopherol

γ-Tocopherol has one less methyl group in its molecule than α-tocopherol. It is the major form of vitamin E found in plants but it is present in LDL at only about one tenth of the level of α-tocopherol. γ-Tocopherol is absorbed from the gut to about the same extent as α-tocopherol but it is not recognized by the α-tocopherol binding protein in the cytosol of the liver and is therefore not incorporated into lipoproteins in the liver to nearly the same extent as α-tocopherol. As there is on average less than one molecule of γ-tocopherol in each LDL particle, it is somewhat doubtful if it plays a major role in protecting against LDL oxidation.

Ubiquinol -10

One molecule of ubiquinol-10 is present in every 10 or so particles of LDL (Esterbauer *et al.*, 1992). This antioxidant is consumed very early in LDL oxidation, even before the α-tocopherol is consumed (Stocker *et al.*, 1991). The level of ubiquinol -10 in LDL correlates negatively with the rate of LDL oxidation during an early part of the lag phase (Kontush *et al.*, 1994). Whether or not ubiquinol levels in the normal diet have much effect on the oxidizability of LDL is debatable, however, because its levels in LDL are usually so low.

Carotenoids

Carotenoids are found in green leafy vegetables, carrots, squash, melons, tomatoes and many other plants. They contain a highly conjugated hydrocarbon backbone and act as antioxidants (Palozza and Krinsky, 1992). One of them, β-carotene, also gives LDL its normal golden orange colour and in fact a simple test of LDL oxidation is to look if it has lost its colour. There is on average about one molecule of β-carotene in every three particles of LDL, with lower levels of the other carotenoids (Esterbauer *et al.*, 1992). The total carotenoid content of LDL amounts to somewhat less than one molecule per LDL particle.

β-Carotene is a difficult molecule to work with in aqueous systems because it is so hydrophobic. It has been shown in one study that enriching LDL in vitro with it decreases its oxidizability (Jialal *et al.*, 1991), but another study could not repeat this (Gaziano *et al.*, 1995). Cocultures of human aortic endothelial cells lying on top of smooth muscle cells oxidize LDL mildly so that it will induce monocytes to migrate through the endothelial cell layer

(Navab *et al.*, 1991). This monocyte migration is prevented if the LDL is preincubated with β-carotene. The reported effects of exogenous β-carotene on LDL oxidation *in vitro* are therefore conflicting. What is clear, however, is that supplementing humans with large amounts of β-carotene does not normally inhibit its oxidation *ex vivo*, even though the levels of it in LDL can be raised about 10–20-fold (Princen *et al.*, 1992; Reaven *et al.*, 1994; Gaziano *et al.*, 1995). An exception to this may be in the case of cystic fibrosis, where the levels of β-carotene in LDL are very low (Winklhoferroob *et al.*, 1995). Supplementing the diets of patients with this disease with β-carotene restored its levels in LDL to the normal levels and made the LDL less oxidizable by copper *ex vivo*.

In an epidemiological study of European populations, it was shown that the mean incidence of coronary heart disease in certain regions is inversely correlated with the mean plasma carotene levels in these regions (Gey *et al.*, 1993*a*). In a prospective study in Basel, Switzerland, it was found that a low plasma level of carotene was associated with a higher mortality from coronary heart disease and strokes (Gey *et al.*, 1993*b*). Gey *et al.* (1993*a*) suggested that lycopene may be an important dietary protective factor against cardiovascular disease in Mediterranean countries and that lutein may be important in the Far East. This remains to be proven, however. Smokers with a low serum β-carotene level were found in a prospective epidemiological study in the USA to be more likely to have a myocardial infarction (Street *et al.*, 1994). In another prospective epidemiological study in the USA, the total serum carotenoid level was inversely related to coronary heart disease in hypercholesterolaemia subjects, although in this study the relationship was stronger for non-smokers than for smokers (Morris *et al.*, 1994). A high dietary intake of carotenoids by Finnish women was associated with a somewhat lower mortality from coronary heart disease (Knekt *et al.*, 1994). There was also an inverse relationship between the intake of fruit and vegetables and coronary mortality in both men and women. The authors concluded that the carotenoids may have been less protective than vitamins E and C but may have strengthened the effect of vitamin E intake. Kardinaal *et al.* (1993) have shown that there is an association between β-carotene levels (but not α-tocopherol levels) in adipose tissue (which is a measure of the long-term intake of fatty acids and lipid-soluble compounds) and acute myocardial infarction in nine European countries, especially if the proportion of polyunsaturated fatty acids in adipose tissue was high (Kardinaal *et al.*, 1995). In the Scottish Health Study, a high intake of carotene was associated with a lower incidence of coronary heart disease in men, but not women, both before and after adjustment for other risk factors (Todd *et al.*, 1995).

In a preliminary study of 333 male physicians suffering from coronary heart disease in the USA, Gaziano *et al.* (1990) showed that the risk of developing future vascular events was reduced by 54% by giving them 50 mg of β-carotene on alternate days for about 5 years. A much larger trial is now under way and its results are eagerly awaited.

Although β-carotene does not usually appear to decrease the oxidizability of LDL, it may conceivably protect against cardiovascular disease in other ways, for instance, by being incorporated into cell membranes and possibly decreasing the ability of cells in atherosclerotic lesions to oxidize LDL or by affecting the expression of certain genes. These possibilities remain to be tested.

Ascorbic acid

Ascorbic acid (vitamin C) is the main water-soluble antioxidant in plasma and is found in many fruits and vegetables. It has been shown to inhibit LDL oxidation *in vitro* by copper (Esterbauer *et al.*, 1989; Jialal *et al.*, 1990), macrophages (Jialal and Grundy, 1991; Stait and Leake, 1994), activated neutrophils (Frei, 1991), a water-soluble azo radical initiator (Frei, 1991; Sato *et al.*, 1990; Ma *et al.*, 1994) or cigarette smoke (Frei, 1991). It sometimes inhibits LDL oxidation effectively in the levels found in plasma (20–55 mM). It may act by scavenging free radicals in the aqueous phase or by converting α-tocopheroxyl radicals in LDL back to α-tocopherol so that it may act as a chain-breaking antioxidant again (Sato *et al.*, 1990; Kagan *et al.*, 1992; Kalyanaraman *et al.*, 1992). In addition, its oxidation product dehydroascorbate, or one of its degradation products, may modify apo B-100 in LDL so that it no longer binds copper ions, thus inhibiting LDL oxidation by copper (Retsky *et al.*, 1993).

Physiological levels of ascorbic acid have recently been shown to increase, rather than decrease, the oxidation of LDL that is already partially oxidized, possibly by reducing iron(III) to iron(II) (Stait and Leake, 1994). Hunt *et al.* (1992) have also shown that ascorbate can sometimes be a double-edged sword, in that it could either increase or decrease ceroid (oxidized and polymerized lipid) accumulation in macrophages.

It is therefore not certain whether ascorbate inhibits or promotes LDL oxidation in atherosclerotic lesions, although it is quite possible its net effect may be to inhibit LDL oxidation, but that this inhibition may not be as strong as it would be if it were not for the prooxidant effect of ascorbate under certain circumstances. It should be noted that ascorbate has been shown to protect against atherosclerosis in cholesterol-fed rabbits, even when the plasma cholesterol level was not affected by it (Verlangieri *et al.*, 1977; Beetens *et al.*, 1986).

The epidemiology of ascorbic acid and coronary heart disease is rather mixed. Some studies suggest that it is protective, for instance, the mean plasma level of ascorbic acid in various European populations is inversely related to the incidence of coronary heart disease (Gey *et al.*, 1993*a*) and in the Basel Prospective Study the mortality from coronary heart disease was higher if the plasma ascorbate was low (Gey *et al.*, 1993*b*). In the NHANES study in the USA, the mortality rate for all causes of death was strongly and inversely related to ascorbate intake in males and weakly in females (Enstrom *et al.*, 1992). For the highest level of intake, the standardized mortality rate from cardiovascular disease was reduced to 0.58 for males and to 0.75 for females. Todd *et al.* (1995) found that in the Scottish Health Study there was an inverse association between ascorbate intake and coronary heart disease in men, both before and after adjustment for other risk factors, but in women the association was removed by adjustment. There was an inverse relationship between plasma ascorbate and the risk of angina in another Scottish study, but this may be partly explained by the fact that smokers tend to have a low plasma level of ascorbate (Riemersma *et al.*, 1991). Khaw and Woodhouse (1995) have recently found that the serum level of ascorbate is strongly inversely correlated to the serum levels of fibrinogen and factor VII. The intake of vitamin C in winter in England was less than in summer and this was reflected in the serum levels. They suggested that the lower plasma levels in winter increased people's susceptibility to infections, which in turn increased the levels of haemostatic factors and increased the tendency to arterial thrombosis. This may have contributed to the considerable excess mortality from cardiovascular disease seen in the winter months. This is an interesting idea but it needs a lot more evidence to prove it. Gale *et al.* (1995) have shown in a prospective study involving a 20 year follow up period in elderly people in the UK that the mortality from stroke was increased if the intake or plasma level of ascorbate were low. The relative risk in the highest tertile of ascorbate intake was 0.5 compared to the lowest tertile, after adjustment for cardiovascular risk factors. There was no association between ascorbate and coronary heart disease. Other studies have found no association between ascorbate intake and coronary heart disease (Stampfer *et al.*, 1993; Rimm *et al.*, 1993; Hodis *et al.*, 1995).

Large trials of the affects of ascorbic acid supplements and cardiovascular disease will be needed to decide if ascorbate protects against cardiovascular disease, or at least against certain types of cardiovascular disease.

Flavonoids

There are literally thousands of different kinds of flavonoids and these compounds are ubiquitous in plants (Harborne, 1988). They are therefore present in our normal diet, although there is uncertainty about the amounts we consume. The daily intake of flavonoids has been estimated to be as great as 1 g (Pierpoint, 1986) or as little as 26 mg (Hertog *et al.*, 1993).

Flavonoids protect against LDL oxidation by macrophages or copper (de

Whalley *et al.*, 1990; Jessup *et al.*, 1990). LDL oxidation by macrophages is typically inhibited by 50% by individual flavonoids at concentrations of 1–5 μM, for instance, quercetin, morin, fisetin, gossypetin or hypolaetin. Flavonoids protect the α-tocopherol in LDL from consumption in oxidizing LDL, either by scavenging free radicals themselves or by rejuvenating α-tocopherol from its radical, and extends the lag phase before the lipid hydroperoxides build up in LDL and it is taken up rapidly by macrophages. These results have been confirmed (Myara *et al.*, 1993; Mangiapane *et al.*, 1992; Miura *et al.*, 1994). Flavonoids also inhibit LDL oxidation induced by ultraviolet light (Negre-Salvayre *et al.*, 1991) and quercetin is capable of protecting cells from the toxic effects of oxidized LDL (Negre-Salvayre and Salvayre, 1992). A mixture of a flavonoid (rutin), α-tocopherol and ascorbic acid inhibited LDL oxidation synergistically and also protected endothelial cells from the toxic effects of previously oxidized LDL in a synergistic manner (Negre-Salvayre *et al.*, 1995).

The relative activities of individual flavonoids in scavenging aqueous peroxyl radicals have been assessed (Rice-Evans *et al.*, 1995). Quercetin and myricetin were highly effective, although almost all the flavonoids tested were more effective than α-tocopherol and ascorbate.

Some flavonoids (myricetin and gossypetin) at high concentrations (100 μM) can modify LDL themselves to greatly increase its uptake by macrophages (Rankin *et al.*, 1993). This is not due to oxidation but to the flavonoids causing the cross-linking of apo B-100 molecules in LDL particles to cause the large-scale aggregation of LDL. These aggregates are presumably taken up by the macrophages by phagocytosis. It is very doubtful, however, if these concentrations of flavonoids are obtained *in vivo* in humans and therefore, if anything, flavonoids are likely to inhibit rather than promote LDL modification.

Hertog *et al.* (1993) have found an epidemiological association between the flavonoid intake of elderly men in the Netherlands and the subsequent risk of coronary heart disease. The relative risk of dying from coronary heart disease in the upper tertile of flavonoid intake was only 0.42 of that in the lower tertile and this was still significant after adjustment for the other risk factors. The main sources of flavonoids were tea, onions and apples, in that order (Hertog *et al.*, 1993). The main flavonoid in the diet is quercetin (Fig. 15.3). In 16 cohorts of the Seven Countries study, Hertog *et al.* (1995) found in a 25 year follow-up period that the average intake of flavonoids at baseline was inversely associated with mortality from coronary heart disease (but not from other causes, including cancer).

There is now a great deal of interest in the 'French paradox', i.e. why people in the south of France and other Mediterranean countries have far less coronary heart disease than people in northern Europe, despite eating a similar amount of fat and smoking as much. This could be due to many causes but one that has attracted a lot of attention is that red wine may possibly protect against coronary heart disease (Renaud and De Lorgeril, 1992).

This idea was given a boost in 1993 when Frankel *et al.* in the USA showed that red wine inhibits LDL oxidation by copper (Frankel *et al.*, 1993*a*). This was presumably due to its flavonoid content (de Whalley *et al.*, 1990). One component of wine, resveratrol (which is not a flavonoid but has 3 phenolic hydroxyl groups), was shown to inhibit LDL oxidation by copper (Frankel *et al.*, 1993*b*), but the amounts present in wine are too low for it to be the main antioxidant against LDL oxidation (Frankel *et al.*, 1995).

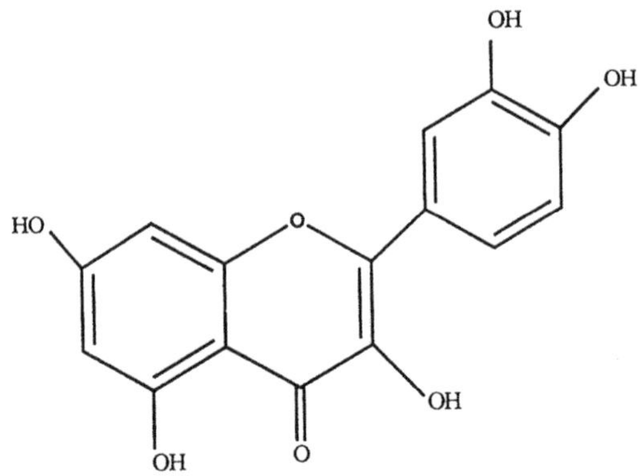

Fig. 15.3. The structure of quercetin, the main flavonoid in the diet.

The concentrations of phenols (most of which are flavonoids) in individual red wines vary quite a lot from about 4 mM to 10 mM and in white wines from about 0.4 mM to 0.7 mM (Vinson and Hontz, 1995). They are higher in red wines than white wines because they are present mainly in the skin of grapes (which is removed in the manufacture of white wine). Vinson and Hontz (1995) reported that the phenols in white wine were somewhat more potent on a molar basis than those in red wine in inhibiting LDL oxidation by copper ions, whereas Frankel *et al.* (1995) found the opposite. Red wine is far more potent than white wine per millilitre in inhibiting LDL oxidation, however, because it contains about ten times the concentrations of phenols. Frankel *et al.* (1995) found that the inhibition of LDL oxidation by various Californian wines correlated with their contents of gallic acid, catechin, myricetin, quercetin, caffeic acid, rutin, epicatechin, cyanidin and malvidin 3-glucoside, in that order. The most abundant monomeric phenolic compound in red wine is the flavonoid catechin, which has a concentration of about 190 mg/l, followed by gallic acid (95 mg/l), epicatechin (82 mg/l), malvidin 3-glucoside (24 mg/l), rutin (9 mg/l), myricetin (8 mg/l), quercetin (8 mg/l), caffeic acid (7 mg/l), cyanidin (3 mg/l) and resveratrol (1.5 mg/l) (Frankel *et al.*, 1995).

We have recently shown that wine inhibits LDL oxidation by macrophages. Red wine diluted 1 to 1000, or sometimes 1 to 10 000, decreased the oxidation of LDL as measured by its rate of macrophage uptake (Fig 15.4) or by thiobarbituric acid-reactive substances. White wine inhibited LDL oxidation only when diluted 1 to 100. Red wine, but not white wine, diluted 1 to 100 actually modified LDL by itself to increase greatly its uptake by macrophages, but this was not due to oxidation as the thiobarbituric acid-reactive substances were not increased. It was due instead to the aggregation of the LDL particles, as they could by sedimented by fairly low speed centrifugation, and was probably analogous to the aggregation of LDL that we have observed previously with high concentrations of certain flavonoids (Rankin *et al.*, 1993).

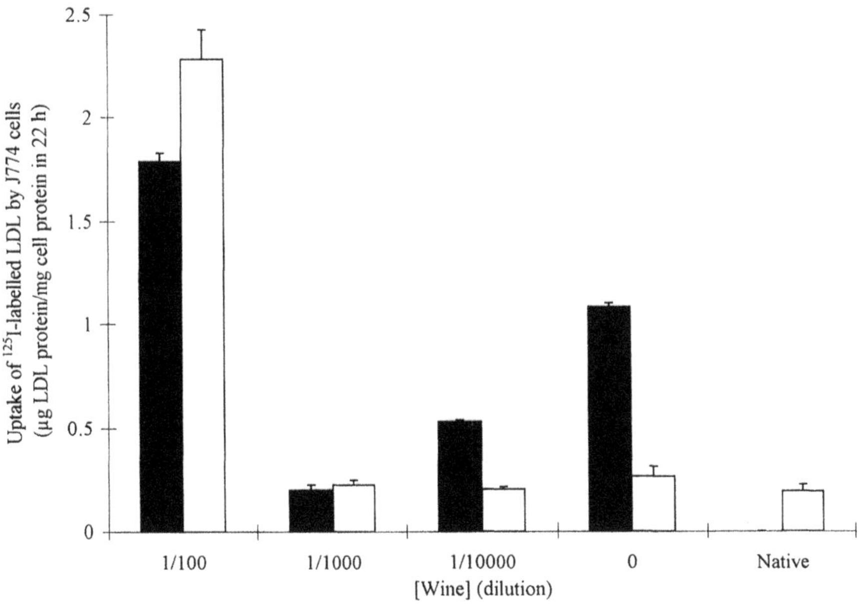

Fig. 15.4. The inhibition of LDL oxidation by red wine. ^{125}I-Labelled LDL (100 μg protein/ml) was incubated for 18 h in triplicate with mouse peritoneal macrophages (macrophage-modified LDL; black bars) or cell-free wells (control LDL; white bars) in Ham's F-10 culture medium in the presence of various concentrations of red wine (Ochoa Tinto Navarra, Spain, 1992). It was then diluted 1 to 10 in serum-containing medium and its uptake by a second set of macrophages was measured. The uptake of native (non-incubated) LDL is also shown. The mean ± SEM of triplicates is shown. Unpublished data of G.M. Wilkins and D.S. Leake (1995).

We have also incubated human plasma overnight with 3 or 10% (v/v) red wine and then isolated LDL from it and shown that the LDL becomes more resistant to oxidation (Fig. 15.5). This shows that antioxidants in red wine become incorporated into LDL in plasma and remained associated with it during a fairly lengthy isolation procedure, which included an overnight dialysis step, and then protected it against oxidation. It also shows that the antioxidant effects of red wine are not due, or at least not due entirely, to low M_r water-soluble components, as these would have been removed during the dialysis step.

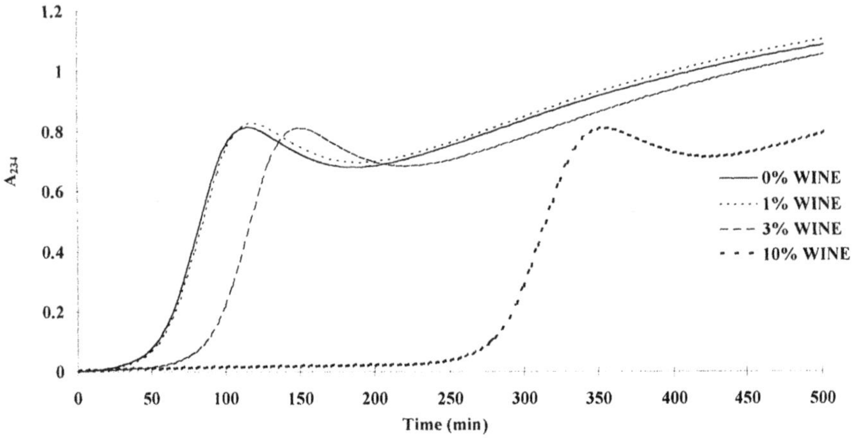

Fig. 15.5. The incubation of plasma with red wine makes its LDL more resistant to oxidation after isolation. Human plasma was incubated for 18 h at 37 C with up to 10% (v/v) red wine (Cabernet Sauvignon, France, 1993). The LDL was isolated by sequential density ultracentrifugation, dialysed overnight and incubated at 37 C at 50 μg protein/ml with copper sulfate (net concentration 2 μM over the EDTA present). The absorbance at 234 nm was monitored as a measure of the conjugated dienes formed in the oxidizing lipid of LDL. Unpublished data of G.M. Wilkins and D.S. Leake (1995).

The important question is whether or not the lipid-soluble antioxidants in red wine are absorbed from the human gut and remain in the body in the appropriate chemical form for long enough to associate with LDL and give it some protection against oxidation *in vivo*? Very little is known about flavonoid absorption, metabolism and excretion in humans (Griffiths, 1982) but there is limited evidence that at least some flavonoids are absorbed (Das, 1971; Hackett, 1986). It is unclear whether or not red wine ingestion by humans makes their LDL less oxidizable. Some workers have shown no effect (de Rijke *et al.*, 1995), whereas others have shown some protection (Kondo *et al.*, 1994; Fuhrman *et al.*, 1995). Kondo *et al.* (1994) showed a small (11%) increase in resistance of LDL to oxidation by a lipid-soluble azo radical initiator after 2 weeks on red wine (about 600 ml/day). Fuhrman *et al.* (1995) have reported that giving men 400 ml of red wine per day with meals for 2 weeks decreased the oxidizability of their LDL *ex vivo* with copper. Little effect was seen after 1 week. White wine consumption was reported to actually increase the oxidizability of LDL. The total polyphenols in LDL, measured using phosphomolybdic phosphotungstic acid, were increased 4-fold after 2 weeks on red wine, up to about 30 mg/g LDL protein. A simple calculation shows that this corresponds to about 50 molecules of phenols per LDL particle, in terms of catechin. This is a very large number and would be an important finding, if it can be confirmed.

More evidence that antioxidants in red wine may actually be absorbed in humans and reach significant concentrations in plasma comes from the work of Whitehead *et al.* (1995), who have shown that the antioxidant capacity of plasma was increased 18% 1 h after drinking red wine. White wine gave no significant increase.

More studies will be needed before a consensus emerges as to whether or not red wine consumption reduces the oxidizability of LDL.

Although the antioxidant properties of flavonoids have attracted most attention in the possible protective effect of these compounds against cardiovascular disease, flavonoids may in theory protect in other ways. For instance, they may possibly protect against thrombosis (Gryglewski *et al.*, 1987) or they may have oestrogenic activity (Miksicek, 1995). The oestrogenic activity of flavonoids may be interesting because coronary heart disease is delayed in women for about a decade compared to men but this protection is lost after the menopause. Male alcoholics sometimes show evidence of feminization. Some alcoholic beverages have recently been shown to have oestrogenic activity in animals *in vivo*, and bourbon and wine contain compounds that bind to the oestrogen receptor in vitro (Gavaler *et al.*, 1995). The idea that flavonoids in wine may possibly protect against coronary heart disease because of their oestrogenic activity may be an interesting one to explore in the future.

Other antioxidants in plants

Plants contain many antioxidants that have the capacity to inhibit LDL oxidation *in vitro*. Thus, micromolar or submicromolar concentrations of chlorogenic acid, caffeic acid, ferulic acid and *p*-coumaric acid, which are hydroxycinnamic acids, inhibit LDL oxidation by metmyoglobin (Castelluccio *et al.*, 1995) or a water-soluble azo initiator (Laranjinha *et al.*, 1994) by scavenging peroxyl radicals. LDL oxidation by a water-soluble azo initiator is also inhibited by protocatechuic acid (a hydroxybenzoic acid derivative) and ellagic acid (a tannic compound) (Larajinha *et al.*, 1994). These compounds are widely distributed in the plant kingdom and the daily intake of ferulic acid had been calculated to be about 500 mg/day.

Other substances in plants that have been shown to inhibit LDL oxidation *in vitro* include oleuropein (present in olive oil) (Visioli and Galli, 1994), hydroxytyrosol and its derivatives (also present in olive oil) (Visioli *et al.*, 1995), garlic (Lewin and Popov, 1994), compounds (maybe phenolics in the tar fraction) in cigarette smoke (Chen and Loo, 1995) (although smoking may increase LDL oxidation *in vivo*), Chinese green tea polyphenols (Zhenhua *et al.*, 1991), Japanese herbal Kampo medicines (Ondriaš *et al.*, 1992) and hydroxyeugenol- and coniferyl alcohol derivatives (Deigner *et al.*, 1994). Oral supplementation with garlic powder has been reported to reduce the oxidizability of human LDL plus VLDL by copper (Phelps and Harris, 1993) but this could not be confirmed in another study, this time in hypercholesterolaemic subjects (Simons *et al.*, 1995).

Conclusions

The number of studies of the possible effects of antioxidants in plant foodstuffs on cardiovascular disease has increased dramatically in recent years and will no doubt continue to grow. This field is a highly complex one, however, one of the main reasons being that atherosclerosis is a disease that takes decades to develop and the clinical symptoms involve not only the atherosclerotic lesions themselves but also the thrombosis that can precipitate a heart attack or stroke and also the electrical disturbances in the heart after a heart attack that can affect its normal rhythmic beat. It is possible that all of these processes may be affected by dietary factors, with vastly different time scales. Reliable clinical trials involving coronary heart disease or strokes require thousands or even tens of thousands of patients and are therefore extremely expensive to carry out. For these reasons and more, few definite recommendations can be made about diet and cardiovascular disease. One of the few recommendations about which there is widespread agreement is the need to reduce the amount of saturated fat in the diet, as saturated fat is the single most important factor in increasing the cholesterol levels in plasma. All other dietary recommendations are either controversial or based on incomplete evidence. We cannot therefore say with certainty that antioxidants in fruit or vegetables are important in protecting against cardiovascular disease; all we can say is that there are epidemiological studies consistent with this possibility and that there are *in vitro* and animal experiments that suggest it. It may be many years before we can prove that antioxidants in plants are important in helping to protect against human disease, but given that definitive proof is especially difficult to obtain in this area it may be prudent in the meantime to increase our intake of fruit and vegetables in case they do.

Acknowledgements

I am grateful to the Wellcome Trust for a Research Leave Fellowship, which gave me the time to write this chapter.

References

Andrews, B., Burnand, K., Paganga, G., Browse, N., Rice-Evans, C., Sommerville, K., Leake, D. and Taub, N. (1995). *Atherosclerosis* **112**, 77–84.
Beetens, J.R., Coene, M.-C., Verheyen, A., Zonnekeyn, L. and Herman, A.G. (1986). *Prostaglandins* **32**, 335–52.
Benz, D.J., Mol, M., Ezaki, M., Moriito, N., Zelan, I., Miyanohara, A., Friedmann, T., Parthasarathy, S., Steinberg, D. and Witztum, J.L. (1995). *J. Biol. Chem.* **270**, 5191–7.
Björkhem, I., Henriksson-Freyschuss, A., Breuer, O., Diczfalusy, U., Berglund, L. and Henriksson, P. (1991). *Arteriosclerosis* **11**, 15–22.
Bowry, V.W. and Stocker, R. (1993). *J. Am. Chem. Soc.* **115**, 6029–44.
Bowry, V.W., Ingold, K.U. and Stocker, R. (1992). *Biochem. J.* **288**, 341–4.
Carew, T.E., Schwenke, D.C. and Steinberg, D. (1987). *Proc. Natl. Acad. Sci. USA* **84**, 7725–9.
Carpenter, K.L.H., Wilkins, G.M., Fussell, B., Ballantine, J.A., Taylor, S.E., Mitchinson, M.J. and Leake, D.S. (1994). *Biochem. J.* **304**, 625–33.
Castelluccio, C., Paganga, G., Melikian, N., Bolwell, G.P., Pridham, J., Sampson, J. and Rice-Evans, C. (1995). *FEBS Lett.* **368**, 188–92.
Cathcart, M.K., McNally, A.K., Morel, D.W. and Chisolm G.M., III (1989). *J. Immunol.* **142**, 1963–9.
Chen, C. and Loo, G. (1995). *Atherosclerosis* **112**, 177–85.
Cominacini, L., Garbin, U., Pastorino, A.M., Davoli, A., Campagnola, M., De Santis, A., Pasini, C., Faccini, G.B., Trevisan, M.T., Bertozzo, L., Pasini, F. and Lo Cascio, V. (1993). *Atherosclerosis* **99**, 63–70.
Das, N.P. (1971). *Biochem. Pharmacol.* **20**, 3435–45.
Davies, M.J. and Thomas, A. (1984). *New. Engl. J. Med.* **310**, 1137–40.
Deigner, H.P., Wolf, G., Ohlenmacher, U. and Reichling, J. (1994). *Arzneimittel-Forschung-Drug Res.* **44**, 956–61.
Dieber-Rotheneder, M., Puhl, H., Waeg, G., Striegl, G. and Esterbauer, H. (1991). *J. Lipid. Res.* **32**, 1325–32.
Ehrenwald, E., Chisolm, G.M. and Fox, P.L. (1994). *J. Clin. Invest.* **93**, 1493–501.
Endemann, G., Stanton, L. W., Madden K.S., Bryant, C.M., White, R.T. and Protter, A.A. (1993). *J. Biol. Chem.* **268**, 11811–16.
Enstrom, J.E., Kanim, L.E. and Klein, M.A. (1992). *Epidemiology* **3**, 194–202.
Esterbauer, H., Jürgens, G., Quehenberger, O. and Koller, E. (1987). *J. Lipid Res.* **28**, 495–509.
Esterbauer, H., Striegl, G., Puhl, H. and Rotheneder, M. (1989). *Free Rad. Res. Comm.* **6**, 67–75.
Esterbauer, H., Gebicki, J., Puhl, H. and Jürgens, G. (1992). *Free Rad. Biol. Med.* **13**, 341–90.

Frankel, E.N., Kanner, J., German, J.B., Parks, E. and Kinsella, J.E. (1993*a*). *Lancet* **341**, 454–7.
Frankel, E.N., Waterhouse, A.L. and Kinsella, J.E. (1993*b*). *Lancet* **341**, 1103–4.
Frankel, E.N., Waterhouse, A.L. and Teissedre, P.L. (1995). *J. Agric. Food Chem.* **43**, 890–4.
Frei, B. (1991). *Am. J. Clin. Nutr.* **54** Suppl., 1113S–1118S.
Fuhrman, B., Lavy, A. and Aviram, M. (1995). *Am. J. Clin. Nutr.* **61**, 549–54.
Gale, C.R., Martyn, C.N., Winter, P.D. and Cooper, C. (1995). *Brit. Med. J.* **310**, 1563–6.
Gavaler, J.S., Rosenblum, E.R., Deal, S.R. and Bowie, B.T. (1995). *Proc. Soc. Exp. Biol. Med.* **208**, 98–102.
Gaziano, J.M., Manson, J.E., Ridker, P.M., Buring, J.E. and Hennekens, C.H. (1990). *Circulation* **82**, III-201.
Gaziano, J.M., Hatta, A., Flynn, M., Johnson, E.J., Krinsky, N.I., Ridker, P.M., Hennekens, C.H. and Frei, B. (1995). *Atherosclerosis* **112**, 187–95.
Gey, K.F., Puska, P., Jordan, P. and Moser, U.K. (1991). *Am. J. Clin. Nutr.* **53** Suppl., 326S–334S.
Gey, K.F., Moser, U.K., Jordan, P., Stähelin, H.B., Eichholzer, M. and Lüdin, E. (1993*a*). *Am. J. Clin. Nutr.* **57** Suppl., 787S–797S.
Gey, K.F., Stähelin, H.B. and Eichholzer, M. (1993*b*). *Clin. Investig.* 71, 3–6.
Graham, A., Wood J.L., O'Leary, V.J. and Stone, D. (1994). *Free Rad. Res.* **21**, 295–308.
Griffiths, L.A. (1982). In *The flavonoids: Advances in research* (ed. J.B. Harborne and T.J. Mabry), pp. 681–718, Chapman & Hall, London.
Gryglewski, R.J., Korbut, R., Robak, J. and Swies, J. (1987). *Biochem. Pharmacol.* **36**, 317–22.
Hackett, A.M. (1986). In *Plant flavonoids in biology and medicine: Biochemical, pharmacological, and structure-activity relationships* (ed. V. Cody, E. Middleton and J.B. Harborne), pp. 177–94. Alan R. Liss, New York.
Harborne, J.B. (1988). *The Flavonoids: Advances in Research Since 1980*, Chapman & Hall, London.
Heinecke, J.W., Rosen, H. and Chait, A. (1984). *J. Clin. Invest.* **74**, 1890–4.
Heinecke, J.W., Baker, L., Rosen, H. and Chait, A. (1986). *J. Clin. Invest.* **77**, 757–61.
Heinecke, J.W., Rosen, H., Suzuki, L.A. and Chait, A. (1987). *J. Biol. Chem.* **262**, 10098–103.
Hertog, M.G.L., Feskens, E.J.M., Hollman, P.C.H., Katan, M.B. and Kromhout, D. (1993). *Lancet* **342**, 1007–11.
Hertog, M.G.L., Kromhout, D., Aravanis, C., Blackburn, H., Buzina, R., Fidanza, F., Giampaoli, S., Jansen, A., Menotti, A., Nedeljkovic, S., Pekkarinen, M., Simic, B.S., Toshima, H., Feskens, E.J.M., Hollman, P.C.H. and Katan, M.B. (1995). *Arch. Int. Med.* **155**, 381–6.
Hiramatsu, K., Rosen, H., Heinecke, J.W., Wolfbauer, G. and Chait, A. (1987). *Arteriosclerosis* **7**, 55–60.
Hodis, H.N., Mack, W.J., Labree, L., Cashinhemphill, L., Sevanian, A., Johnson, R. and Azen, S.P. (1995). *J. Am. Med. Ass.* **273**, 1849–54.
Hunt, J.V., Bottoms, M.A. and Mitchinson, M.J. (1992). *FEBS Lett.* **311**, 161–4.
Jessup, W., Rankin, S.M., de Whalley, C.V., Hoult, J.R.S., Scott, J. and Leake, D.S. (1990). *Biochem. J.* **265**, 399–405.
Jessup, W., Darley-Usmar, V., O'Leary, V. and Bedwell, S. (1991). *Biochem. J.* **278**, 163–9.

Jessup, W., Simpson, J.A. and Dean, R.T. (1993). *Atherosclerosis* **99**, 107–20.
Jialal, I. and Grundy, S.M. (1991). *J. Clin. Invest.* **87**, 597–601.
Jialal, I. and Grundy, S.M. (1992). *J. Lipid Res.* **33**, 899–906.
Jialal, I., Vega, G.L. and Grundy, S.M. (1990). *Atherosclerosis* **82**, 185–91.
Jialal, I., Norkus, E.P., Cristol, L. and Grundy, S.M. (1991). *Biochim. Biophys. Acta* **1086**, 134–8.
Kagan, V.E., Serbinova, E.A., Forte, T., Scita, G. and Packer, L. (1992). *J. Lipid Res.* **33**, 385–97.
Kalyanaraman, B., Darley-Usmar, V.M., Wood, J., Joseph, J. and Parthasarathy, S. (1992). *J. Biol. Chem.* **267**, 6789–95.
Kardinaal, A.F.M., Kok, F.J., Ringstad, J., Gomez-Aracena, J., Mazaev, V.P., Kohlmeier, L., Martin, B.C., Aro, A., Kark, J.D., Delgado-Rodriguez, M., Riemersma, R.A., van't Veer, P., Huttunen, J.K. and Martin-Moreno, J.M. (1993). *Lancet* **342**, 1379–84.
Kardinaal, A.F.M., Aro, A., Kark, J.D., Riemersma, R.A., van't Veer, P., Gomez-Aracena, J., Kohlmeier, L., Ringstad, J., Martin, B.C., Mazaev, V.P., Delgado-Rodriguez, M., Thamm, M., Huttunen, J.K., Martin-Moreno, J.M. and Kok, F.J. (1995). *Aterioscler. Thromb. Vasc. Biol.* **15**, 726–32.
Khaw, K.-T. and Woodhouse, P. (1995). *Brit. Med. J.* **310**, 1559–63.
Knekt, P., Reunanen, A., Järviven, R., Seppänen, R., Heliövaara, M. and Aromaa, A. (1994). *Am. J. Epidemiol.* **139**, 1180–9.
Kodama, T., Freeman, M., Rohrer, L., Zabrecky, J., Matsudaira, P. and Krieger, M. (1990). *Nature* **343**, 531–5.
Kondo, K., Matsumoto, A., Kurata, H., Tanahashi, H., Koda, H., Amachi, T. and Itakura, H. (1994). *Lancet* **344**, 1152.
Kontush, A., Hübner,C., Finckh, B., Kohlschütter, A. and Beisiegel, U. (1994). *FEBS Lett.* **341**, 69–73.
Lamb, D.J. and Leake, D.S. (1994*a*). *FEBS Lett.* **338**, 122–6.
Lamb, D.J. and Leake, D.S. (1994*b*). *FEBS Lett.* **352**, 15–18.
Lamb, D.J., Wilkins, G.M. and Leake, D.S. (1992). *Atherosclerosis* **92**, 187–92.
Laranjinha, J.A.N., Almeida, L.M. and Madeira, V.M.C. (1994). *Biochem. Pharmacol.* **48**, 487–94.
Leake, D.S. (1993). *Brit. Heart J.* **69**, 476–8.
Leake, D.S. and Rankin, S.M. (1990). *Biochem. J.* **270**, 741–8.
Lewin, G. and Popov, I. (1994). *Arzneimittel-Forschung. Drug Res.* **44**, 604–7.
Ma, Y.-S., Stone, W.L. and LeClair, I.O. (1994). *Proc. Soc. Exp. Biol. Med.* **206**, 53–9.
Mabile, L., Fitoussi, G., Periquet, B., Schmitt, A., Salvayre, R. and Negre-Salvayre, A. (1995). *Free Rad. Biol. Med.* **19**, 177–87.
Mangiapane, H., Thomson, J., Salter, A., Brown, S., Bell, G.D. and White, D.A. (1992). *Biochem. Pharmacol.* **43**, 445–50.
Miksicek, R.J. (1995). *Proc. Soc. Exp. Biol. Med.* **208**, 44–50.
Miura, S., Watanabe, J., Tomita, T., Sano, M. and Tomita, I. (1994). *Biol. Pharm. Bull.* **17**, 1567–72.
Morgan, J. and Leake, D.S. (1993). *FEBS Lett.* **333**, 275–9.
Morris, D.L., Kritchevsky, S.B. and Davis, C.E. (1994). *J. Am. Med. Ass.* **272**, 1439–41.
Myara, I., Pico, I., Vedie, B. and Moatti, N. (1993). *J. Pharmacol. Toxicol. Method.* **30**, 69–73.

Navab, M., Imes, S.S., Hama, S.Y., Hough, G.P., Ross, L.A., Bork, R.W., Valente, A.J., Berliner, J.A., Drinkwater, D.C., Laks, H. and Fogelman, A.M. (1991). *J. Clin. Invest.* **88**, 2039–46.

Negre-Salvayre, A. and Salvayre, R. (1992). *Free Rad. Biol. Med.* **12**, 101–6.

Negre-Salvayre, A., Alomar, Y., Troly, M. and Salvayre, R. (1991). *Biochim. Biophys. Acta* **1096**, 291–300.

Negre-Salvayre, A., Mabile, L., Delchambre, J. and Salvayre, R. (1995). *Biol. Trace Elem. Res.* **47**, 81–91.

Ondriaš, K., Staško, A., Gergel, D., Hromadová, M. and Beně, L. (1992). *Free Rad. Res. Comm.* **16**, 227–37.

Ottnad, E., Parthasarathy, S., Sambrano, G.R., Ramprasad, M.P., Quehenberger, O., Kondratenko, N., Green, S. and Steinberg, D. (1995). *Proc. Natl. Acad. Sci. USA* **92**, 1391–5.

Palinski, W., Rosenfeld, M.E., Ylä-Herttuala, S., Gurtner, G.C., Socher, S.S., Butler, S.W., Parthasarathy, S., Carew, T.E., Steinberg, D. and Witztum, J.L. (1989). *Proc. Natl. Acad. Sci. USA* **86**, 1372–6.

Palozza, P. and Krinsky, N.I. (1992). *Meth. Enzymol.* **213**, 403–20.

Parker, R. A., Sabrah, T., Cap, M. and Gill, B.T. (1995). *Arterioscler. Thromb. Vasc. Biol.* **15**, 349–58.

Parthasarathy, S., Wieland, E. and Steinberg, D. (1989). *Proc. Natl. Acad. Sci. USA* **86**, 1046–50.

Phelps, S. and Harris, W.S. (1993). *Lipids* **28**, 475–7.

Pierpoint, W.S. (1986). In *Plant flavonoids in biology and medicine: Biochemical, pharmacological, and structure-activity relationships* (ed. V. Cody, E. Middleton, Jr. and J.B. Harborne), pp. 125–140. Alan R. Liss, New York.

Princen, H.M.G., Van Poppel, G., Vogelezang, C., Buytenhek, R. and Kok, F.J. (1992). *Arterioscler. Thromb.* **12**, 554–62.

Princen, H.M.G., van Duyvenvoorde, W., Buytenhek, R., van der Laarse, A., van Poppel, G., Gevers Leuven, J.A. and van Hinsbergh, V.W.M. (1995). *Aterioscler. Thromb. Vasc. Biol.* **15**, 325–33.

Rankin, S.M., Parthasarathy, S. and Steinberg, D. (1991). *J. Lipid Res.* **32**, 449–56.

Rankin, S.M., de Whalley, C.V., Hoult, J.R.S., Jessup, W., Wilkins, G.M., Collard, J. and Leake, D.S. (1993). *Biochem. Pharmacol.* **45**, 67–75.

Reaven, P.D., Khouw, A., Beltz, W.F., Parthasarathy, S. and Witztum, J.L. (1993). *Aterioscler. Thromb.* **13**, 590–600.

Reaven, P.D., Ferguson, E., Navab, M. and Powell, F.L. (1994). *Arterioscler. Thromb.* **14**, 1162–9.

Regnström, J., Nilsson, J., Tornvall, P., Landou, C. and Hamsten, A. (1992). *Lancet* **339**, 1183–6.

Reid, V.C. and Mitchinson, M.J. (1993). *Atherosclerosis* **98**, 17–24.

Renaud, S. and De Lorgeril, M. (1992). *Lancet* **339**, 1523–6.

Retsky, K.L., Freeman, M.W. and Frei, B. (1993). *J. Biol. Chem.* **268**, 1304–9.

Rice-Evans, C. and Bruckdorfer, K.R. (1995). *Oxidative stress, lipoproteins and cardiovascular dysfunction.* Portland Press, London.

Rice-Evans, C.A., Miller, N.J., Bolwell, P.G., Bramley, P.M. and Pridham, J.B. (1995). *Free Rad. Res.* **22**, 375–83.

Riemersma, R.A., Wood, D.A., Macintyre, C.C.A., Elton, R.A., Gey, K.F. and Oliver, M.F. (1991). *Lancet* **337**, 1–5.

de Rijke, Y.B., Demacker, P.N.M., Assen, N.A., Sloots, L.M., Katan, M.B. and Stalenhoef, A.F.H. (1995). *Atherosclerosis* **115** Suppl., S15.
Rimm, E.B., Stampfer, M.J., Ascherio, A., Giovannucci, E., Golditz, G.A. and Willett, W.C. (1993). *N. Engl. J. Med.* **328**, 1450–6.
Sato, K., Niki, E. and Shimasaki, H. (1990). *Arch. Biochem. Biophys.* **279**, 402–5.
Shaikh, M., Martini, S., Quiney, J.R., Baskerville, P., LaVille, A.E., Browse, N.L., Duffield, R., Turner, P.R. and Lewis, B. (1988). *Atherosclerosis* **69**, 165–72.
Simons, L.A., Balasubramaniam, S., von Konigsmark, M., Parfitt, A., Simons, J. and Peters, W. (1995). *Atherosclerosis* **113**, 219–25.
Smith, C., Mitchinson, M.J., Aruoma, O.I. and Halliwell, B. (1992). *Biochem. J.* **286**, 901–5.
Smith, D., O'Leary, V.J. and Darley-Usmar, V.M. (1993). *Biochem. Pharmacol.* **45**, 2195–2201.
Sparrow, C.P. and Olszewski, J. (1992). *Proc. Natl. Acad. Sci. USA* **89**, 128–31.
Sparrow, C.P. and Olszewski, J. (1993). *J. Lipid Res.* **34**, 1219–28.
Stait, S.E. and Leake, D.S. (1994). *FEBS Lett.* **341**, 263–7.
Stampfer, M.J., Hennekens, C.H., Manson, J.E., Colditz, G.A., Rosner, B. and Willett, W.C. (1993). *N. Engl. J.Med.* **328**, 1444–9.
Stanton, L.W., White, R.T., Bryant, C.M., Protter, A.A. and Endemann, G. (1992). *J. Biol. Chem.* **267**, 22446–51.
Steinberg, D., Parthasarathy, S., Carew, T.E., Khoo, J.C. and Witztum, J.L. (1989). *New Engl. J.Med.* **320**, 915–24.
Steinberg, D. and Workshop Participants (1992). *Circulation* **85**, 2337–44.
Steinbrecher, U.P. (1988). *Biochim. Biophys. Acta* **959**, 20–30.
Stocker, R., Bowry, V.W. and Frei, B. (1991). *Proc. Natl. Acad. Sci. USA* **88**, 1646–50.
Street, D.A., Comstock, G.W., Salkeld, R.M., Schüep, W. and Klag, M.J. (1994). *Circulation* **90**, 1154–61.
Swain, J.A., Darley-Usmar, V. and Gutteridge, J.M.C. (1994). *FEBS Lett.* **342**, 49–52.
Todd, S., Woodward, M. and Bolton-Smith, C. (1995). *J. Clin. Epidemiol.* **48**, 307–16.
Verlangieri, A.J. and Bush, M.J. (1992). *J. Am. Coll. Nutr.* **11**, 131–8.
Verlangieri, A.J., Hollis, T.M. and Mumma, R.O. (1977). *Blood Vessels* **14**, 157–74.
Vinson, J.A. and Hontz, B.A. (1995). *J. Agric. Food Chem.* **43**, 401–3.
Visioli, F. and Galli, C. (1994). *Life Sci.* **55**, 1965–71.
Visioli, F., Bellomo, G., Montedoro, G. and Galli, C. (1995). *Atherosclerosis* **117**, 25–32.
de Whalley, C.V., Rankin, S.M., Hoult, J.R.S., Jessup, W. and Leake, D.S. (1990). *Biochem. Pharmacol.* **39**, 1743–50.
Whitehead, T.P., Robinson, D., Allaway, S., Syms, J. and Hale, A. (1995). *Clin. Chem.* **41**, 32–5.
Wilkins, G.M. and Leake, D.S. (1994a). *Biochim. Biophys. Acta* **1215**, 250–8.
Wilkins, G.M. and Leake, D.S. (1994b). *Biochim. Biophys. Acta* **1211**, 69–78.
Wilkins, G.M., Segal, A.W. and Leake, D.S. (1994). *Biochem. Biophys. Res. Comm.* **202**, 1300–7.
Winklkoferroob, B.M., Puhl, H., Khoschsorur, G., Vanthof, M.A., Esterbauer, H. and Shmerling, D.H. (1995). *Free Rad. Biol.Med.* **18**, 849–59.
Witztum, J.L. and Steinberg, D. (1991). *J. Clin. Invest.* **88**, 1785–92.
Ylä-Herttuala, S., Palinski, W., Rosenfeld, M.E., Parthasarathy, S., Carew, T.E., Butler, S., Witztum, J.L. and Steinberg, D. (1989). *J. Clin. Invest.* **84**, 1086–95.
Zhenhua, D., Yuan, C., Mei, Z. and Yunzhong, F. (1991). *Med. Sci. Res.* **19**, 767–8.

16. Potentially anticarcinogenic secondary metabolites from fruit and vegetables

MICHAËL G.L. HERTOG[1], GEERT VAN POPPEL[2] and DORETTE VERHOEVEN[2]

[1]Department of Chronic Diseases and Environmental Epidemiology, National Institute of Public Health and Environmental Protection, PO Box 1, 3720 BA Bilthoven, The Netherlands and [2]TNO Nutrition and Food Research Institute, PO Box 360, 3700 AJ Zeist, The Netherlands

Introduction

Cancer embraces a group of diseases which have in common that a cell or a group of cells is altered in such a way that it escapes growth control and replicates, thus prohibiting a normal functioning of the organ in which these cells are located. There is overwhelming evidence from experimental and epidemiological studies that most common forms of human cancers are related to environmental life-style factors such as diet, in combination with genetic factors that influence the susceptibility of individuals to these factors (Doll and Peto, 1981; Steinmetz and Potter, 1991*a*; Marshall, 1991). A large number of epidemiologic studies have shown that the consumption of vegetables and fruit is inversely associated with risk of cancer, particularly cancers of the alimentary and respiratory tract (Block *et al.*, 1992). Vegetables and fruit contain a large number of components that could, by a variety of mechanisms, explain these cancer-protective effects. Until recently, most attention has been paid to vitamins such as vitamin C, pro-vitamin A, and dietary fibre in these foods but in recent years it has been shown that these plant tissues contain a whole variety of potentially anticarcinogenic secondary metabolites. These include, amongst others, flavonoids, phenolic acids, terpenes, glucosinolates, and carotenoids (Table 16.1). The term 'secondary metabolite' refers to compounds that are not essential to the survival of the whole plant or certain parts of the plant. Some of these compounds have been called non-nutritive components because they do not have any apparent nutritive value for humans. More recently the term 'bioactive components' has been coined for the group of components in foods that have, or are supposed to have, biological effects in the human body. In this overview, three important groups of potentially anticarcinogenic bioactive components—carotenoids, flavonoids and glucosinolates—

and their significance in cancer prevention are discussed. Firstly a short overview of carcinogenesis and anticarcinogenesis is given. Additional information on structure and occurrence in foods is presented for flavonoids and glucosinolates.

Table 16.1. Selected bioactive food components and their occurrence in foods

Component	Representative	Food	Reference
Carotenoids	β-Carotene	Green–yellow vegetables, fruits	Mangels *et al.*, 1993
Flavonoids	Quercetin	Onions, apples, tea, red wine	Hertog *et al.*, 1992, 1993*a*
Phenolic acids	Cinnamic acid	Coffee, fruits	Stöhr and Herrman, 1975
Glucosinolates	Indole-3-carbinol	Cruciferous vegetables e.g. Brussels sprouts, broccoli	Fenwick *et al.*, 1983
Terpenes	D-limonene	Orange oil	Fiala *et al.*, 1985
Organosulfur compounds	Diallylsulfide	*Allium* vegetables e.g. leek, onions, garlic	Fiala *et al.*, 1985
Phytoestrogens, isoflavonoids	Genistein	Soya products, legumes	Franke *et al.*, 1994

Carcinogenesis and anticarcinogenesis

There is considerable evidence that carcinogenesis is a multistage process which can be initiated and modulated by chemical agents present in the environment (e.g. diet), but also by viruses and ultraviolet radiation (IARC, 1990). Most experimental evidence on the multistage process of carcinogenesis is derived from the mouse-skin model. In general, three distinct steps in chemical carcinogenesis are considered: initiation, promotion, and malignant conversion or progression. In the initiation phase, irreversible damage to the DNA of a cell is introduced, which provides the cell with a relative growth advantage, possibly due to the activation of proto-oncogenes (Guyton and Kenseler, 1993). In the promotion phase, proliferation of the initiated cell is stimulated, possibly by binding to and activation of, cellular receptors which are involved in growth control, or by inhibition of intracellular communication (Weisburger, 1992). Promotion leads to benign growth that can evolve in the third stage into cancer. Malignant conversion requires, as with initiation, genetic alterations in which growth control is further inhibited, possibly by inhibition of tumour suppression genes (Guyton and Kensler, 1993). It is probable that multiple initiating and promoting events are involved in tumour development and growth. Vogelstein and co-workers presented a multistage genetic model for human colorectal tumour development. In this model at least five genetic mutations in affected cells are needed for cancer formation (Vogelstein *et al.*, 1988; Fearon and Vogelstein, 1990). In experimental animal studies it was shown that initiation, promotion, and malignant conversion can be modulated, e.g. stimulated or inhibited, by chemical compounds in the diet. Compounds that inhibit chemically induced carcinogenesis are called 'anticarcinogens'. Due to the pioneering work of Wattenberg (1992), anticarcinogens have been classified—mainly upon the moment in the process of carcinogenesis in which they are active—into:

- inhibitors of carcinogen formation
- carcinogen scavengers
- agents that inhibit promotion and/or malignant conversion
- agents that stimulate DNA repair mechanisms.

Most compounds show several mechanisms of action. Inhibitors of carcinogen formation are substances that stimulate the activity of phase I and phase II enzymes, which are important in deactivating potential toxins, including carcinogens. Phase I enzymes are microsomal mono-oxygenases such as the P450 family. Phase II enzymes conjugate products from phase I transformation with for instance glutathione, thus making carcinogens more hydrophilic and more easily excreted. For instance, the activation of the precarcinogen benzo(*a*)pyrene to its active metabolite (ultimate carcinogen) is inhibited and its excretion is enhanced by induction of phase II conjugating enzymes (Wattenberg, 1987). Carcinogen scavengers trap ultimate carcinogens by binding and forming inactive complexes. Polycyclic aromatic hydrocarbons are thus inactivated (Wattenberg and Loub, 1978). Antioxidants trap reactive oxygen species which are thought to be involved in promotion and malignant conversion (Guyton and Kensler, 1993). In addition, products from the arachidonic acid cascade such as prostaglandins and leucotrienes stimulate promotion and, possibly, progression (Fürstenberger *et al.*, 1989). Anti-inflammatory agents which inhibit the arachidonic acid cascade may thus inhibit promotion and malignant conversion (Fischer *et al.*, 1982).

Carotenoids

General

Almost 600 carotenoids from natural sources have been characterized. Green leafy vegetables and many coloured fruits are rich in carotenoids. Mangels and colleagues (1993) generated an extensive database giving the content of five major carotenoids of a large variety of foods. Carotenoids in humans are absorbed from the intestine and appear in the blood. Human serum contains β-carotene, α-carotene, cryptoxanthin, lycopene and lutein as major components (Parker, 1989). A number of biological activities of carotenoids have been described that may pertain to their putative cancer preventive potential. Less than 10% of all carotenoids serve as precursor for vitamin A. In humans, β-carotene, α-carotene, and cryptoxanthin are converted to retinol, whereas lutein and lycopene are not. Retinoids have an important role in control of cell differentiation, cell adhesion and cell membrane permeability (Sporn and Roberts, 1983; DiGiovanni, 1990). Blood levels of retinol are under homeostatic control but De Vet (1989) argued that β-carotene, especially, may exert an influence through local conversion into retinol, i.e. after reaching posthepatic tissues. Also, it has been hypothesized that β-carotene is capable of rapidly compensating localized vitamin A deficiencies induced by carcinogens (Van Poppel, 1993).

Antioxidant functions of carotenoids would enable them to neutralize free radicals, intermediates of metabolism that contain a non-paired electron and that can damage biomolecules like DNA and proteins (Krinsky, 1989). Burton and Ingold (1984) postulated that β-carotene may exert its antioxidant function through its ability to interact with a radical (e.g. a peroxyl radical), yielding a resonance-stabilized carbon-centred radical species. Very little is as yet known, however, about the chemistry of reactions of carotenoids with radicals. Carotenoids function as an antioxidant in many, but not all *in vitro* systems (Krinsky, 1991). Several animal studies have shown that carotenoids may exert immunomodulatory functions. For example, T and B lymphocytes in the spleens of rats showed enhanced proliferative responses after supplementation with either β-carotene or canthaxanthin (Bendich and Shapiro, 1986). The postulated immunomodulatory effects of carotenoids have summarized by Bendich (1989). Immunoenhancing mechanisms have been postulated to involve the quenching of free radicals, which could lower the level of immunosuppressing lipid peroxides, alter arachidonic acid metabolism, stabilize lysosome membranes or protect nuclear structures. Alternatively, retinol-mediated mechanisms could be involved, since retinol has several influences on the immune system (Tomkins and Hussey, 1989).

In addition to the above mentioned mechanisms, other effects of carotenoids have been described, including modulation of P450-mediated carcinogen metabolism (Edes *et al.*, 1989; Tan and Chu, 1991) and enhancement of gap-junction communication (Zhang *et al.*, 1991). Enhanced cell-to-cell communication would restrict clonal expansion of initiated cells and has been described to be brought on both by retinoids (Rutten *et al.*, 1988) and by carotenoids independent of their provitamin A activity (Zhang *et al.*, 1991).

Experimental research

The possible mechanisms of action for carotenoids given above have mainly been derived from *in vitro* experiments. The interest in carotenoids and cancer has also prompted a number of animal studies, which have been reviewed by Krinsky (1991). Many studies, though certainly not all, indicate that carotenoids can prevent the appearance of skin tumours induced by ultraviolet light with or without chemical carcinogens. In addition, carotenoids may also prevent internal tumours induced by carcinogens. The results of these studies are not always consistent and many workers have used systems unique to their own laboratory. Moreover, the metabolism of carotenoids in most laboratory animals differs notably from that in humans.

Human intervention studies on the effect of carotenoids and retinol on DNA fragments or micronuclei in buccal mucosa of tobacco chewers have been performed by Stich and co-workers. Strong reductions in micronucleated cells were observed after Indian and Filipino villagers had been supplemented with β-carotene, β-carotene combined with retinol, or retinol alone; cathaxanthin was not effective (Stich *et al.*, 1984*a*,*b*, 1988). β-Carotene was also effective in a population of tobacco chewers having normal plasma retinol levels (Stich *et al.*, 1985). Van Poppel and coworkers (1992) also observed a reduction of micronuclei in lung sputum of cigarette smokers after β-carotene treatment. Immunological studies in humans on the effect of β-carotene have yielded inconsistent results. Earlier studies reported increases in the numbers of lymphocytes expressing CD4+ (indicating T-helper function) (Alexander *et al.*, 1985; Prabhala *et al.* 1991; Watson *et al.*, 1991), Leu-11b (indicating natural killer cell function) (Prabhala *et al.*, 1991) and interleukin-2 (Prabhala *et al.*, 1991; Watson *et al.*, 1991). In contrast to these studies, Ringer *et al.* (1991) and Van Poppel (1993) did not observe any effects on lymphocyte subpopulations in placebo-controlled studies.

Oral leukoplakia is a potentially premalignant lesion in oral carcinogenesis (Stich *et al.*, 1988). The combination of β-carotene and vitamin A resulted in more frequent regression of established leukoplakias, and less frequent appearance of new leukoplakias. β-Carotene alone showed some effects, but these were not significant. Results by Garewal *et al.* (1990) indicate (partial) regression of oral leukoplakias in 17 of 24 patients after 3–6 months of β-carotene supplementation (30 mg/day). De Vet *et al.* (1991) studied the effect of β-carotene on regression or progression of cervical dysplasia, a putative precursor lesion of cervical carcinoma. Their carefully designed placebo-controlled study showed no effect of β-carotene.

Epidemiological research

Epidemiological studies on carotenoid intake and cancer incidence have focused mainly on β-carotene. Van Poppel and Goldbohm (1995) recently reviewed almost 100 observational case-control and cohort studies. The results show a remarkable consistency for the association of increased risk of lung cancer with low levels of dietary β-carotene or low plasma β-carotene levels in 14 case-control and 11 cohort studies. For stomach cancer, the evidence for a protective effect of β-carotene is also consistent, through the number of studies is more modest. For breast and prostate cancer, both hormone-dependent cancers, the studies indicate no consistent association of plasma or dietary β-carotene and reduced risk. For colorectal cancer, the possible effect of β-carotene appears to be moderate, if it exists. For a number of other cancer sites, e.g. oesophageal cancer, data are promising but the number of studies and the power of these studies is limited. The observational epidemiological studies cannot resolve whether the associations are indeed due to β-carotene, or to other compounds in β-carotene-rich fruit and vegetables, e.g. other carotenoids. Helzlsouer and colleagues (1989) present data that suggest that serum lycopene may be more strongly associated with subsequent risk of bladder cancer than is serum β-carotene. Likewise, the same cohort study (Burney *et al.*, 1989) showed that pancreatic cancer risk was significantly negatively associated with high serum lycopene, whereas β-carotene showed a weak, nonsignificant association.

The first results of human intervention studies on cancer incidence are now becoming available. A combination of β-carotene, vitamin E and selenium reduced stomach cancer mortality in China (Blot *et al.*, 1993). On the other hand, the results indicate a lack of protective effect of β-carotene on lung cancer in smokers with 30–35 years smoking history in Finland (ATBC, 1994), colorectal adenoma in the USA (Greenberg *et al.*, 1994), and second skin cancers in the USA (Greenberg *et al.*, 1990). In the Finnish trial lung cancer mortality was even statistically significantly higher in the β-carotene-supplemented group compared to the placebo group. These trials, however, had a relatively short follow up period (5–7 years) and may only provide information on absence of efficacy of β-carotene in very late stages of carcinogenesis. Intervention studies may need a long follow up to establish beneficial effects in earlier stages of carcinogenesis and other ongoing trials and continued follow up will provide more experimental data on the efficacy of β-carotene.

Flavonoids

General

Flavonoids comprise one of the largest groups of secondary plant metabolites, occurring widely throughout the plant kingdom. Flavonoids consist of flavonols, flavones, flavanones, anthocyanidins, catechins, and biflavans. Over 4000 different types of flavonoids have been described and this number is still increasing (Markham, 1989). Flavonoids share the common skeleton of diphenylpyrans (C_6-C_3-C_6), e.g. two benzene rings (A and B) linked through a heterocyclic pyran or pyrone ring (C) (Fig. 16.1) (Kühnau, 1976). Several functions of flavonoids in plants have been either demonstrated or proposed. These include: protection of plants from ultraviolet light, insects, fungi, viruses, and bacteria; pollinator attractors; plant hormone controllers; and enzyme inhibitors (Markham, 1989). Flavonoids usually occur in food as glycosides with sugars attached at C-3 and, less commonly, at the C-7position. The sugar-free part of the flavonoid molecule is referred to as the aglycone. Of particular importance in this review are flavonols and flavones because of their ubiquitous occurrence in plant-derived foodstuffs and their potentially anticarcinogenic effects. Catechins, especially those occurring in (green) tea, have in recent years attracted much attention because of their potent anticarcinogenic activities. Little attention will be paid to those compounds in this review, however, because of their rather specific occurrence in tea. The combined intake of flavonols and flavones is approximately 23 mg/day in the Netherlands, mainly provided by tea, onions and apples (Hertog *et al.*, 1993*b*). In Mediterranean countries, red wine is a major source of flavonoids. Intake of flavonols and flavones around 1960 in a variety of countries ranged from an average of 6 mg/day in Finland to 64 mg/day in Japan (Hertog *et al.*, 1995). Quercetin is the most important flavonoid, at 16 mg/day in the Netherlands. Approximate levels of quercetin and four related flavonoids in a number of selected foods are shown in Table 16.2.

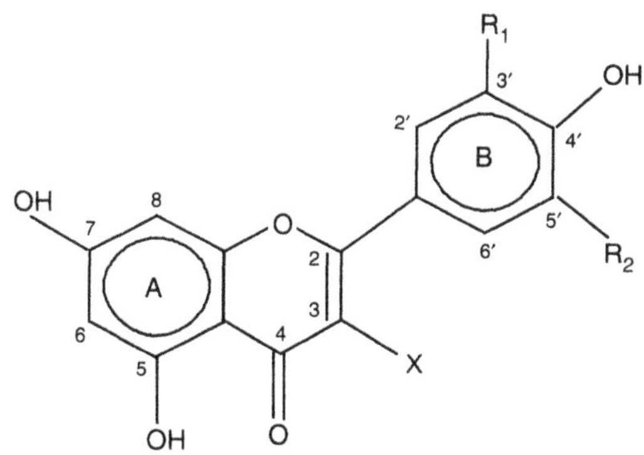

Fig. 16.1. Structure of flavonoids. Flavonols: X = OH; quercetin: R_1 = OH, R_2 = H; kaempferol: R_1 = H, R_2 = H; myricetin: R_1 = OH, R_2 = OH. Flavones: X = H; apigenin: R_1 = H, R_2 = H; luteolin: R_1 = OH, R_2 = H.

The extent of absorption of quercetin has long been an unsolved problem. It was reported that only flavonoid aglycones could be absorbed and that no enzymes cleaving the sugar residues were present in the gut wall. Gugler *et al.* (1975) reported, however, that quercetin aglycone was not absorbed either.

Nieder (1981) did suggest that flavonol glycosides from *Ginkgo biloba* were absorbed in human subjects, but no information on the extent of absorption was given. Hollman *et al.* (1995) recently showed, in a study involving ileostomy patients, that the absorption of quercetin was dependent of the type of sugar residue in the glycoside and that the absorption of quercetin from onions amounted to 50%. Quercetin is thus absorbed in humans and appears in plasma, although its metabolic fate is unknown.

Table 16.2. Flavonol and flavone content (mg/100 g fresh weight) of selected vegetables, fruits and beverages (from Hertog *et al.*, 1992, 1993*a*)

Food	Quercetin	Kaempferol	Myricetin	Luteolin
Lettuce*	0.7–3.0	<0.2	<0.1	<0.1
Onion[a]	28.4–48.6	<0.2	<0.1	<0.1
Endive[a]	<0.1	1.5–9.5	<0.1	<0.1
Red pepper[a]	<0.1	<0.2	<0.05	0.7–1.4
Broad beans	2.0	<0.2	2.6	<0.1
Apples[b]	2.1–7.2	<0.2	<0.1	<0.1
Strawberry[a]	0.8–1.0	12	<0.1	<0.1
Black tea (bags)[cd]	1.7–2.5	1.3–1.7	0.3–0.5	<0.1
Red wine [cd]	0.4–1.6	<0.1	0.7–0.9	<0.1
Apple juice[c]	0.3	<0.1	<0.05	<0.1

<0.05, <0.1, <0.2 below limit of detection.
[a] Variation in three seasons.
[b] Five varieties.
[c] mg/100 ml.
[d] Five brands.

Experimental research

Flavonoids such as quercetin are strong antioxidants (Rice-Evans *et al.*, 1995) and scavengers of free radicals which are possibly involved in cell damage and tumour promotion (Fujiki *et al.*, 1986; Guyton and Kenseler, 1993). Quercetin and other related flavonoids inhibited the formation of carcinogen-induced tumours in rats and mice. Topical application of quercetin and other flavonoids inhibited rat skin tumour promotion induced by 12-*O*-tetradecanoylphorbol-13-acetate (TPA) (Kato *et al.*, 1983; Wei *et al.*, 1990), possibly by inhibition of epidermal ornithine decarboxylase activity. Quercetin and other flavonoids also inhibited 7,12-dimethylbenz(*a*)anthracene- (DMBA), benzo(*a*)pyrene- (BP), 3-methylcholanthrene-, and *N*-methyl-*N*-nitrosourea-induced skin tumourigenesis in mice (Mukhtar *et al.*, 1988). The effect of flavonols administered in the diet was investigated in a number of animal studies. Verma and co-workers (1988) reported that dietary quercetin (2%)

inhibited mammary tumour initiation by DMBA and tumour promotion with TPA. Deschner *et al.* (1991, 1993) showed that both under low and high fat intake, dietary quercetin and rutin suppressed hyperproliferation of colonic epithelial cells and, ultimately, colon tumour incidence induced by azoxymethanol. Quercetin also inhibited proliferation of human colon cancer cells, breast cancer cells, and leukemic cells (Scambia *et al.*, 1991; Ranelleti *et al.*, 1992; Yoshida *et al.*, 1992). Some flavonoids, e.g. quercetin, were also mutagenic in bacterial test systems (Nagao *et al.*, 1991) but quercetin was reported not to be mutagenic *in vivo* (Aeschbacher *et al.*, 1982). Quercetin at a level of 2% in the diet was claimed to induce bladder tumours in rats (Pamucku *et al.*, 1980), although these results were disputed in other studies which used quercetin levels up to 10% of the diet (Saito *et al.*, 1980; Hirono *et al.*, 1981; Morino *et al.*, 1982).

Epidemiological research

So far the association between flavonol and flavone intake and cancer risk has been investigated in three epidemiological studies. Two are individual-based prospective cohort studies (the Zutphen Elderly Study and the Netherlands Cohort Study), both carried out in the Netherlands, and one study was an ecological comparison between cohorts from a variety of countries (the Seven Countries Study). In these studies flavonoid intake was calculated using flavonoid data determined in Dutch foods (Hertog *et al.*, 1992;1993*a*). In the Zutphen Elderly Study, flavonol and flavone intake at baseline in 1985 of 800 men aged 65–85 years was determined using the cross-check dietary history. After 5 years of follow up, 57 men had developed cancer of the alimentary and respiratory tract, but intake of quercetin and four other flavonols and flavones was not associated with these cancers (Hertog *et al.*, 1994). Goldbohm and co-workers (1995) also did not find an association between quercetin consumption and subsequent cancer incidence in the Netherlands Cohort Study. This a study among 120 852 men and women aged 55–69 years, who filled in a semi-quantitative food frequency questionnaire in 1986. During 4.3 years of follow up 200 cases of stomach cancer, 650 cases of colonic cancer, and 764 cases of lung cancer occurred. Flavonoid intake at baseline was not independent of other risk factors associated with any of these cancers (Goldbohm *et al.*, 1995). Also in the Seven Countries Study, a study on risk factors for chronic diseases that started around 1960, flavonoid intake at baseline was not related to differences in site-specific cancer mortality rates after 25 years of follow up (Hertog *et al.*, 1995). No human intervention study on the health effects of quercetin have been reported to date.

Glucosinolates

General

Glucosinolates occur predominantly, though not exclusively, in vegetables of the family Cruciferae. So far, over 100 glucosinolates have been identified. All glucosinolates share a common basic skeleton containing a β-D-thioglucose grouping, a side chain and a sulfonated oxime moiety, but differ in the side chain (R) group (Fig. 16.2) (Fenwick *et al.*, 1983). Several effects of glucosinolates and their hydrolysis products have been proposed: goitrogenicity, toxicity, antifungal and antibacterial activities, and inhibition of chemical carcinogens. Glucosinolates are usually broken down through hydrolysis catalysed by myrosinase, an enzyme that is released from damaged plant cells. The hydrolysis products consist of equimolar amounts of an aglycone, glucose moiety and sulfate. The aglycones are unstable and undergo further reactions to form thiocyanates, nitriles, isothiocyanates or indoles (Fenwick *et al.*, 1983). In the present review attention is paid to isothiocyanates and indoles because of their anticarcinogenic properties.

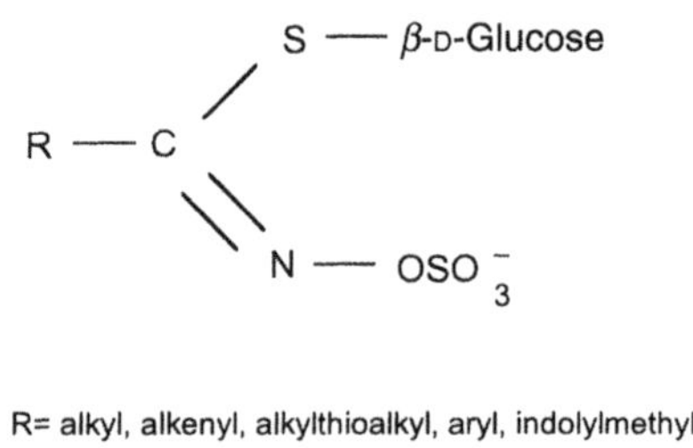

Fig. 16.2. General structure of glucosinolates.

Very little is known about the daily intake of glucosinolates and their hydrolysis products. Brassicas, vegetables of a genus of the Cruciferae including, among others, cabbages, broccoli, Brussels sprouts and cauliflower, are the principle contributors to our intake of glucosinolates. Sones *et al.* (1984) have estimated the consumption of total glucosinolates in the UK to be 46 mg/day from fresh foods and 29 mg from cooked foods. The intake of indole glucosinolates, often studied because of their anticarcinogenic potencies, was estimated to be approximately 15 mg from fresh foods. The intake of glucosinolates in Japan could amount to up to 112 mg/day (Steinmetz and Potter, 1991*b*) The total glucosinolate contents of a variety of brassica vegetables are reported in Table 16.3.

Table 16.3. Contents of total glucosinolates in selected brassica vegetables (from: Fenwick *et al.*, 1983; Sones *et al.*, 1984)

Brassica vegetable	Total glucosinolate content (mg/100 g)	
	Fresh vegetable	Cooked vegetable
Cabbage		
Red	41–109	55
White	33–84	
Brussels sprouts	60–390	60–254
Cauliflower	14–208	9–111
Broccoli	61 and 28	37

Experimental research

Both isothiocyanates and indoles are considered to be able to influence phase I and phase II biotransformation enzyme activities (McDanell and McLean, 1988), thereby possibly influencing several processes related to chemical carcinogenesis. Indoles affect estradiol metabolism through induction of estradiol-2-hydroxylase, a phase I enzyme, and may reduce the risk of oestrogen-dependent diseases such as mammary cancer (Davis *et al.*, 1993). Besides being modulators of biotransformation enzymes isothiocyanates are seen as suppressing agents (Wattenberg, 1992), acting during the earliest stages of the neoplastic process (promotion phase) via an inhibition of the expression of the malignant phenotype.

Many experimental studies, mostly *in vivo* studies on animals, have been carried out to examine the possible anticarcinogenic effects of isothiocyanates and indoles. In various test systems, isothiocyanates and indoles induced both phase I enzymes, e.g. 7-pentoxyresorufin *O*-dealkylase (Guo *et al.*, 1992; Park and Bjeldanes, 1992; Wortelboer *et al.*, 1992; Smith *et al.*, 1993) and 7-ethoxyresorufin *O*-deethylase (Jongen *et al.*, 1989; Smith *et al.*, 1993), and Phase II enzymes, e.g. NAD(P)H-quinone reductase (Benson *et al.*, 1986; Shertzer and Sainsbury, 1991; Guo *et al.*, 1993) and glutathione-*S*-transferase (Sparnins *et al.*, 1982; Benson *et al.*, 1986; Stresser *et al.*, 1994), although the induction patterns differed between tissues. Isothiocyanates decreased the activation of mutagenic compounds, mostly nitrosamines (Guo *et al.*, 1992; Smith *et al.*, 1993), reduced formation of *N*-nitrosobenzylmethylamine (NBMA)-DNA adducts (Stoner *et al.*, 1991) but not of BP-DNA adducts (Adam-Rodwell *et al.*, 1993) and reduced DNA methylation (Morse *et al.*, 1989; Guo *et al.*, 1991). Indoles were inhibitors of the DNA-binding of Aflatoxin B1 (AFB_1), BP and NDMA (Shertzer *et al.*, 1983, 1984) and reduced the number of sister chromatid exchanges induced by mutagens (Jongen *et al.*, 1989). Both isothiocyanates and indoles reduced tumour formation in miscellaneous tissues, test systems and when various carcinogenic compounds were used (Morse *et al.*, 1991; Sugie *et al.*, 1993).

In an experimental study with humans, supplementation of indole-3-carbinol induced estradiol 2-hydroxylase activities (Michnovicz and Bradlow, 1990). Supplementation of 300 g of Brussels sprouts to the diet of human volunteers during 3 months significantly reduced the urinary excretion of 8OHdG, a marker of oxidative DNA damage (Verhagen *et al.*, 1995) and increased glutathione *S*-transferase levels (Nijhoff *et al.*, 1995). Besides anticarcinogenic properties, a few isothiocyanates and indoles caused genetic mutations in bacteria (Fekadu *et al.*, 1994; Musk *et al.*, 1994) and a few isothiocyanates caused chromosome mutations in cultured mammalian cells (Musk and Johnson, 1993).

Epidemiological research

The association between intake of glucosinolates or their hydrolysis products and risk of cancer has not been specifically investigated yet. On the other hand, many epidemiological studies have examined the association between consumption of brassica vegetables and cancer risk (Bidoli *et al.*, 1992; Steinmetz and Potter, 1991*a*; Gao *et al.*, 1993). The results of the case-control studies suggest inverse associations between the consumption of brassica vegetables and the risk of cancer. These associations appear to be most consistent for lung, stomach, colonic and rectal cancers, and least consistent for prostate and ovarian cancer. The cohort studies conducted so far confirm the inverse association, in particular for lung cancer (Steinmetz *et al.*, 1993). The results of these epidemiological studies should, however, be interpreted with caution, because the protective effects of brassica vegetables cannot be clearly attributed to glucosinolates as other bioactive components present may be contributing also. In addition, the consumption of several types of vegetables are usually highly correlated and it is difficult to conclude whether the protective effect is attributable to brassica vegetables *per se* or to vegetables in general.

Conclusion

In this overview, a number of bioactive components of plant foods and their association with cancer have been discussed. Most of these compounds conferred in experimental (animal) studies a certain amount of protection against chemically-induced tumours in animals. A number of these compounds such as β-carotene, have been also shown to be related to certain markers of tumour growth in humans. The potential importance of these compounds in reducing cancer risk is also supported by the overwhelming epidemiologic evidence showing a protective effect of the consumption of vegetables and fruit on cancer risk. The importance and relevance of the single compounds in preventing human cancer is, however, still unclear. The unexpected results from the Finnish intervention trial on β-carotene and lung cancer in which a higher rate of lung cancer was observed in the supplemented group, suggests that β-carotene is not the important cancer preventive compound in green-leafy vegetables it was supposed to be. Possibly other carotenoids are more important in explaining these inverse associations. In addition, the extent of absorption of β-carotene from green leafy vegetables has been questioned recently by a number of investigators (De Pee and West, 1995), again suggesting that β-carotene may not be the most important active component in these foods. Similarly, the flavonoid quercetin had strong anticarcinogenic effects in *in vitro* and *in vivo* studies but did not show an effect in three epidemiological studies. Foods associated with lower cancer rates, such as green–yellow vegetables, are not important sources of quercetin in the human diet. It seems therefore that quercetin does not play an important role in explaining the inverse association between the consumption of vegetables and fruit, and cancer risk. These two examples nicely demonstrate the weaknesses of both epidemiological and experimental research in proving associations between single food components and cancer. Epidemiological research is hampered by the fact that it is impossible to pinpoint the exact compound in a food or diet responsible for the observed association, because of the high correlation with other food compounds and possibly other confounding factors. On the other hand the results from experimental work on single compounds in usually very controlled environments are not always easily to extrapolate to the human situation. What seems evident is that the combination of various compounds with anticarcinogenic potential, present in vegetables and fruit, account for their protective effects. Future research should, therefore, in addition to the study of single compounds, also examine the cancer preventive effects of whole foods. In addition, more attention should be paid when studying diet-cancer relations to the genetic susceptibilities of individuals and to their interactions with life-style factors such as diet.

In the present overview, only a limited number of potentially anticarcinogenic bioactive components were discussed and our choice was limited to components occurring in a large variety of foods. Other compounds in specific foods that have attracted much interest are for instance the terpenoids in citrus oils, phytoestrogens in legumes and organosulfur compounds in *Allium* vegetables. All of these compounds showed anticarcinogenic activities in animal experiments but in general their effects in humans are unknown. The study of potentially anticarcinogenic bioactive food components presents therefore a new and exciting area of research.

References

Adam-Rodwell, G., Morse, M.A. and Stoner, G.D. (1993). *Cancer Lett.* **71**, 35–42.
Aeschbacher H.U. Meier H. and Ruch E. (1982). *Nutrition and Cancer* **4**, 90–8.
Alexander, M., Newmark, H. and Miler, R.G. (1985). *Immunol. Lett.* **9**, 221–4.
ATBC (1994). *N. Engl. J. Med.* **30**, 1029–35.
Bendich, A. (1989). *J. Nutr.* **119**, 112–15.
Bendich, A. and Shapiro, S.S. (1986). *J. Nutr.* **116**, 2254–62.
Benson, A.M., Barretto, P.B. and Stanley, J.S. (1986). *J. Natl. Cancer Inst.* **76**, 467–73.
Bidoli, E., Franceschi, S., Talamini, R., Barra, S. and La Vecchia, C. (1992). *Int. J. Cancer* **50**, 223–9.
Block, G., Patterson, B. and Subar, A. (1992). *Nutr. Cancer* **17**, 1–29.
Blot, W.J., Li, J., Taylor, P.R., Guo, W., Dawsey, S., Wang, G-Q., Yang, C.S., Zheng, S-F., Gail, M., Li, G-Y., Yu, Y., Liu, B.Q., Tangrea, J., Sun, Y.H., Liu, F., Fraumeni, J.F., Zhang, Y-H. and Li, B. (1993). *J. Natl. Cancer Inst.* **85**, 1483–92.
Burney, P.G.J., Comstock, G.W. and Morris, J.S. (1989). *Am. J. Clin. Nutr.* **49**, 895–900.
Burton, G.W. and Ingold, K.U. (1984). *Science* **224**, 569–73.
Davis, D.L., Bradlow, H.L., Wolff, M., Woodruff, T., Hoel, D.G. and Anton-Culver, H. (1993). *Environ. Health Perspect.* **101**, 372–7.
De Pee, S., West, C., Muhilal, K.D. and Hantvast J.G.A.J. (1995). *Lancet* **346**, 75–8.
De Vet, H.C.W. (1989). *Anticancer Res.* **9**, 145–52.
De Vet, H.C.W., Knipschild, P.G. and Willebrand, D.(1991). *J. Clin. Epidemiol.* **44**, 273–83.
Deschner, E.E., Ruperto, J.F., Wong, G.Y. and Newmark, H.L. (1991). *Carcinogenesis* **7**, 1193–6.
Deschner, E.E., Ruperto, J.F., Wong, G.Y. and Newmark, H.L. (1993). *Nutr. Cancer* **20**, 199–204.
DiGiovanni, J. (1990). In *Chemical carcinogenesis and mutagenisis* II. (ed. C.S. Cooper and P.L. Grover), pp. 159–202. Springer-Verlag, Berlin.
Doll, R. and Peto, R. (1981). *J. Natl. Cancer Inst.* **66**, 1196–305.
Edes, T.E., Thornton, W. and Shah, J.(1989). *J. Nutr.* **19**, 796–9.
Fearon, E.R. and Vogelstein, B. (1990). *Cell* **61**, 759–67.
Fekadu, K., Musk, S., Parzefall, W., Johnson, I., Sontag, G., Schulte-Hermann, R. and Knasmüller, S. (1994). In *Proceedings of the International Conference Euro Food Tox IV 'Bioactive substances in food of plant origin'.* (ed. H. Kozlowska, J. Fornal and Z. Zduczyk), Vol.2.
Fenwick, G.R., Heany, R.K. and Mullin, W.J. (1983). *CRC Crit. Rev. Food Sci. Nutr.***18**, 123–201.
Fiala, E.S., Reddy, B.S. and Weisburger, J.H. (1985). *Ann. Rev. Nutr.* **5**, 445.
Fischer, M., Mills, G.D. and Slaga, T.J. (1982). *Carcinogenesis* **3**, 1243–5.
Franke, A.A, Custer, L.J., Cerna, C.M. and Narala, K.K. (1994). *J. Agric. Food Chem.* 42, 1905–13.
Fujiki, H., Horiuchi, T. and Yamashita, K. (1986). In *Plant flavonoids in biology and medicine: Biochemical, pharmacological and structure-activity relationships.* (ed. V. Cody, E. Middleton and J.B. Harborne), pp. 429–40. Alan R. Liss, New York.

Fürstenberger, G., Gross, M. and Marks, F. (1989). *Carcinogenesis* **10**(1), 91–6.
Gao, C-M., Tajima, K., Kuroishi, T., Hirose, K. and Inoue, M. (1993). *Jpn. J. Cancer Res.* **84**, 594–600.
Garewal, H.S., Meyskens, F.L. and Killen, D. (1990). *J. Clin. Oncol.* **8**, 1715–20.
Goldbohm, RA., Brandt P.A., Hertog M.G.L., Brants, H.A.M. and van Poppel G. (1995). *Am. J. Epidemiol.* (Suppl) **141**, S61.
Greenberg, E.R., Baron, J.A. and Stukel, T.A.(1990). *New Engl. J. Med.* **323**, 789–95.
Greenberg, E.R., Baron, J.A. and Tosteson, T.D. (1994). *New Engl. J. Med.* **331**, 141–7.
Gugler, R., Leschik, M. and Dengler, H.J. (1975). *Europ. J. Clin. Parmacol.* **9**, 229–34.
Guo, Z., Smith, T.J., Thomas, P.E. and Yang, C.S. (1991). *Cancer Res.* **51**, 4798–803.
Guo, Z., Smith, T.J., Wang, E., Sadrieh, N., Ma, Q., Thomas, P.E. and Yang, C.S. (1992). *Carcinogenesis* **13**, 2205–10.
Guo, Z., Smith, T.J., Wang, E., Eklind, K.I., Chung, F.-L. and Yang, C.S. (1993). *Carcinogenesis* **14**, 1167–73.
Guyton, K.Z. and Kensler, T.W. (1993). *Br. Med. Bull.* **49**, 523–44
Helzlsouer, K.J., Comstock, G.W. and Morris, J.S. (1989). *Cancer Res.* **49**, 6144–8.
Hertog, M.G.L. Hollman, P.C.H. and Katan, M.B. (1992). *J. Agric. Food Chem.* **40**, 2379–83.
Hertog, M.G.L. Hollman, P.C.H. and Putte van de, B. (1993a). *J. Agric. Food Chem.* **41**, 1242–6.
Hertog, M.G.L., Hollman, P.C.H., Katan, M.B. and Kromhout, D. (1993b). *Nutr. Cancer* **20**, 21–9.
Hertog, M.G.L., Feskens, E.J.M., Hollman, P.C.H., Katan, M.B. and Kromhout, D. (1994). *Nutr. Cancer* **22**,175–84.
Hertog, M.G.L., Kromhout, D., Aravanis, C., Blackburn, H., Buzina, R., Fidanza, F., Giampaoli, S., Jansen, A., Menotti, A., Nedeljkovc, S., Pekkarinen, M., Simic, B.S., Toshima, H., Feskens, E.J.M., Hollman, P.C.H. and Katan, M.B. (1995). *Arch. Intern. Med.* **155**, 381–6.
Hirono, I., Ueno, I., Hosaka, S., Takanashi, H., Matsushima, T., Sugimura, T. and Natori, S. (1981). *Cancer Lett.* **13**, 15–21.
Hollman, P.C.H., deVries, J.H.M. and Katan, M.B. (1995). *Atheroscleosis* **115**, S49.
IARC (1990). In *Cancer: Causes, occurrence, and control.* (ed. L. Tomatis, A. Aitio, N.E. Day, E. Heseltine, J. Kaldor, A.B. Miller, D.M. Parkin and E. Riboli). International Agency for Research on Cancer, Lyon.
Jongen, W.M.F., Topp, R.J., Wienk, K.J.H. and Homan, E.C. (1989). *Mutat. Res.* **222**, 263–9.
Kato, R., Nakadate, T., Yamamoto, S. and Sugimura, T. (1983). *Carcinogenesis* **4** (10), 1301–5.
Krinsky, N.I. (1989). *Prev. Med.* **18**, 592–602.
Krinsky, N.I. (1991). *Am. J. Clin. Nutr.* **53**, 238S-246S.
Kühnau, J. (1976). *World Rev. Nutr. Diet.* **24**, 117–20.
Mangels, A.R., Holden, J.M., Beecher, G.R., Forman, M.R. and Lanza, E. (1993). *J. Am. Diet Assoc.* **93**, 284–96.
Markham, K.R. (1989). *Meth. Plant Biochem.* **1**, 197–235.
Marshall, C.J. (1991). *Cell* **64**, 313–26.
McDanell, R. and McLean, A.E.M. (1988). *Food Chem. Toxicol.* **26**, 59–70.
Michnovicz, J.J. and Bradlow, H.L. (1990). *J. Natl. Cancer Inst.* **82**, 947–9.
Morino, K., Matsukura, N., Ohgaki, H., Kawachi, T., Sugimura, T. and Hirono, I. (1982). *Carcinogenesis* **3**, 93–7.

Morse, M.A., Amin, S.G., Hecht, S.S. and Chung, F.-L. (1989). *Cancer Res.* **49**, 2894–7.
Morse, M.A., Eklind, K.I., Hecht, S.S. and Chung, F.-L. (1991). *IARC Sci. Publ.* **105**, 529–34.
Mukhtar, H., Das, M., Khan, W.A., Wang, Z.Y., Bik, D.P. and Bickers, D.R. (1988). *Cancer Res.* **48**, 2361–5.
Musk, S.R.R. and Johnson, I.T. (1993). *Mutat. Res.* **300**, 111–17.
Musk, S.R.R., Preobrazhenskaya, M.N., Belitsky, G.A., Korolev, A.M., Lytcheva, T.A., Khitrovo, I.A., and Johnson, I.T. (1994). *Mutat. Res.* **323**, 69–74.
Nagao, M., Morita, N. and Yahagi, T. (1991). *Environ. Mutagenesis* **3**, 401–19.
Nieder, M. (1991). *Münch. Med. Wschr.* **133**, S61–2.
Nijhoff, W.A., Nagengast, F.M., Grubben, M.J.A.L., Jansen, J.B.M.J., Verhagen, H., Van Poppel, G. and Peters, W.H.M. (1995). *Carcinogenesis* **16**, 2125–8.
Pamucku, A.M., Yalçiner, S., Hatcher, J.F. and Bryan, G.T. (1980). *Cancer Res.* **40**, 3468–72.
Park, J.-Y. and Bjeldanes, L.F.(1992). *Chem. Biol. Interact.* **83**, 235–47.
Parker, R.S. (1989). *J. Nutr.* **119**, 101–4.
Prabhala, R.H., Garewal, H.S. and Hicks, M.J. (1991). *Cancer* **67**, 1556–60.
Ranelletti, F.O., Ricci, R., Larocca, L.M., Maggiano, N., Capelli, A., Scambia, G., Benedetti-Panici, P., Mancuso, S., Rumi, C. and Piantelli, M. (1992). *Int. J. Cancer* **50**, 486–92.
Rice-Evans, C.A., Miller, N.J., Bolwell, P.G. Bramley, P.M. and Pridham, J.B. (1995). *Free Rad. Res.* **22**, 275–83.
Ringer, T.V., De Loof, M.J. and Winterrowd, G.E. (1991). *Am. J. Clin. Nutr.* **53**, 688–94.
Rutten, A.A.J.J.L., Jongen, W.F.M. and de Haan, L.H.J. (1988). *Carcinogenesis* **9**, 315–20.
Saito, D., Shirai, A., Matsushima, T. Sugimura, T. and Hirono, I. (1980). *Teratog. Carcinog. Mutagen.* **1**, 213–21.
Scambia, G., Ranelletti, F.O., Benedetti-Panici, P., Piantelli, M., Bonanno, G,. DeVincenzo, R., Ferrandina, G., Pierelli, L., Capelli, A. and Mancuso, S. (1991). *Cancer Chemother. Pharmacol.* **28**, 255–8.
Shertzer, H.G. (1983). *Food Chem. Toxicol.* **21**, 31–5.
Shertzer, H.G. (1984). *Chem. Biol. Interact.* **48**, 81–90.
Shertzer, H.G. and Sainsbury, M. (1991). *Food Chem. Toxicol.* **29**, 237–42.
Smith, T.J., Guo, Z., Li, C., Ning, S.M., Thomas, P.E. and Yang, C.S. (1993). *Cancer Res.* **53**, 3276–82.
Sones, K., Heany, R.K. and Fenwick, G.R. (1984). *J. Sci. Food Agric.* **35**, 712–20.
Sparnins, V.L., Venegas, P.L. and Wattenberg, L.W. (1982). *J. Natl. Cancer Inst.* **68**, 493–6.
Sporn, M.B. and Roberts, A.B. (1983). *Cancer Res.* **43**, 3034–40.
Steinmetz, K.A. and Potter, J.D. (1991*a*). *Cancer Causes Cont.* **2**, 325–57.
Steinmetz, K.A. and Potter, J.D. (1991*b*). *Cancer Causes Cont.* **2**, 427–42.
Steinmetz, K.A., Potter, J.D. and Folsom, A.R. (1993). *Cancer Res.* **53**, 536–43.
Stich, H.F., Rosin, M.P. and Vallejera, M.O. (1984*a*). *Lancet* **1**, 1204–6.
Stich, H.F., Stich, W., Rosin, M.P. and Vallejera, M.O. (1984*b*). *Int. J. Cancer* **34**, 745–50.
Stich, H.F., Hornby, A.P. and Dunn, B.P. (1985). *Int. J. Cancer* **36**, 321–7.
Stich, H.F., Rosin, M.P. and Hornby, A.P. (1988). *Int. J. Cancer* **42**, 195–9.

Stöhr, H. and Herrmann, K. (1975). *Z. Lebensm. Unters.-Forsch.* **159**, 305–12.
Stoner, G.D., Morrissey, D.T., Heur, Y.-H., Daniel, E.M., Galati, A.J. and Wagner, S.A. (1991). *Cancer Res.* **51**, 2063–8.
Stresser, D.M., Bailey, G.S. and Williams, D.E. (1994). *Drug Metab. Dispos.* **22**, 383–91.
Sugie, S., Okumura, A., Tanaka, T. and Mori, H. (1993). *Jpn. J. Cancer Res.* **84**, 865–70.
Tan, B. and Chu, F.L. (1991). *Am. J. Clin. Nutr.* **53**, 1071S-1075S.
Tomkins, A. and Hussey, G.(1989). *Nutr. Res. Rev.* **2**, 17–28.
Van Poppel, G. (1993). *Eur. J. Cancer* **29A**, 1335–55.
Van Poppel, G. and Goldbohm, R.A. (1995). *Am. J. Clin. Nutr.* **62**, 13935–14025.
Van Poppel, G., Kok, F.J. and Hermus, R.J.J. (1992). *Br. J. Cancer* **66**, 1164–8.
Van Poppel, G., Spanhaak, S. and Ockhuizen, T. (1993). *Am. J. Clin. Nutr.* **57**, 404–7.
Verhagen, H., Poulsen, H.E., Loft, S., Van Poppel, G., Willems, M.I. and Van Bladeren, P. J. (1995). *Carcinogenesis* **16**, 969–70.
Verma, A.K., Johnson, J.A., Gould, M.N. and Tanner, M.A. (1988). *Cancer Res.* **48**, 5754–88.
Vogelstein, B., Fearon, E.R. and Hamliton, S.R. (1988). *N. Engl. J. Med.* **319**, 525–32.
Watson, R.R., Prabhala, R.H., Plezia, P.M. and Alberts, D.S. (1991). *Am. J. Clin. Nutr.* **53**, 90–94.
Wattenberg, L.W. (1987). *Carcinogenesis* **8**, 1971–3.
Wattenberg, L.W. (1992). *Cancer Res.* **52**, 2085S–2091S.
Wattenberg, L.W. and Loub W.D. (1978). *Cancer Research* **38**, 1410–13.
Wei, H., Tye, L., Bresnick, E. and Birt, D.F. (1990). *Cancer Res.* **50**, 499–502.
Weisburger, J.H. (1992). In *Phenolic compounds in food and their effects on health* II. (ed. M.T. Huang, C.H. Ho and C.Y. Lee). American Chemical Society, Washington, DC.
Wortelboer, H.M., Van Der Linden, E.C.M., De Kruif, C.A., Noordhoek, J., Blaauboer, B.J., Van Bladeren, P.J. and Falke, H.E. (1992). *Food Chem. Toxicol.* **30**, 589–99.
Yoshida, M., Yamamoto, N. and Nikaido, T. (1992). *Cancer Res.* **52**, 6676–81.
Zhang, L.X., Cooney, R.V. and Bertram, J.S. (1991). *Carcinogenesis* **12**, 2109–14.

17. Toxic compounds from fruit and vegetables

J.P. FELIX D'MELLO

The Scottish Agricultural College, Department of Crop Science and Technology, West Mains Road, Edinburgh EH9 3JG, UK

Introduction

It is possible to adopt a very wide interpretation of this title, based on conventional meanings of the terms 'fruit' and 'vegetables'. The Oxford dictionary defines fruit as 'the edible product of a plant or tree, consisting of the seed and its envelope'. The definition for vegetables includes plants 'cultivated for food', particularly edible herbs or roots. If one also applies a broad definition for the phrase 'toxic compounds', then it is immediately apparent that we would be considering an almost infinite array of deleterious substances in the context of a wide variety of plant products. The constraints of space do not allow such an approach. This chapter therefore focuses on a limited number of toxic constituents in selected fruits and vegetables. In fruits, the adverse effects of allergens, hypoglycin and patulin are considered. In vegetables, the existence of a wider range of toxicants is well recognized, including lectins, isoflavones, glucosinolates, *S*-methylcysteine sulfoxide, glycoalkaloids, nitrate, oxalates and furanocoumarins. The deleterious effects of these compounds are reviewed, and the risks for human health considered. In many cases the assessment and management of risk necessitates reference to animal studies, particularly where the underlying basis for toxicity is of relevance. The toxicology of these compounds is addressed in a global context in order to illustrate the diversity of these substances. This review excludes issues arising from contamination of fruit and vegetables with pesticide residues.

Toxicants from fruits

Allergens

Allergies to apples, other fruits and even vegetables, particularly celery, are very common in people with hay fever caused by birch and other pollens (Ortolani *et al.* 1988; Vieths *et al.* 1992; 1993). The cross-reactivities are due to structural similarities between food and pollen allergens. Both local and systemic manifestations may be seen in sensitive individuals. Initially, the symptoms are of the immediate type, occurring within a few minutes of consuming the fruits or vegetables. These symptoms are usually localized in the mouth and consist of intraoral and lip irritation and angioedema of the throat. In sensitized individuals one bite of Golden Delicious apples, for example, is sufficient to evoke an allergic response (Vieths *et al.* 1993). Although these manifestations are not serious and often disappear within a short period of time, other symptoms may ensue, such as rhinoconjunctivitis, urticaria–angioedema, asthma and anaphylactic shock.

In a study with 262 patients suffering from hay fever and food allergy, Ortolani *et al.* (1988) showed that 53% were sensitive to apples, 40% to peach, 23% to apricot and 22% to pears. The results of another investigation (Vieths *et al.* 1993) indicated that the allergenic potency increased markedly during maturation of Golden Delicious apples and to a smaller extent during ripening of Boskoop apples. The allergenic potency was attributed to the occurrence of an 18 kD allergen in Golden Delicious apples, whereas even in mature Boskoop apples this protein was present in very low concentrations.

Hypoglycin

The toxic amino acid, hypoglycin A, and its glutamyl dipeptide derivative, hypoglycin B, occur in the fruit of *Blighia sapida,* a relatively small tree native to Africa, but now also found in parts of Florida and in Jamaica. The fruit is edible when ripe, but in its unripe form contains a comparatively high quantity of hypoglycin. The tissue covering the seed (arillus) contains about 0.08 g/kg dry weight but in the arillus of the unripened fruit this concentration can be almost 15 times higher (Rosenthal, 1991). In times of food shortages, the unripened fruit is in much demand, but consumption leads to severe manifestations of 'vomiting sickness', particularly in children and undernourished adults. It is now acknowledged that the induction of vomiting sickness is linked with the consumption of hypoglycin. The main features of this debilitating condition, which may be apparent within a few hours, include violent retching, vomiting, convulsions, with death occurring within 2–3 days. In addition to the gross manifestations, acute internal changes also occur, the primary effect being the precipitation of a severe hypoglycaemia which can totally deplete hepatic reserves of glycogen. This particular feature is readily reproduced in animal models of the condition. Thus the intravenous administration of hypoglycin A (20 mg/kg body weight) to rabbits reduces blood glucose concentrations from 90 to 30 mg/100 ml within 6–7 h (Rosenthal, 1991). Organ damage and functional disruption of mitochondria are also evident in animals given the amino acid. The mode of action of hypoglycin is thought to be via an inhibition of fatty acid oxidation, although there is some dispute as to whether the disruption is associated with long-chain or short-chain fatty acids. The overall consequence, however, is that such an inhibition causes an inordinate reliance on energy production from glucose, and depletion of this limited nutrient culminates in the characteristic hypoglycaemia observed in affected subjects. The vomiting syndrome, however, has been associated with a structural antagonism between hypoglycin and the branched-chain amino acid, leucine (Cheeke and Shull, 1985).

Patulin

Patulin is a hemiacetal lactone mycotoxin produced by a number of *Penicillium* and *Aspergillus* species of fungi. Of these species, *Penicillium expansum* is notable for its synthesis of patulin in decaying apples. Indeed, this fungus is considered to be the primary cause of rot of apples, and a variety of other fruit. The occurrence of patulin in fruit juice is, therefore, to be expected and has been a cause for concern in the UK and elsewhere in Europe. The presence of patulin in apple juice is generally attributed to the use of mouldy fruit (MAFF, 1993). The mycotoxin may also occur as a contaminant of other products derived from fruit, e.g. home-made jam. The disquiet arises from reports that the mycotoxin is acutely toxic to mice, is mutagenic to yeast and induces tumours in rats, although its carcinogenicity for humans is an issue which requires resolution. However, the UK Committee on Carcinogenicity and the Committee on Mutagenicity have designated patulin as a mutagen and current recommendations (MAFF, 1993) are that the level of this mycotoxin should be 'reduced to the lowest level that is technologically achievable and that appropriate measures are taken to ensure that this occurs'.

Levels of patulin as high as 250 mg/kg have been found in damaged or infected parts of fruit. Due to the potential health hazards and the possible use of patulin concentrations in quality assurance in the manufacture of fruit products, a limit of 50 μg/kg (ppb) has been adopted in a number of European countries.

In recent years, there has been a considerable increase in the UK in the production of cloudy apple juices, prepared by pressing the fruit, stabilizing with vitamin C and subsequently pasteurizing the juice. The reduction in processing steps, as compared with the procedure employed in the preparation of clear juices, means that patulin losses during fining and filtration are restricted, with higher residual levels of the mycotoxin in the cloudy juices. Table 17.1 provides a comparison of clear and cloudy juices based on recent data in the UK and Spain (MAFF, 1993; Prieta *et al.*, 1994). Although the UK data were derived from a relatively small number of samples, it is apparent that the incidence of patulin was higher in cloudy juices, with a median value of 28 μg/kg, compared with 0–10 μg/kg for clear juices. Four cloudy samples contained patulin concentrations in excess of 50 μg/kg, compared with only one of the clear juice samples. Highest concentrations of patulin occurred in the cloudy juices. In a more extensive study in Spain, patulin was detected in 82% of commercial samples (assumed to be clear juices for the purpose of this comparison). However, 70% of these juices contained levels of less than 10 μg/l, although 22% had levels of 10–50 μg/l and in two samples, concentrations of 164 and 170 μg/l were recorded (Prieta *et al.*, 1994). On a more reassuring note, this study reported 12 samples of apple food for children to be free of patulin.

In the UK survey (MAFF, 1993) estimates of patulin intakes for mean and extreme consumers of apple juice were calculated, based on the median values of 28 μg/kg and 5 μg/kg (i.e. half the detection limit) for cloudy and clear juices. Individuals in the age group of 2–5 years were calculated to ingest patulin at rates of 0.5 μg/kg body weight per week if they were normal consumers of cloudy apple juice and three times that level if they were extreme consumers. For adults, the respective values were 0.2 and 0.9 μg/kg per week. However, if the highest value of 434 μg/kg for cloudy juice is used, then intakes of extreme consumers would be 22.8 and 14.7 μg/kg body weight per week respectively for children aged 2–5 years and for adults in the UK. One may assume that such intakes are undesirable. It is fortuitous that the incidence of high levels of patulin in apple juice is low, but the case for vigilance is not diminished.

The occurrence of patulin in apple juice can be controlled by ensuring that mouldy fruit is not used in the production of apple juice, in accordance with the best aims of good manufacturing practice. Apple juice manufacturers are now complying with this recommendation and the prospects for reducing patulin levels further would appear to be promising, particularly in view of recent developments in the analytical methodology for the mycotoxin. It is worth noting here that patulin disappears on alcoholic fermentation of fruit juices (Stinson *et al.*, 1978).

Toxicants from vegetables

Lectins

Lectins are proteins with a characteristic affinity for certain sugar residues or glycoproteins present in the membranes of animal cells, including those of the intestinal mucosa (D'Mello, 1995). Since lectins react with red blood cells *in vitro* to induce agglutination, the alternative term haemagglutinins is often used. Lectins are present in the seeds of a variety of plants, including *Phaseolus vulgaris* (kidney bean), *Phaseolus lunatus* (lima bean), *Psophocarpus tetragonolobus* (guar) and *Canavalia ensiformis* (jack bean). Lectins from

Table 17.1. Distribution of patulin in retail apple juices

Type of apple juice	Number of samples analysed	Number of samples containing patulin in the range (μg/kg)						References
		< 10	10–15	51–100	101–150	151–200	> 200	
Clear	17	10	6	0	1	0	0	MAFF (1993)
Cloudy	15	4	7	2	0	1	1	MAFF (1993)
Commercial	82	57	18	5	0	2	0	Prieta *et al.* (1994)

these beans are capable of precipitating severe gastrointestinal disturbances, rapid weight loss and, in certain animal species, a high incidence of mortality (Grant, 1991). In contrast to most dietary proteins, lectins resist proteolytic breakdown *in vivo* and, consequently, significant quantities of ingested lectins may be found in the digestive tract and faeces. The classical view of lectins, based on studies with raw kidney beans, is that they exert their adverse effects by causing extensive structural and functional disruption of the intestinal mucosa, thereby reducing nutrient absorption. However, the lethal effects cannot be explained on this basis. It is now apparent that lectins are absorbed intact into the systemic circulation to elicit profound effects on organ size and a depletion of body reserves of lipid, glycogen and protein. In addition, an immunological response may also occur (D'Mello, 1995).

Beans are an important part of the diet for humans in virtually every country, including western Europe where the recent expansion of vegetarianism heightens the need for awareness of the potential hazards in consuming raw forms of these legume seeds. In the more affluent countries, the consumption of commercially processed beans poses no threat, since the cooking temperatures employed are designed to inactivate the heat-labile lectins. However, the increased availability of raw beans in the retail market has led to a higher incidence of gastrointestinal problems in people eating raw or partially cooked kidney beans (Noah *et al.*, 1980; Bender and Reaidi, 1982). The use of 'slow cookers' employing temperatures of less than 75 °C have compounded the problem. The symptoms include nausea and vomiting 1–3 h after consumption of the beans and in addition abdominal distension and diarrhoea may also occur. In developing countries, there is potentially greater risk in areas where fuel is in short supply and also because commercially processed beans may not be available or for economic reasons are not part of the weekly food purchases. Under these conditions, the consumption of toxic lectins over a protracted period of time is possible, the long-term effects of which remain uncertain.

Since the earlier cases of kidney bean poisoning, public awareness has been enhanced, particularly by the food industry and retail outlets which now provide special labelling and instructions for cooking. It is particularly important to ensure that prior to consumption or addition to other foods, beans are fully hydrated and boiled for at least 10–20 min. The use of slow cookers is also actively discouraged.

Isoflavones

Soybeans have been an important food item for people in Asia for many centuries (Golbitz, 1995), but over the past 15–20 years soya products have made a significant impact on Western diets. In America, these soya foods include commodities such as veggieburgers, tempeh burgers, soyamilk cheeses and soyaflour pancake mix. It is well recognized that soybeans contain relatively high quantities of the glycosides of the isoflavones, daidzein, genistein and glycitein. For example, concentrations of daidzein can vary from 28 to 233 mg/kg dry soybean seed, depending on sowing time and location (Tsukamoto *et al.*, 1995). Values for genistein were 7–149 g/kg dry seed. Isoflavones also occur in numerous other plants. Subsequent to consumption, these isoflavones undergo metabolic changes in the gut resulting in the synthesis of products with oestrogenic activity and the ability to bind with the hormone receptors. These phytoestrogens have been shown to influence not only sex hormone metabolism and physiological activity but also protein synthesis, malignant cell proliferation, differentiation and angiogenesis. Their role as natural cancer-protective agents has been proposed and the existing epidemiological evidence supports this view (Herman *et al.,* 1995). However, studies with farm animals grazing subterranean clover, another rich source of isoflavones, indicate the potential risks of these substances in reproductive function. Consumption of this herbage has been associated with a pronounced decline in fertility of sheep. In laboratory animals it is possible to demonstrate a dose-related increase in uterine weight following the administration of a range of phytoestrogens, including genistein and daidzein (Cheeke and Shull, 1985). Although phytoestrogens have a low potency in this respect, in comparison with oestrone, marked effects may ensue through an additive action with endogenous oestrogen. Recent investigations with rats showed that the phytoestrogen, coumestrol, stimulated the onset of oestrous cycles when provided pre-pubertally to immature females but inhibited ovarian cycles when provided to cycling adult females (Whitten *et al.,* 1995). There is currently considerable interest in dietary oestrogenic load in humans. Whether or not isoflavone intake can account for the reduced mean sperm count of men and for the irregularities of the menstrual cycle in women consuming high levels of plant foods remains an open question.

Glucosinolates

Glucosinolates are thioglucosides which are widely distributed in plants, particularly among members of the Cruciferae. Their universal presence in brassica vegetables and forage crops has attracted much attention because of the potential to induce toxicity in animals (Duncan, 1991). In addition, there is some evidence of the value of the glucosinolates in the prevention of cancer in humans (Nugon-Baudon and Rabot, 1994). Over 100 different glucosinolates have been identified. They all share a common structure, being composed of a thioglucose group attached to an R-group. A wide range of R-groups have been identified. In cabbage, for example, the main glucosinolates are sinigrin, progoitrin and glucobrassicin, contributing to a total glucosinolate content of between 360 and 2754 mg/kg fresh weight. These three glucosinolates also predominate in a variety of other brassica vegetables, including brussels sprouts, broccoli and cauliflower (Nugon-Baudon and Rabot, 1994). These authors summarize data which indicate contrasting intakes of total glucosinolates, of the order of 323 mg/person in the UK and only 56 mg/person in Canada, assuming all brassica vegetables are eaten raw. Upon hydrolysis by plant or microbial thioglucosidase (myrosinase), a diverse range of aglycone moeities are released from the glucosinolates. The aglycone residues undergo further degradation to yield a number of toxic metabolites. The most common breakdown products are the isothiocyanates and nitriles, but depending on various conditions such as pH, temperature and metallic ion concentrations, a number of other metabolites may also be synthesized. The isothiocyanate residues are known to possess goitrogenic properties. However, the goitrogenic potential is less than that of thiouracil. The goitrogenic effect of the isothiocyanates may be related to their metabolism to thiocyanate ion which inhibits iodine uptake by the thyroid gland. The isothiocyanates also reduce total and free plasma concentrations of T4 in a dose-related manner (Duncan, 1991). In addition, it has been suggested that isothiocyanates have the capacity to cleave the disulfide bonds of key proteins, thus disrupting their tertiary structure. The toxicity of nitriles derived from glucosinolate degradation is generally regarded to emanate from their capacity to precipitate organ damage, principally in the liver and kidney of experimental animals. They are the most toxic of the usual breakdown products, with LD_{50} values of around 175 mg/kg body weight in mice (Heaney and Fenwick, 1987). Liver damage in rats dosed with crude nitrile extracts includes disruption of the normal lobular architecture of tissues together with irregular bile duct proliferation (VanEtten *et al.*, 1969). In the kidney, such extracts induce hypertrophy and enlarged nuclei among the epithelial cells lining the convoluted tubules of rats. All of these changes are also seen in animals dosed with the pure forms of the nitriles.

Despite the foregoing, there is no evidence that the glucosinolates present in brassica vegetables are detrimental to human health. If this were the case, epidemiological studies with people consuming high levels of cabbage (e.g. central and eastern Europeans) would have exposed such evidence. In any case cooking usually reduces glucosinolate levels by 30–60%, depending upon the vegetable, and this should ensure that intakes are not excessive. Although goitre in humans has been associated with high intakes of cabbage, it is generally acknowledged that 96% of all goitre is caused by simple iodine deficiency and consequently the role of glucosinolates in the incidence of this condition is thought to be minor (Cheeke and Shull, 1985).

Far from being toxic, there is evidence that the glucosinolates may confer advantages as regards protection from certain types of cancer. On the basis of epidemiological studies, the Committee on Diet, Nutrition and Cancer (1982) in the USA concluded that the consumption of brassica vegetables was associated with reductions in the incidence of cancer at several sites in humans. It is important to recognize, however, that these vegetables contain a variety of other substances in addition to the glucosinolates (e.g. *S*-methylcysteine sulfoxide, see below) and these may also confer benefits in cancer prevention. A number of studies with animals models have indicated the advantages of glucosinolate consumption in protection against chemically induced cancers. Thus, rats given a diet with freeze-dried cauliflower showed reduced toxic effects of the carcinogen, aflatoxin B_1 (Stoewsand *et al.*, 1978). Other evidence in a similar vein has been reviewed by Nugon-Baudon and Rabot (1994).

S-Methylcysteine sulfoxide (SMCO)

In addition to the glucosinolates, brassica vegetables and forage also contain a non-protein sulfur amino acid in the form of *S*-methylcysteine sulfoxide. Brassica forage crops such as kale and rape are grown as winter feed for cattle and sheep, but the presence of SMCO in these plants is considered to be a significant factor limiting their use as animal fodder (D'Mello, 1991). Values for SMCO in four brassica vegetables are shown in Table 17.2 (Marks *et al.*, 1992) which illustrate considerable differences between the vegetables. Even higher concentrations have been reported in kale and other forage brassicas. Highest levels generally occur in the young leaves and growing shoots and increases have also been observed as the brassica crop reaches maturity.

Table 17.2. Concentrations of *S*-methylcysteine sulfoxide (mg/100 g fresh weight) in brassica vegetables (Marks *et al.*, 1992)

Vegetable	Concentration
Broccoli	19
Brussels sprouts	68
Cabbage	18
Cauliflower	14

The toxicity of SMCO in cattle and sheep emanates from its metabolism by the rumen bacteria to dimethyl disulfide. The resulting adverse effects are characterized by a severe haemolytic anaemia which appears within one to three weeks in animals fed mainly or exclusively on brassica forage. This condition is also referred to as 'kale anaemia'. Overt indications of this disorder consist of a loss of appetite, reduced milk production in dairy cows, while internally, the changes include the appearance of refractile, stainable granules (Heinz–Ehrlich bodies) within the erythrocytes and a decline in blood haemoglobin concentrations. In addition, extensive organ damage may occur. The liver, for example, may appear hypertrophic, pale and necrotic (Smith 1980). Critical daily intakes of SMCO range from 15 to 19 g/kg body weight, irrespective of whether the amino acid is supplied in crystalline form or as a component of brassica forage. Those animals which survive may show spontaneous but incomplete recovery if they continue to graze the crop, but further episodes of haemolytic anaemia may follow. If brassica forage is withdrawn, normal blood chemistry is restored within 3–4 weeks. Although much needs to be elucidated with regard to underlying mechanisms, current views are that dimethyl disulfide inactivates key proteins through blockage of sulfhydryl groups (Barry *et al.*, 1985). Affected animals appear to compensate for this protein inactivation by increasing synthesis of growth hormone and thyroxine which then stimulate the production of replacement proteins. The risk of SMCO toxicity can be reduced or even eliminated by management practices, for example by controlling brassica intake, particularly in the winter when SMCO concentrations in the forage are maximal (Smith, 1980). In addition, further control is possible by growing brassica crops in low-sulfur soils. Under these conditions SMCO concentrations are markedly reduced (Barry *et al.*, 1984).

Although the toxicity of SMCO to cattle and sheep is undisputed, the incidence of adverse effects in humans has not been reported. This would be consistent with the view that its toxicity requires the intervention of microbial activity and this is limited in humans compared with ruminants. In addition, intermittent consumption of brassica vegetables as part of a balanced diet ensures that SMCO intakes are not excessive. In any event, moderate intakes of the amino acid may be beneficial in the prevention of cancers of the digestive tract, a property also attributed to the other major sulfur-containing constituents, the glucosinolates (see above). The biological activities of the glucosinolates, however, do not fully account for the protective effects of brassica vegetables and a role for SMCO in cancer prevention has therefore been proposed. Evidence for this role is still somewhat sketchy (Marks *et al.*, 1992) and considerably more work is required in this area.

Glycoalkaloids

The glycoalkaloids of relevance here are those occurring in the potato, since even in commercially available tubers destined for human consumption there is always a residual level of these substances. These toxic steroidal glycosides occur predominantly in the forms of α-solanine and α-chaconine. In both compounds, the aglycone moeity is solanidine and in both cases two of the sugar residues are glucose and rhamnose. The differences reside in the third sugar molecule which is galactose in the case of solanine and a second molecule of rhamnose in the case of chaconine. A further point of difference arises from the attachment of the galactose residue to the aglycone in solanine whereas in chaconine the attaching sugar is glucose (Petterson *et al.*, 1991). In the UK the total glycoalkaloid concentration in tubers available for human consumption is normally within the range 25 to 150 mg/kg fresh weight (Morgan and Coxon, 1987), with a generally agreed upper limit of 200 mg/kg, but considerably higher levels have been recorded elsewhere for commercial varieties. Indeed, unacceptably high levels of glycoalkaloids in tubers led to the withdrawal of the Lenape potato variety from commercial growing in Canada and USA (Anonymous, 1970). In Sweden, tubers of the established variety Magnum Bonum, harvested in 1986, were found to contain potentially toxic levels of glycoalkaloids, resulting in the imposition of a conditional sales ban on such potatoes (Hellenas *et al.*, 1995). Total glycoalkaloid levels ranged from 61 to 665 mg/kg fresh weight, with a mean value of 254 mg/kg.

Tuber levels of glycoalkaloids are dependent upon a variety of factors. Large variations have been reported for different cultivars, particularly when comparisons are made between commercial and wild varieties which may differ not only with regard to concentrations of α-solanine and α-chaconine but also in the occurrence of unusual glycoalkaloids in the wild types. In addition, agronomic factors and geographical location (Sinden and Webb, 1972) may have a significant effect on these concentrations. Exposure to light induces marked effects on the concentrations of both total and individual glycoalkaloids, which may have important implications for food safety. For example, during marketing and sale in retail outlets, tubers may be displayed under fluorescent light and this can increase glycoalkaloid concentrations. The glycoalkaloid concentrations in tubers exposed for up to 13 days to continuous light from different sources have recently been reported by Percival *et al.* (1994). The sources of illumination were high-pressure mercury, high-pressure sodium and fluorescent tube (warm white). Glycoalkaloid concentrations increased over this period regardless of source of illumination, but marked differences were observed between the mercury source and the other two forms of illumination. Thus, at 13 days, concentrations of 211, 558 and 698 mg/kg fresh weight, respectively, were recorded for tubers exposed to mercury, fluorescent and sodium sources of light. Thus replacement of fluorescent with mercury lighting for potatoes on display would significantly reduce glycoalkaloid concentrations in tubers and produce safety would be ensured. Furthermore, the ratio of α-chaconine to α-solanine also varied with exposure to light such that this ratio decreased significantly from 69 : 31 at the beginning of the study to 52 : 48 after 13 days of exposure to the sodium light source. At that point in time, the ratios for tubers exposed to fluorescent and mercury light sources were 54 : 46 and 59 : 41 respectively. The increase in α-solanine levels on exposure of tubers to light may be important in view of the higher LD_{50} value of this alkaloid relative to that of α-chaconine for rats (Morgan and Coxon, 1987).

It might be assumed that peeling would reduce glycoalkaloid concentrations in the tuber. The results of Hellenas *et al.* (1995) confirm this expectation. However, the efficacy of peeling was dependent upon the initial level of the alkaloids prior to peeling. Thus, the proportion left in peeled tubers was 38% in those containing 121 g/kg fresh weight, but 86% in tubers with an initial glycoalkaloid concentration of 785 g/kg fresh weight. Indeed, a positive linear relationship was observed between the alkaloid concentrations of peeled tubers and the original levels in whole tubers. It would appear that in tubers containing inordinately high levels of glycoalkaloids, these compounds diffuse deeper into the cortex tissue from the periderm and skin where it is usually concentrated.

The consumption of tubers with levels of glycoalkaloids in excess of normal values has long been associated with toxicity in humans. Cooking of potatoes provides no protection as both alkaloids are heat-stable. A summary of the various cases of toxicity in humans, based largely on a review by McMillan and Thompson (1979), is presented in Table 17.3. A notable instance of toxicity in recent times concerns an outbreak of suspected poisoning at a school in South East London (McMillan and Thompson, 1979). On the second day of the autumn term 78 schoolboys became ill after eating potatoes stored over the summer period. Seventeen of the boys were admitted to hospital, with gastrointestinal, circulatory, neurological and dermatological symptoms which were consistent with glycoalkaloid poisoning. Similar symptoms were reported in Sweden in an outbreak alleged to have been caused by a high-alkaloid batch of potatoes from the variety Magnum Bonum (Hellenas *et al.*, 1995). McMillan and Thompson (1979) cited earlier cases of potato glycoalkaloid poisoning which resulted in the death of a number of subjects. The glycoalkaloids have, in addition, been associated with teratogenic properties in animals. On the basis of epidemiological evidence, damaged or diseased potatoes were implicated in the incidence of congenital deformities in humans. However, avoidance of potatoes shortly before and during pregnancy failed to reduce the incidence of these deformities (Morgan and Coxon, 1987).

Apart from causing toxicity, glycoalkaloids may also impart adverse organoleptic properties in that they are associated with a bitter taste and a burning sensation in the throat. Levels exceeding 110 g/kg fresh weight have been suggested as being undesirable from this point of view (Morgan and Coxon, 1987).

The monitoring of glycoalkaloid levels in potatoes is desirable on an ongoing basis, particularly as new varieties are developed. It is salutary to recall an often-quoted statement that if potatoes were introduced at the present time as a new food commodity, they would probably not satisfy current food standard requirements and regulations for natural toxicants.

Table 17.3. Cases of glycoalkaloid poisoning in humans (based partly on the review of McMillan and Thompson, 1979)

Year of report	Details	Adverse effects
1899	56 soldiers in Berlin affected after consuming potatoes.	Gastroenteritis; fever; headache; drowsiness; apathy
1918	61 people in eastern Glasgow	Headache, vomiting; diarrhoea few hours after eating potatoes
1925	7 members of a family consumed greened potatoes	Extreme exhaustion; restlessness; accelerated respiration; loss of consciousness prior to death of 2 of this family.
1933	60 persons in Cyprus consumed young potato shoots and leaves as vegetables.	Headache; colic; vomiting or nausea; diarrhoea; fever; irritation of throat; one death at 44 hours
1952–53	382 North Koreans affected after eating rotten potatoes.	52 entered hospital, 22 died (3 within 24 h); pain; nausea; vomiting; feeble pulse; oedema of the face; respiratory failure; cardiac arrest
1979	78 London schoolboys consumed potatoes left over from the previous term	17 required hospital admission; diarrhoea; vomiting; circulatory, neurological and dermatological symptoms
1986	6 separate incidents in Sweden, involving a total of 11 people who consumed potatoes of the variety Magnum Bonum	Nausea; vomiting; epigastric pain; headache

Nitrate

The level of nitrate in vegetables has been a matter of some concern in many countries, to the extent that in the UK a working party was set up by the then Steering Group on Food Surveillance, resulting in the publication of two reports (MAFF, 1987, 1992). The disquiet arose primarily from the finding that a proportion of ingested nitrate may be converted to the nitrite ion which in excess may contribute to the condition of infantile methaemoglobinaemia. In this syndrome, the nitrite ion reacts with haemoglobin to impair its oxygen-carrying capacity. When the nitrite ion combines with secondary or tertiary amines, *N*-

nitroso compounds may be formed, some of which are recognized to be potent carcinogens in laboratory animals, and this is another source of concern.

There is little doubt now that vegetables are the single greatest source of nitrate in the human diet, accounting for 76% of the total intake (MAFF, 1992). A summary of recent data relating to nitrate contents of different vegetables is presented in Table 17.4. The wide variation in concentrations for each vegetable is clear, possibly emanating from differences in nitrogen status of the soil and fertilizer application rates (Lyons *et al.*, 1994). Equally, there are marked variations between the three data sources listed in Table 17.4. There are also wide differences among the vegetables in nitrate content. One of the reports (MAFF, 1992) grouped various vegetables into three categories according to nitrate content. Those with a relatively high content included lettuce, celery and beetroot. For some unexplained reason, spinach was placed in the second category with moderate concentrations of nitrate, which included potatoes and cabbage. Tomatoes were placed in the third group with low concentrations of nitrate. In a recent assessment of geographical variations (MAFF, 1992), it was observed that nitrate levels in samples of Dutch and English lettuce were similar, but lettuce from Spain had significantly lower concentrations (600 vs 3900 mg/kg fresh weight). These differences may be partly due to extensive trimming of the outer leaves of lettuce imported into the UK. The outer leaves are known to contain higher concentrations of nitrate. Furthermore, environmental factors may exert an effect on nitrate content. These levels tend to be negatively correlated with light intensity and temperature. Consequently, local conditions prior to harvesting and the extent of trimming of outer leaves may be important determinants of nitrate concentrations in vegetables.

Table 17.4. Nitrate content of vegetables (mg/kg fresh weight)

Vegetable	Nitrate-N	Nitrate	
Cabbage	70–370		390
Lettuce	65–330	800–3500	1200
Parsley	10–330	1360–5780	2100
Spinach		3000–8000	2100
Beetroot	290–650		1500
Radish	145–790		1100
Broccoli	30–85		410
Brussels sprout	10–15		7–12
Cauliflower	40–75		37
Celery	295		1200
Data source	Lyons *et al.* (1994)	Lima *et al.* (1995)	MAFF (1992)

Data of Lima *et al.* (1995) recalculated using dry matter values taken from MAFF (1992).

The effects of fertilizer application rates and crop maturity on nitrate concentrations of summer cabbages, carrots, leeks, onions, beetroot and potatoes were considered by MAFF (1992). In the absence of nitrogen fertilizers, nitrate concentrations remained relatively low. After treatment with fertilizer, there was an initial phase when plant nitrate levels remained high followed subsequently by a period of rapid decline and finally a phase coinciding with normal harvest when nitrate levels remained low.

It is worth noting at this juncture that cooking has a marked effect on nitrate content of vegetables, and this is clearly an element which needs to be considered in any assessment of risk to humans. However, as might be expected, the effects are variable, depending on the vegetable in question. For cauliflower, 100% losses occurred on boiling, but with spinach losses of only 43% were recorded (MAFF, 1992). In the case of beetroot, on the other hand, there appeared to be an increase in nitrate content on boiling, possibly arising from changes in dry matter levels and in nitrate release.

The UK survey (MAFF, 1992) also considered the risk posed to special groups of people by nitrate in vegetables. In this context, the intakes by vegetarians would be expected to be higher than those of the average population. This is borne out in the MAFF (1992) study. Vegetarians consume six times more lettuce and four times more celery than the general population. Consequently, the intake of nitrate in the former group is higher. This was estimated to be 185–194 mg/day among vegetarians, compared with 54 mg/day for the general population. However, the acceptable daily intake (ADI) for nitrates is greater than both these sets of values, being 219 mg/day for a 60 kg person (MAFF, 1992). It was concluded that there was no cause for concern regarding current dietary intakes of nitrate (and nitrite) by the general population and indeed by vegetarians. Nevertheless, it was recommended that efforts to reduce levels of nitrate in foods and to monitor nitrate intakes by vegetarians should continue.

Oxalates

Oxalates are widely distributed in plants in soluble and insoluble forms and have been the cause of some concern as regards human and animal health since some plants accumulate relatively high quantities. The paper of Libert and Franceschi (1987) provides a comprehensive survey of the distribution of oxalates in different plants. Total oxalate concentrations in petioles of rhubarb (*Rheum* spp.) range between 18 and 216 g/kg dry weight and in spinach leaf (*Spinacia oleracea*) values of 44 to 157 g/kg dry weight have been recorded. In both cases a substantial proportion of the oxalate content may be in water-soluble forms (Libert and Franceschi, 1987). Thus in one set of values listed by these authors, proportions of water-soluble oxalates in spinach ranged from 39 to 68% of total values. The wide range of concentrations of total and water-soluble oxalates reflect the influence of several factors including stage of growth, genotype/cultivar, mineral nutrition and, in addition, post-harvest treatment may exert a considerable effect. Nitrate fertilizers increase oxalate accumulation in spinach, for example, whereas ammonium nitrogen has little effect. In general, phosphate fertilizers decrease oxalate levels in spinach whereas the effect of potassium is small. The most obvious method of reducing the risk of oxalate poisoning is to cook the leaves or petioles. Blanching, for example, is commonly used for spinach and other vegetables, and the effect is to reduce the levels of water-soluble oxalates.

An important factor to be considered in assessing risk is the extent of absorption of oxalates from the digestive system. Up to 40% of dietary oxalate may be absorbed, depending upon form of oxalates, overall nutritional status and variation from one individual to another (Libert and Franceschi, 1987). These authors also highlighted recent evidence indicating absorption of calcium oxalate, previously regarded as being unavailable for uptake from the gut.

The adverse effects of oxalates are attributable to a variety of causes. In the acute toxicity of oxalates three overlapping phases have been recognized (Hodgkinson, 1977). Initially, a local corrosive action may occur in the mouth or in the alimentary tract, preceding symptoms such as vomiting and diarrhoea. This phase may be followed by effects on the nervous system (causing muscle cramp and tetany) and cardiovascular system, causing for example, reduced blood coagulability and other symptoms associated with low levels of calcium in body fluids. In the final phase, renal insufficiency may be induced as a result of the direct toxicity of oxalic acid or precipitates of calcium oxalate. In humans, the lethal dose varies from 2 to 30 g depending upon a number of physiological and exogenous factors. Lethal poisoning of a child after excessive ingestion of rhubarb leaves has been imputed to high oxalate intake, but the aetiology of this particular case is somewhat complicated by the presence of other toxicants in the plant (Libert and Franceschi, 1987).

Since oxalic acid is a relatively strong chelating agent, the major chronic effects emanate from this property. Complexation of calcium by oxalate may result in calcium deficiency, although it has been calculated that in UK diets the oxalate to calcium ratio is satisfactory (Hodgkinson, 1977) but this may not be the case in regional diets in India, for example. The majority of urinary stones formed in humans are derived from calcium oxalate and in this case dietary sources of oxalate may be of relevance, particularly for those in specific risk groups. This highlights the importance of eating balanced diets especially in respect of calcium-rich foods such as milk and cheese and of ensuring adequate vitamin D status.

Furanocoumarins

Linear furanocoumarins occur widely in plants, including commonly consumed grocery vegetables such as celery and parsnips (Diawara *et al.*, 1995). These compounds are photoactivated plant secondary metabolites which have long been used as therapeutic agents in the treatment of skin disorders. However, the use of fuanocoumarins for this purpose has been accompanied by higher incidence of skin cancer in treated subjects. In celery, the major furanocoumarins include psoralen, bergapten and xanthotoxin. Of these, bergapten is present in higher concentrations, particularly in the outer leaves of the celery plant where levels of up to 28 μg/g have been found (Diawara *et al.*, 1995). The distribution of linear furanocoumarins varies according to the different parts of the celery plant. Thus, Diawara *et al.* (1995) reported values for total furanocoumarins (on fresh weight basis) of 44.9 μg/g in the outer leaves, 9.9 μg/g in the inner leaves, with significantly lower quantities in the petioles and root (ranging from 0.9 to 1.5 μg/g). Furanocoumarin levels also vary with other factors such as infection of the celery plant with *Sclerotina sclerotorum*, ultraviolet radiation and atmospheric pollution.

Since the linear furanocoumarins are heat-stable there is added concern about their potential to induce toxicity in humans. The data of Diawara *et al.* (1995) show that in the outer and inner leaves, levels of these substances can exceed those reported to cause both acute and chronic dermatitis, 18 and 7–9 μg/g, respectively (Austad and Kavli, 1983; Seligman *et al.*, 1987). The furanocoumarins can reach the skin through direct contact with vegetables such as celery, causing dermatitis. In an epidemiological study of grocery workers in Oregon, it was established that the dermatitis in these individuals arose through handling of celery, and exposure to UVA irradiation exacerbated this condition (Berkley *et al.*, 1986). Another more recent study confirmed the incidence of photodermatitis in workers handling celery containing total furanocoumarins at levels of up to 84 μg/g fresh weight (Finkelstein *et al.*, 1994). The potential for toxicity is greater when celery plants become infected with disease organisms. Linear furanocoumarins may also reach the skin through dietary consumption of celery. Thus, Ljunggren (1990) reported a case of severe phototoxic burn in a person who had consumed large quantities of celery root. The furanocoumarins are also associated with carcinogenic and mutagenic properties.

Positive steps can be taken in the management of risk from linear furanocoumarins. A key factor in this strategy would be to avoid contact with or consumption of the leaves of mature petioles. This is readily achieved by trimming off the leaves at harvest while wearing protective clothing. Diseased celery plants should be discarded and prolonged cold storage avoided as both factors increase levels of furanocoumarins.

Conclusions

This review illustrates the diversity of compounds, present naturally in fruit and vegetables, with the potential to induce adverse effects in humans. No attempt has been made to present a comprehensive catalogue of such compounds.

The toxic potential of these substances should be weighed against the overwhelming benefits to human health which accrue from the intake of vitamins and fibre in fruit and vegetables. Furthermore, there is no evidence from epidemiological studies to indicate any major hazards to human health when fruit and vegetables form part of a balanced diet. Indeed, under these conditions, the glucosinolates and SMCO present in brassica vegetables and the isoflavones which occur in soyabean products may confer benefits in terms of reduced incidence of several types of cancer in humans. Other advantages of secondary plant metabolites should also be considered. Thus, in the processing of apples for juice, it has recently been found that the presence of oxalic acid in commercial pectins prevents enzymatic browning (Tong *et al.*, 1995). However, the use of pure oxalic acid as an additive is unlikely owing to its toxicity. The putative role of many of these compounds in plants as part of a defence system against predators and fungal diseases should not be ignored.

However, there are circumstances when hazards to human health may occur, particularly in certain risk groups, and lethalities have been recorded in the recent past. By the very nature of their diets, vegetarians may be at greater risk from excessive nitrate consumption than the general population, and the possible effects of oxalates from vegetables on the calcium status of these individuals need to be considered. Some hay-fever sufferers may be vulnerable to fruit and vegetable allergens. In developing countries, undernourished people are at risk from a variety of plant toxicants in addition to those reviewed here, especially when fuel and water are in short supply.

The potato alkaloids and the linear furanocoumarins in celery do present a potential problem for the general population, but the risk can be reduced substantially through monitoring, enhanced public awareness, and careful post-harvest processing, storage and display of these vegetables. Rejection of decayed apples prior to processing appears to be prudent advice for reducing patulin levels in the juice. Adequate cooking of kidney beans should ensure inactivation of the powerful lectins, while boiling of spinach effectively leaches out the water-soluble oxalates.

Many issues remain unresolved. Particularly with regard to the benefits of glucosinolates and SMCO in brassica vegetables and the hazards of isoflavones in human reproduction. A major question arises over possible additive or synergistic efforts caused by different secondary compounds in the normal intake of fruit and vegetables. For example, a role for lectins has been proposed in the aetiology of fruit allergies, and one might speculate about a possible link between furanocoumarins and allergy to celery. Much still needs to be elucidated as regards the underlying mechanisms of toxicity. However, a unifying feature of recent observations is the demonstration that the activity of ornithine decarboxylase is an important focal point for the action of a number of natural toxicants including glycoalkaloids, lectins and at least one non-protein amino acid (D'Mello, 1993). This enzyme plays a critical role in polyamine synthesis. These are challenging issues which can only be resolved with concerted efforts from phytochemists, toxicologists and medical researchers.

References

Anonymous (1970). *Amer. Potato J.* **47**, 103.

Austad, J. and Kavli, G. (1983). *Contact Dermatitis* **9**, 448–51.

Barry, T. N., Manley, T.R., Millar, K.R. and Smith, R.H. (1984). *J. Agric. Sci.* **102**, 635–43.

Barry, T.N., Manley, T.R., Redekopp, C. and Allsop, T.F. (1985). *Br. J. Nutr.* 54, 165–73.

Bender, A.E. and Reaidi, G.B. (1982). *J. Plant Foods* **4**, 15–22.

Berkley, S.F., Hightower, A.W., Beier, R.C., Fleming, D.W., Brokopp, C.D., Ivie, G.W. and Broome, C.V. (1986). *Ann. Intern. Med.* **105**, 351–5.

Cheeke, P.R. and Shull, L.R. (1985). In *Natural toxicants in feeds and poisonous plants.* Avi Publishing Company, Connecticut.

Committee on Diet, Nutrition and Cancer. (1982). *National Research Council.* National Research Council, Washington DC.

Diawara, M., Trumble, J.T., Quiros, C.F. and Hansen, R. (1995). *J. Agric. Food Chem.* **43**, 723–7.

D'Mello, J.P.F. (1991). In *Toxic substances in crop plants* (ed. J. P.F. D'Mello, C.M. Duffus and J.H. Duffus), pp. 21–48. Royal Society of Chemistry, Cambridge.

D'Mello, J.P.F. (1993). *Amino Acids* **5**, 212–13.

D'Mello, J.P.F. (1995). In *Tropical legumes in animal nutrition* (ed. J.P.F. D'Mello and C. Devendra), pp. 135–72. CAB International, Wallingford.

Duncan, A.J. (1991). In *Toxic substances in crop plants* (ed. J.P.F. D'Mello, C.M. Duffus and J.H. Duffus), pp. 126–47. Royal Society of Chemistry, Cambridge.

Finkelstein, E., Afek, U., Gross, E., Aharoni, N., Rosenberg, L. and Havely, S. (1994). *Int. J. Dermatol.* **33**, 116–18.

Golbitz, P. (1995). *J. Nutr.* **125**, 5702–72S.

Grant, G. (1991). In *Toxic substances in crop plants* (ed. J.P.F. D'Mello, C.M. Duffus and J.H. Duffus), pp. 49–67. Royal Society of Chemistry, Cambridge.

Heaney, R.K. and Fenwick, G.R. (1987). In *Natural toxicants in food: progress and prospects* (ed. D.H. Watson), pp. 76–108. Ellis Horwood, Chichester.

Hellenas, K-E., Branzell, C., Johnsson, H. and Slanina, P. (1995). *J. Sci. Food Agric.* **68,** 249–55.

Herman, C., Aldercreutz, T., Goldin, B.R., Gorbach, S.L. Hockerstedt, K.A.V., Wanatabe, S., Hamalainen, E.K., Markkanen, M.H., Makela, T.H., Wahala, K.T., Hase, T.A. and Fotsis, T. (1995). *J. Nutr.* **125,** 757S–70S.

Hodgkinson, A. (1977). *Oxalic acid in biology and medicine*. Academic Press, London.

Libert, B. and Franceschi, R. (1987). *J. Agric. Food Chem.* **35,** 926–38.

Lima, J.L.F.C., Rangel, A.O.S.S. and Souto, M.R.S. (1995). *J. Agric. Food Chem.* **43,** 704–7.

Ljunggren, B. (1990). *Arch. Dermatol.* **126,** 1334–6.

Lyons, D.J., Rayment, G.E., Nobbs, P.E. and McCallum, L.E. (1994). *J. Agric. Food Chem.* **64,** 279–81.

Marks, H.S., Hilson, J.A., Leichtweis, H.C. and Stoewsand, G.S. (1992). *J. Agric. Food Chem.* **40,** 2098–101.

McMillan, M. and Thompson, J.C. (1979). *Quart. J. Med.* **48,** 227–43.

Ministry of Agriculture, Fisheries and Food (1987). *Food Surveillance Paper No.* **20,** HMSO, London.

Ministry of Agriculture, Fisheries and Food (1992). *Food Surveillance paper No.* **32,** HMSO, London.

Ministry of Agriculture, Fisheries and Food (1993). *Food Surveillance paper No.* **36,** HMSO, London.

Morgan, M.R.A. and Coxon, D.T. (1987). In *Natural toxicants in food* (ed. D.H. Watson), pp. 221–30. Ellis Horwood, Chichester.

Noah, D.N., Bender, A.E., Reaidi, G.B. and Gilbert, R.J. (1980). *Br. Med. J.* **281,** 236–7.

Nugon-Baudon, L. and Rabot, S. (1994). *Nutr. Res. Rev.* **7,** 205–31.

Ortolani, C., Ispano, M., Pastorello, E., Bigi, B. and Ansaloin, R. (1988). *Ann. Allergy* **61,** 47–52.

Percival, G., Dixon, G. and Sword, A. (1994). *J. Sci. Food Agric.* **66,** 139–44.

Petterson, D.S., Harris, D.J. and Allen, D.G. (1991). In *Toxic substances in crop plants* (ed. J.P.F. D'Mello, C.M. Duffus and J. H. Duffus), pp. 148–79. Royal Society of Chemistry, Cambridge.

Prieta, J., Moreno, M.A., Diaz, S., Suarez, G. and Dominguez, L. (1994). *J. Agric. Food Chem.* **42,** 1701–3.

Rosenthal, G.A. (1991). In *Herbivores: Their interactions with secondary plant metabolites, volume 1: The chemical participants* (ed. G.A. Rosenthal and M.R. Berenbaum), pp. 1–34. Academic Press, San Diego.

Seligman, P.J., Mathias, C.G., O'Malley, M.A., Beier, R.C., Fehrs, L.J., Serril, W.S. and Halperin, W.E. (1987). *Arch. Dermatol.* **123,** 1478–82.

Sinden, S.L. and Webb, R.E. (1972). *Amer. Potato J.* **49,** 334–8.

Smith, R.H. (1980). *Vet. Rec.* **107,** 12–15.

Stinson, E.E., Osman, S.F., Huhtanen, C.N. and Bills, D.D. (1978). *Appl. Environ. Microbiol.* **36,** 620–2.

Stoewsand, G.S., Babish, J.B. and Wimberly, H.C. (1978). *J. Environ. Path. Toxicol.* **2,** 399–406.

Tong, C.B.S., Hicks, K.B., Osman, S.F., Hotchkiss, A.T. and Haines, R.M. (1995). *J. Agric. Food Chem.* **43,** 592–7.

Tsukamoto, C., Shimada, S., Igita, K., Kudou, S., Kokubun, M., Okubo, K. and Kitamura, K. (1995). *J. Agric. Food Chem.* **43,** 1184–92.

VanEtten, C.H., Daxenbichler, M.E. and Wolff, I.A. (1969). *J. Agric. Food Chem.* **17**, 483–91.

Vieths, S., Schoning, B. and Baltes, W. (1992). *Food Agric. Immunol.* **4**, 181–97.

Vieths, S., Schoning, B. and Jankiewicz, A. (1993). *Food Agric. Immunol.* **5**, 93–105.

Whitten, P., Lewis, C., Russell, E. and Naftolin, F. (1995). *J. Nutr.* **125**, 771S–6S.

18. Phytochemistry of fruits and vegetables: an ecological overview

JEFFREY B. HARBORNE

Department of Botany, School of Plant Sciences, University of Reading, Whiteknights, Reading RG6 6AS, UK

Introduction

The ripe fruit is the one part of the plant which is likely to be undefended chemically, since it is provided for animals in return for the widespread dispersal of the seed that lies within that fruit. By contrast, the seed and seedcoat usually possess some chemical toxins, although they are often protected physically as well. Nevertheless, as Janzen (1978) has pointed out, ripe fruit tissue may still contain some chemical constituents which have a harmful effect on the herbivore because of the need to deter the 'wrong' seed dispersers, namely those animals which either damage the seed in eating the fruit or which fail to distribute the seed away from the parent plant. For example, ripe *Andira inermis* fruits appear to have a potent antibiotic in the juicy pulp. This has no effect on the Costa Rican fruit bats which avidly eat the fruit and disperse the seed. However, the fruits are rejected as food by cattle and pigs, because the antibiotic inhibits the gut flora required for proper food digestion; such animals would feed under the parent tree and the seed might not be moved far enough away from the parent to germinate and develop successfully.

By contrast with the ripe fruit, the unripe fruit will differ not only in nutritional content but also in its secondary chemistry. Thus it is likely to be defended chemically in the same way that foliage is. The green chlorophylls of the leaf will camouflage the unripe fruit and reduce the extent of mammalian herbivory. Some phytophagous insects however are known to home in on green tissues and have the ability to transfer from foliage to green fruit with little difficulty. This is true of the insect *Helicoverpa zea*, which attacks green tomato fruits. Herbivores which have overcome the chemical defence of unripe fruits will tend to benefit more, since the nutritional value (and especially the nitrogen content) is often greater than that of the leaf.

The fact that most unripe fruits will have some chemical traits (acidity, bitterness, astringency) which deter the majority of herbivores means that chemical changes inevitably take place during the ripening process. Just as there are increases in the nutritional value, there will be significant changes in secondary chemistry and typically toxic constituents will either disappear or become reduced in amount. During ripening, other changes may occur. The fruit may develop a colour (yellow, red, or black) which advertises it against a green background. Attractive odours and taste constituents may also accumulate, with a decrease in organic acid content.

Although our knowledge of the chemistry of cultivated fruits (and vegetables) is considerable, we know much less about that of wild species. The chemical ecology of fruits and their seeds has not been widely studied in natural habitats and we are dependent on the occasional observations of ecologists, studying other aspects of vegetation, for much of our information. In this review, it is intended to consider briefly the attractive chemistry of fruits and then consider their protective chemistry, in conjunction with that of the seed. Illustrations will be based on fruit eaten by birds and by primates. The third and final section of this review considers the protective chemistry of vegetative tissues (i.e. leaves and roots) which are used as vegetables.

Attractive chemistry of fruits

The ripe fruit is usually exposed to herbivores by its attractive and distinctive colour, as well as a normally pleasant odour. It will equally be attractive to the herbivore on tasting it, by its sweetness and nutritional value. Colour is provided by chlorophylls, carotenoids and anthocyanins. Green fruits will contain chlorophylls, yellow, orange and red fruits carotenoids and red to purple black fruits anthocyanins. Certain fruits such as the apple can have all three classes of pigment: chlorophylls in green varieties, carotenoids in yellow varieties and anthocyanins in red varieties. Much information is available on the fruit pigments of cultivated plants (Gross, 1987) but less is known about the fruit pigments of wild plant species.

The ecological importance of fruit colour for attracting herbivores is illustrated here with reference to examples from temperate and tropical habitats. Some years ago, we surveyed the fruit anthocyanins of two closely related species of the shrub *Empetrum*, namely *E. nigrum* and *E. rubrum*. *E. nigrum* is a Northern hemisphere plant, which as the name implies, has purple black fruit. The pigments are based on a mixture of delphinidin, petunidin, malvidin, peonidin and cyanidin 3-galactosides. By contrast, *E. rubrum* which replaces *E. nigrum* in the southern hemisphere has red fruits and the pigments are peonidin and cyanidin 3-galactosides (Moore *et al.*, 1970). It would appear therefore that what is usually a recessive mutation has become fixed in the southern hemisphere plants because of the need to switch to a fruit colour attractive to the local bird populations. One assumes that the purple black berries eaten by northern temperate bird frugivores, failed to attract the frugivores of the southern hemisphere and hence the production of red fruits in these plants.

The second example refers to the iridescent blue colour of fruits of *Elaeocarpus*. Blue fruit colour is relatively uncommon among those plant species where the seed is dispersed by birds. Only some 5–7% of such fruits are blue, the majority being red or black. Most of those which are blue are expectedly pigmented by anthocyanin and this has been confirmed by a survey of 26 species in 18 genera in Costa Rica, India, Florida and Malaysia (Lee, 1991). Hence the blue fruits of *Elaeocarpus angustifolius* appear to be unique in lacking anthocyanin, but instead being structurally coloured. This was first noted by Corner (1952) and confirmed by more recent investigation (Lee, 1991). Colour is produced by the presence of an iridosome structure, beneath the outer cell wall of the adaxial epidermis. This iridescent blue colour appears to be of selective advantage in ensuring that the fruit is readily apparent to the cassowary bird (*Casuarius casuarius*) and fruit-eating pigeons. Not only do the iridescent fruits present a brilliant colour against the green foliage of the forest, but also the colour persists even when the mesocarp is almost completely senescent or has been consumed by beetles. Evolution here has produced a structural colour in the fruit which is superior to the more usual anthocyanin pigmentation.

Many fruits have attractive odours which are pleasant to humans. This is not surprising, considering that humans and other primates have an ecological role in dispersing the seed of such plants. Chimpanzees living in forest reserves depend on fruits of trees such as figs for much of their diet. Through their faeces, they distribute the seeds along the forest floor and their food choice may determine those tree species which are able to regenerate through seed and those which cannot (Reynolds, 1993). The odour principles of most fruits consumed by humans have been identified and much is known of the chemistry involved. Some fruits have a single aroma principle (e.g. citral in lemon), whereas others have mixtures of many components (Harborne, 1993).

Some fruits may have odours repellent to humans, simply reflecting the fact that other animals may be responsible for eating the fruit and correctly dispersing the seed. This would explain why the durian fruit, *Duria zibethium,* which is regarded as a delicacy by some humans, is rejected by others because of its offensive sewage-like smell. The odour is largely associated with the outer flesh of the rind and the aril around the seeds has a delicious taste, with a fruity flavour and an undescribable creaminess. Many animals consume the fruit in the Malaysian rainforest, but elephants are particularly attracted to the foul odour and are probably good seed dispersers, since after eating the fruit they carry the seed away from the parent plant before defecating (Corner, 1964). Similarly, human beings avoid growing female trees of *Ginkgo biloba* because the fruits smell of rancid butter due to the presence of butanoic and hexanoic acids in large amount. The seed disperser of the fruit of *Ginkgo* is more problematical since this plant is a relict survival from a past age in evolution. The most likely seed disperser was probably a reptile.

Protective chemistry in fruits

As Janzen (1978) has pointed out, one would expect a differential distribution of toxins within the fruit, with a concentration in the seed and especially the seed coat, and a general lack of toxin from the fruit pulp, which has to be attractive to the herbivore. There are certainly a number of examples where this is so. The serendipity berry, *Dioscoreophyllum cumminsii*, is noted for its intensely sweet protein, which is 3000 times as sweet as sucrose. This sweetness is entirely confined to the pulp. Analysis of the seed of *Dioscoreophyllum* revealed the presence of two intensely bitter diterpenoids, columbin and isocolumbin (Ramstad *et al.*, 1975), which would appear to protect the seed from predation. This fact prevents the whole fruit being used for sweetening purposes.

That toxic seed coats slow down the feeding of a seed predator is exemplified by red squirrels which eat the cones of the white cedar, *Thuja occidentalis*. Each seed has a winged and membraneous seed coat and the squirrel takes time to carefully remove this flimsy seed coat before eating the rest of the tissue. The reason for this caution in eating the whole cone becomes apparent when chemical analysis has revealed a pungent mixture of monoterpenoids secreted in the oval blisters of the seed coat. One component, a-thujone, which makes up 26% of the oil mixture, is the well known toxin of absinth. This is known to cause convulsions associated with lesions of the cerebral cortex in humans (Janzen, 1978).

Another fruit with a toxic seed coat is the guarana, *Paullinia cupana*. The fruits are black and white, with the appearance of a human eye when resting in the fruit case. They are dispersed by toucans and guans in the Amazonian rainforest. Both the seed kernel (4.28% dry weight) and the seed coat (1.64% dry weight) contain the purine alkaloid caffeine. The aril, by contrast, is completely free of alkaloid and contains up to 70% dry weight of glucose, fructose and sucrose. Experiments by Baumann *et al.* (1995) show that caffeine is not released from the seed while the fruit is being digested by the bird, there being a powerful diffusion barrier preventing its intoxication. Evidence suggests that the high concentrations in the seed, which are 3–5 times the level of caffeine in the coffee bean, would be toxic to the bird if the seed coat became damaged during eating the fruit. The seed is therefore adequately protected during its passage through the bird's digestive system and it will be able to germinate when it eventually reaches the soil.

The differential restriction of plant toxins to the seed may not necessarily apply in the case of unripe fruits. In such cases, there may be deterrent substances in the pulp to prevent them from being eaten before they are ripe. This is so in the case of the fruit of ackee, *Blighia sapida*, where the aril around the seed contains about 0.11% of the non-protein amino acid, hypoglycin A. Human consumption of unripe fruits in Jamaica leads to vomiting sickness and can even be fatal. And yet the ripe fruits can be eaten safely, because the concentration of toxin has dropped dramatically during ripening to 0.008%. Such reductions in fruit toxins with ripening also apply to many solanaceous plants. Steroidal alkaloids make unripe fruits of *Solanum dulcamara* and *S. nigrum* poisonous, but the levels of alkaloid in the pulp decrease significantly during the ripening. The alkaloid tends to be concentrated in the seed in the ripe fruit and birds are well known to feed safely on the berries and subsequently disperse the seed.

One interesting point about fleshy fruits eaten by frugivorous birds is whether or not they are mildly toxic, even when ripe. Studies carried out by Barnea *et al.* (1993) suggest that ripe fruits of ivy, holly and hawthorn do contain secondary constituents (saponins, flavonoids and cyanogens), which are mildly distasteful to foraging blackbirds, starlings and redwings. This mild toxicity prevents the consumption of too many fruits in any one foraging bout and hence regulates seed retention time. This in turn ensures better seed dispersal by the bird, since only a few seeds will be deposited in any one site at one time. Observations on the feeding behaviour of these birds support this hypothesis, since feeding bouts were limited to 1.3–5.3 min and seeds eaten per bout to 4–6 seeds.

Frugivorous birds may incidentally be restricted in their choice of fruits to eat by primary constituents. Thus starlings lack the digestive enzyme sucrase which hydrolyses sucrose to glucose and fructose and hence exhibit a feeding aversion to sucrose-rich fruits. They are liable to develop osmotic diarrhoea if they eat such fruits, and this can be fatal. They can safely eat cherries, figs and grapes, which have glucose and fructose as the major reducing sugars, but would have difficulty with the sucrose-containing peach and apricot (Whiting, 1970).

One kind of protective chemical in fruits that has been the subject of detailed investigation is the presence of condensed tannins. These compounds occur fairly regularly in the fruits of trees eaten by primates, such as wild figs, and because of their astringency represent a possible barrier to feeding. Tannins also occur widely in leaves of most woody plants and the evidence here indicates that leaves with a concentration of tannin above 4% dry weight are rejected; this is true for example of feeding on leaves by Colobus monkeys (Oates *et al.*, 1977). The situation regarding the tannins of fruits is more complex. There is evidence of reductions in extractable tannin levels with ripening (Wrangham and Waterman, 1983), so that it is possible that astringent unripe fruits will be rejected while blander-tasting ripe fruits are eaten. Our own experiments at Reading with tannin-containing *Ficus* species eaten by chimpanzees in the Budongo Forest (Tanzania) suggest that the presence of tannins does not determine feeding preference (Table 18.1). Furthermore ripe and unripe fruits seems to be eaten irrespective of the tannin levels. One explanation of our results is that the tannins are concentrated in the fig seeds and hence do not have a role in the taste of the fruit and we are currently investigating this possibility. Indeed, Wrangham *et al.* (1993) indicate that nine fig species eaten by chimpanzees in the Kibale Forest (Zaire) have two or three times more tannin in the seed than in the pulp. We have also observed the chimpanzees of Budongo consuming the bark of *Khaya* tree, where the tannin levels reach 3.3% dry weight; this is surprising considering that there is no free sugar in the bark. What is also curious in our forest reserve is that chimpanzees can reject ripe fruits of one *Ficus sur*, and eat those of an adjacent tree, which would appear to be identical in terms of nutritional status and secondary chemistry. It is apparent that other as yet unknown factors determine whether ripe fruits are consumed or not by these chimpanzees.

Table 18.1. Tannin and free sugar content of *Ficus* species available to chimpanzees in the Budongo Forest reserve

Species	Whether eaten	Tannin[a] (mg/g dry wt)	Free sugars[a] (mg/g dry wt)
Ficus sur	Ripe fruit, not eaten[b]	33	26
Ficus sur	Ripe fruit, not eaten[b]	19	nd
Ficus sur	Ripe fruit, wadge spat out[c]	20	80
Ficus sur	Unripe fruit, juice spat out[c]	24	41
F. matalensis	Ripe fruit, eaten[c]	24	20
F. mucosa	Unripe fruit, eaten[c]	23	50

[a]Data from unpublished results of V. Reynolds, J.B. Harborne and J. Greenham; only dried material was available for analysis.
[b]Fruit from adjacent trees
[c]Analysis was carried out on identical fruits which were handled by the chimpanzees in the manner shown.
nd, not determined

The protective chemistry of the fruit should also extend to protection from microbial infection, since it is important that the fruit ripens and the seeds mature within, before any infection takes place. The chemistry and location of such protection is likely to differ from that produced against herbivory, since the surface and skin of the fruit are the first points where microbial attack occurs. The waxy coating of fruits like the apple would appear to be important in reducing the humidity around the fruit and hence limiting fungal spore germination. Also, concentrated in the fruit coat, one might expect the presence of antifungal agents and their distribution might well include the pulp as well. A range of antifungal constituents have been reported in recent years from several taxonomically different fruits (Table 18.2) and there is evidence in some cases that the concentrations are highest in unripe fruit and they drop dramatically during ripening. This is well known for the tomatine content of tomato, which is effective against *Botrytis cinerea* and *Fusarium solani* in the green fruit (Defago and Kern, 1988). Here the amount of tomatine present is reduced severalfold during ripening. In the mango, *Mangifera indica*, 5-pentadecylresorcinol (see Table 18.2) together with related structures, occur in the peel of the unripe fruit and protect it from infection by the black spot disease, *Alternaria alternata*; this only develops after the fruit is ripened and the antifungal agents have disappeared. This resorcinol is only present in trace amounts in the flesh, but if the unripe fruit is peeled, there is massive synthesis of this antifungal material. The increase in the whole fruit rises from 30 μg/g fresh weight to 160 μg/g fresh weight after four days (Droby *et al.*, 1986).

Table 18.2. Constitutive antifungal constituents of fruits

Plant source	Antifungal agent	Class of chemical
Mangifera indica Anacardiaceae	5-Pentadecylresorcinol	Alkylphenol
Ecballium elaterium Cucurbitaceae	Cucurbitacin I	Nortriterpenoid
Persea americana Lauraceae	1,2,4-Trihydroxyhepta-dec-16-yne	Unsaturated alcohol
Chisocheton paniculatus Meliaceae	1,2-Dihydroxy-6α-acetoxyazadirone	Triterpenoid
Lycopersicon esculentum Solanaceae	Tomatine	Steroidal alkaloid

For references, see Grayer and Harborne (1994).

Phytoalexins are well known to be induced in leaves of many plants following infection as a second line of defence against microbial attack. The question may be posed: is phytoalexin induction a feature of unripe fruits? Since unripe fruits may be actively photosynthesizing, a dynamic response to infection via the synthesis of phytoalexins might well be expected. In fact, a search of the literature suggests that phytoalexin synthesis is rarely a feature of fruit tissue. One clear example appears to be the apple fruit, where two cultivars Bramley's Seedling and Cox's Orange Pippin, respond to infection by *Nectria galligena* by the synthesis of benzoic acid. This compound is the simplest of all known phytoalexins and its fungitoxic effects are pH-dependent (Brown and Swinburne, 1971). However, this seems to be a very specialized case of phytoalexin production, at least within the family Rosaceae. A survey of immature fruits at Reading in a representative selection of species from the family, including *Malus, Prunus* and *Rubus,* failed to yield any further examples where phytoalexins were formed (Kokubun, 1994). Instead, these tissues contained constitutive antifungal compounds in bound form, and these were released in the free state during the inoculation experiments. Fruits of *Malus* species contain phloridzin and phloretin its breakdown product was detected as the antifungal agent in infected fruit. In particular no benzoic acid could be detected after infection by *Botrytis cinerea* of unripe fruit of the apple cultivars Gala, Greensleeves and Kent. The conclusion from these results is that phytoalexin response is a very rare phenomenon in fruits of the family Rosaceae.

In other plant families, there is some evidence of phytoalexin production in fruit tissues. The pod of the pea, *Pisum sativum*, for example, produces pisatin the characteristic phytoalexin of the pea leaf, if challenged with a suitable fungus. Similarly, the fruit of *Capsicum annuum* or pepper produces the sesquiterpenoid capsidiol on fungal challenge. In general, though, relatively little study has been devoted to phytoalexin induction of fruits, or to comparisons of the responses in unripe and ripe tissues.

Protective chemistry of vegetables

Apart from avocados and tomatoes, which are fruits, the vegetables of the European dinner table can be divided botanically into those that are leaf tissue (lettuce, endive, cabbage and related crucifers, spinach, etc.) and those that are storage tissue (potato, radish, parsnip, onion, etc.). From the ecological viewpoint, both types of tissue would be expected to contain some secondary metabolites protective against herbivory and microbial infection. More investment into secondary metabolism might be expected in the storage tissue, because it is usually exposed to herbivory and infection for longer periods than the leaves. Chemical comparison of cultivated vegetables with the wild ancestors or relatives, however, usually indicates a reduction or loss of the toxins which characterize the wild relative. This is the reason why the vegetable is palatable and acceptable in the human diet. It also partly explains why crop plants in general are more prone to microbial disease and pest attack than wild plant species; their protective chemistry has been diluted as a consequence of plant breeding.

The protective toxins may occasionally remain at the same level as in the wild species, in those cases where the toxin may be inactivated or changed by cooking, leaching or otherwise processing the plant tissue. This is true in chicory, where the fresh root is roasted to provide a coffee substitute. It is also true in the tropical crop, cassava, *Manihot esculenta*, where the toxic cyanogenic glycoside linamarin is largely removed by processing and leaching with water. It is only when this is done inefficiently, as in some African communities, that a cassava diet can represent a threat to human health.

In further discussion of the chemical ecology of our common vegetables here, three examples will be used—the potato, the cabbage and the chicory—to indicate the continuing role that secondary metabolism has in the development of these crops.

The alkaloids of the potato

The potato, *Solanum tuberosum*, and the wild related tuber-bearing species of this genus contain steroidal alkaloids, notably solanine and α-chaconine, which are potentially poisonous to humans and also have antimicrobial properties (Roddick, 1986). As Simmonds (1976) has pointed out, human selection of tubers with safe alkaloid levels occurred in the South American home of the potato (see Fig. 18.1). When the cultivated South American tetraploids were introduced into Europe and became *Solanum tuberosum*, the domestic potato was low in alkaloid and any remaining base was readily lost during cooking.

The low alkaloid content of the potato, however, refers only to the edible tuber and other parts of the plant still contain significant levels of alkaloid. This is true of the sprouts, the stem and leaves, the flowers, the fruit and the seed. The potential for producing toxic amounts of alkaloids in the tuber remains within the plant, and this may be expressed in the progeny of hybrids between *Solanum tuberosum* and its wild relatives such as *S. demissum* or *S. chacoense*. Such breeding programmes have been developed throughout this century to produce potato plants which are resistant to Colorado beetle attack or blight infection (see Fig. 18.1). As a result, new varieties may be produced with poisonous levels of alkaloid present and it is essential to test for alkaloid content before putting them on to the market. One new variety called Lenape almost reached the farmer before it was realized that it had dangerously high levels of steroidal alkaloid present (Zikmak and Johnston, 1970).

Until recently, a little known aspect of the human consumption of potato was that the Ahmara Indians in Peru continue to cultivate and eat tubers which have potentially lethal amounts of alkaloid present (Johns, 1990). Steroidal alkaloids taste bitter and the normal level of rejection of a bitter-tasting tuber is about 14 mg alkaloid/100 g tuber and yet the Ahmara Indian potatoes contain up to 17.4 mg/100 g tuber. How have they adapted to eating these bitter-tasting potatoes? The answer lies partly in the development of a considerable sensitivity to bitterness, so that they can recognize a really poisonous tuber from a reasonably safe one. But, they have also learnt to detoxify the alkaloids in the potato either by leaching or by geophagy, the consumption of the potato with clay, which absorbs the alkaloid so that it passes safely through the human system. In this way it is possible to eat a potato which is still protected to a considerable extent by the alkaloids originally present in the wild ancestor.

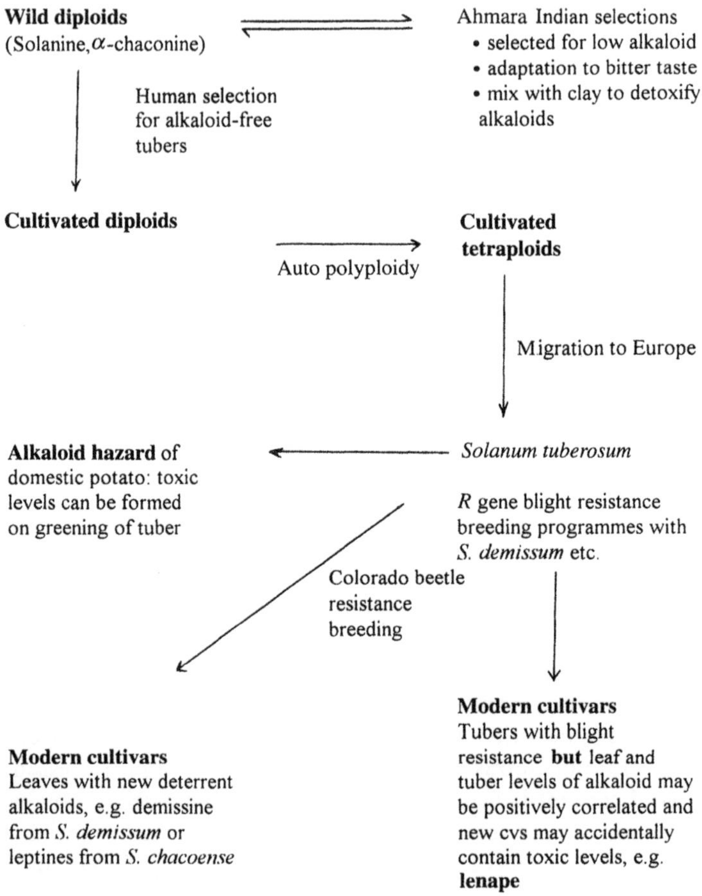

Fig. 18.1. Origin of alkaloids in modern potato varieties.

The sesquiterpene lactones of chicory

Chicory, *Cichorium intybus*, is an interesting plant, because although it is cultivated, cultivation has had little effect on its secondary chemistry. Chicory root, when dried and roasted, is used as a coffee substitute because it imparts a bitter flavour. Hence the bitter principles, which protect the plant when alive, are valued in the food product. Chicory root is also used as a source of the fructan inulin, and again, the bitter principles are readily removed during the isolation process so they do not interfere with human use. It may be noted however that several related plants including the endive, *Cichorium endivia*, and the lettuce, *Lactuca sativa* also have the same bitter principles. However, the concentrations have been drastically reduced in order to make these salad crops palatable (but also more susceptible to herbivory).

The bitter principles of these composite plants are three sesquiterpene lactones, lactucin, lactupicrin and 8-deoxylactucin. These bitter principles are secreted in the latex and occur throughout the plant, except for the flower. A study of the concentrations in chicory throughout the growing season showed the largest amounts in the upper leaves, which are more closely exposed to potential herbivores than the lower leaves, and in the roots (Table 18.3). The distribution of toxins within the plant in general conforms with the idea that those parts that are most vulnerable to herbivory have the highest concentrations. This plant is noticeably free from herbivore attack in natural populations, and the extreme bitter taste of the lactones is a clear deterrent to mammalian feeding. The plant is also distasteful to insects, at least to three coleoptera and one lepidoptera that were tested. Our own experiments with the locust *Schistocerca gregaria* showed that a crude plant extract was an effective antifeedant and that the antifeedant activity was centred on the two major sesquiterpene lactones present. When tested in no-choice and two-choice experiments, lactupicrin was antifeedant at a concentration of 0.21% dry weight and 8-deoxylactucin at a concentration of 0.24%. These correspond precisely with the levels of lactone (between 0.22 and 0.46%) that occur in the plant (see Table 18.3). There is a phenolic, cichoric acid, which occurs throughout the plant in concentrations of about 0.5%, but since it only significantly deterred locust feeding at a concentration of 1.1%, it does not appear to be an antifeedant (Rees and Harborne, 1985). In summary, it does appear that the sesquiterpene lactones of chicory are a significant barrier to herbivory in this unusual food plant.

Glucosinolates of crucifer crops

The toxins of cabbage, kale, turnip, rape and mustard are the glucosinolates or mustard oil glycosides which occur universally throughout the family Cruciferae. A typical structure is sinigrin which on enzymic hydrolysis with myrosinase produces allyl isothiocyanate. Such glucosinolates, or their breakdown products, are toxic to animals. However, they are also important flavour components of crucifer vegetables. At high concentrations in mustard, they provide the biting taste, while at lower concentrations in cabbage leaf, they provide important flavour components, together with their breakdown products. Unlike the situation with the potato alkaloids (see above), the toxins of wild crucifers have been retained in the cultivated forms.

Table 18.3. Sesquiterpene lactone content of different parts of the chicory plant

Part of plant	Lactone content[a] (% dry wt)
Petal	Nil
Top-stem leaves	0.34
Mid-stem leaves	0.26
Base leaves	0.22
Roots	0.42

[a]Mixture of 8-deoxylactucin, lactupicrin and lactucin; average values of measurements made from March to September (Rees and Harborne, 1985).

Evidence that the cultivated cabbage *Brassica oleracea* has lower levels of glucosinolates than the wild species has come from the studies of resistance and susceptibility to the mildew pathogen *Peronospora parasitica.* Allyl isothiocyanate is highly toxic to this pathogen, so that the levels of glucosinolate appear to determine resistance or susceptibility. Most cultivars have betweeen 450 and 21 μg/g dry weight and are susceptible, while a resistant cultivar has 630 μg/g dry weight. By contrast, the wild populations have higher levels and up to 47% of seedlings grown from wild seed are resistant to infection. Hence, selective breeding for milder flavoured *Brassica* vegetables has been a contributory factor in lowering the resistance of modern varieties to powdery mildew (Greenhalgh and Mitchell, 1976).

The interaction between crucifer crops such as cabbage, turnip and rape and phytophagous insects is mediated by the glucosinolates present in these plants (Fig. 18.2). Specialist insects such as the cabbage white butterfly, *Pieris brassicae* and the small white *Pieris rapae* use the glucosinolates both as feeding attractants and oviposition cues. Even so, the concentration of glucosinolate in the plant at a particular stage in plant growth may determine when it feeds successfully on its own host plant. Thus, experiments with the specialized flea beetle *Phyllotetra cruciferae* feeding on *Sinapis alba* have shown that the cotyledons and young leaves are not eaten, because of the high concentrations (20 mM and 10 mM respectively) of glucosinolates. It is only the mature leaves, with levels of 2–3 mM, that are fed upon. These levels, however, still protect the plant from feeding by a generalist, such as the armyworm *Mamestra configurata* (Bodnaryk, 1991). In this case, natural selection would appear to have evolved a mechanism by which the most vulnerable tissues of the plant are protected even from specialist feeders.

A more remarkable response in defense chemistry has been observed in oil seed rape *Brassica napus* that has been infested by the flea beetle *Psylliodes chrysocephala*. This plant switches from the synthesis of aliphatic glucosinolates to the production of indole glucosinolates, which are more toxic on a molar basis than the aliphatic derivatives (Koritsas *et al.*, 1989). In laboratory grown plants, the aliphatic glucosinolate content of the petiole fell from 3.56 to 2.22 mg/g dry weight, while the indole glucosinolate content rose from 0.36 to 12.88 mg/g dry weight. The plant thus responds to insect attack by synthesizing glucosinolates which are more effective in defense than the constitutive mustard oils.

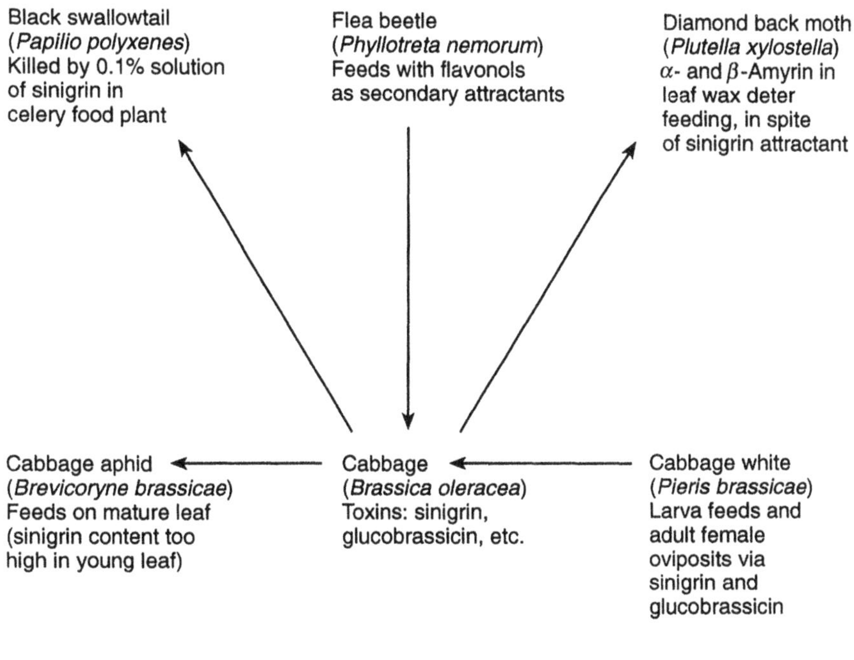

Fig. 18.2. Interactions of cabbage with various insects influenced by the glucosinolates present.

Conclusions

While the chemical ecology of cultivated vegetables is reasonably well advanced, as illustrated here with respect to the cabbage, chicory and potato, the chemical ecology of fruits is still a neglected area. This is especially true of the fruits of tropical wild plant species. We know relatively little about those secondary substances that may be present in the pulp of many fruits, with a dual purpose. Thus they may be unattractive or positively harmful to those animals in the habitat which are poor at seed dispersal. At the same time, they may be attractive or at least neutral to those animals which disperse the seed correctly and without damage.

The chemistry of seeds is reasonably well known but the protective chemistry has been little investigated as part of an ecological programme. The distribution of toxins, such as alkaloids or non-protein amino acids within the different parts of the seed has rarely been surveyed. We need further studies of the type mentioned above where Baumann *et al.* (1995) analysed the caffeine content of the bird-dispersed seed of the guarana. Finally, the roles of fruit volatiles and fruit colours in attracting the right seed dispersal agents to eat ripe fruit require much further study.

References

Barnea, A., Harborne, J.B. and Pannell, C. (1993). *Biochem. Syst. Ecol.* **21**, 421–9.
Baumann, T.W., Schulthess, B.H. and Hanni, K. (1995). *Phytochemistry* **39**, 1063–70.
Bodnaryk, R.P. (1991). *J. Chem. Ecol.* **17**, 1543–56.
Brown, A.E. and Swinburne, T.R. (1971). *Physiol. Plant Pathol.* **1**, 469–75.
Corner, E.J.H. (1952). *Wayside trees of Malaya.* Government Printing Office, Singapore.
Corner, E.J.H. (1964). *The life of plants.* Weidenfeld and Nicholson, London.
Defago, G. and Kern, H. (1988). *Physiol. Plant Path.* **22**, 29.
Droby, S., Prusky, D., Jacoby, B. and Goldman, A. (1986). *Physiol. Mol. Plant Pathol.* **29**, 173–83.
Grayer, R.J. and Harborne, J.B. (1994). *Phytochemistry* **37**, 19–42.
Greenhalgh, J.R. and Mitchell, N.D. (1976). *New Phytologist* **77**, 391–8.
Gross, J. (1987). *Pigments in fruits.* Academic Press, London.
Harborne, J.B. (1993). *Introduction to ecological biochemistry*, 4th edn. Academic Press, London.
Janzen, D.H. (1978). In *Biochemical aspects of plant and animal coevolution* (ed. J.B. Harborne), pp. 163–206. Academic Press, London.
Johns, T. (1990). *With bitter herbs they shall eat it.* Arizona University Press, Tucson.
Kokubun, T. (1994). Ph.D. thesis, University of Reading.
Koritsas, V.M., Lewis, J.A. and Fenwick, G.R. (1989). *Experientia* **45**, 493–5.
Lee, D.W. (1991). *Nature* **349**, 260–2.
Moore, D.M., Harborne, J.B. and Williams, C.A. (1970). *Bot. J. Linn. Soc.* **63**, 277–93.
Oates, J.F., Swain, T. and Zantovska, J. (1977). *Biochem. Syst. Ecol.* **5**, 317–21.
Ramstad, E., Powell, J.W., Wilson, B.J., Adesina, S.K., Higginbotham, J.D. and Harborne, J.B. (1975). *Phytochemistry* **14**, 2719–20.
Rees, S.B. and Harborne, J.B. (1985). *Phytochemistry* **24**, 2225–31.
Reynolds, V. (1993). *Swara*, 13–17.

Roddick, J.G. (1986). In *Solanaceae, biology and systematics* (ed. W.G. D'arcy), pp. 201–22. Columbia University Press, New York.
Simmonds, N.W. (ed.) (1976). *Evolution of crop plants*. Longman, London.
Whiting, G.C. (1970). In *The biochemistry of fruits and their products*, Vol. 1 (ed. A.C. Hulem), pp. 1–32. Academic Press, London.
Wrangham, R.W. and Waterman, P.G. (1983). *Biotropica* **15**, 217–22.
Wrangham, R.W., Conklin, N.L., Etot, G., Obna, J., Hunt, K.D., Hauser, M.D. and Clark, A.P. (1993). *Internat. J. Primatology* **14**, 243–56.
Zikmak, A. and Johnston, R. (1970). *Amer. Potato J.* **47**, 256.

Index

ACO1, ACO2, and ACO3 244
aflatoxin 183–4
aglycone 33–4, 36–7, 42, 113, 117, 319, 338, 340
Agrobacterium rhizogenes 255
  tumefaciens 247
alfalfa 180
allergen 331–2
Allium spp. 138–40, 325
almond, *see Prunus amygdalus*
Alternaria alternata 232, 234, 359
aluminium 45
amino acid 90, 92, 96, 217, 245, 254
  proline 96–7
  metabolism 132–43
  nitrogen and sulphur compounds 137–43
1-amino-cyclopropane-1-carboxylic acid (ACC) 243–9
Ammi majus 180, 258–9
Andira inermis 353
Angelica angelica 180
  archangelica 187
  dahurica 180
annatto, *see Bixa orellana*
antheraxin 12, 13
anthocyanin 6, 11, 29–46, 56, 251, 318, 355
anticarcinogenic 313–25
anticopigmentation 43
antimicrobial 221–3, 236–8
  preformed compounds 231–3
antinutritional 251–61
  metabolites 253–4 (*table*)
antioxidant 24, 74–8
  heart disease, protection against 287–306
  synthetic 54
apocarotenoids 15
apple
  aroma 128, 130, 137
  antioxidant 301
  colour 33
  flavonols 319
  gibberellins 279
  patulin 332–5
  phenolic 40, 56, 58–60, 63–74
  physical damage 205, 211
  phytoalexins 223–4, 231
  phytochemistry 354
apricot, *see Prunus armeniaca*
Arabidopsis 21
  thaliana 272–3, 276
arctic bramble 146
aroma
  amino acid analysis 133–7
  carbohydrate metabolism 126
  chemical analysis 111, 25–49
  fruit 109–21
  fatty acid metabolism 126
  gas chromatography 113–17
  gell chromatography 117
  mass spectrometry 113–17
  β-oxidation 126–32
  volatile compounds 110
ascorbic acid, *see* vitamin C
Aspergillus spp. 174, 333
  niger 112, 118
astringency 87–104, 251–2, 353
atherosclerosis 287
avocado 57, 216, 231–2, 234, 237

banana 56, 109, 112–13, 126, 130, 133–5, 147–9, 228
bean 234
  broad 57–60
  faba 211, 252
  jack, *see Canavalia ensiformis*
  kidney 336
beer 96
bioactive components 314 (*table*), 325
biosynthesis 183, 212
birdsfoot trefoil, *see Lotus corniculatus*
Bixa orellana 11, 15
black pepper 76
Blighia sapida 332, 357

blueberry, *see Vaccinium angustifolium*
boron 5
Botrytis cinerea 221, 224–30, 232–3, 359–60
Brassica spp. 138, 140, 337, 364
broccoli 322–3, 338–9, 344
browning, *see* enzymatic browning; oxidation
brussels sprout 322–3, 338–9, 344

cabbage
  carotenoids 13–14
  glucosinolates 322–3, 337–9
  nitrates 344–5
  physical damage 205–6
  phytochemistry 365
  red 34, 40
caffeic acid 63
caffeine 37, 41, 102, 356
caffeoyl 38–9, 64
calcium 5, 104
callistephin 32–3
Campanula 173
Canavalia ensiformis 334
capsanthin 17
Capsicum annuum
  aroma 136, 144
  carotenoids 15, 21
  gibberillins 279
  physical damage 211
  phytoalexins 223, 225, 229, 235–6
  phytochemistry 360
capsorubin 17
carbohydrate metabolism 143–9
carbon dioxide 4–5
carcinogenesis 77, 313–25
carotenes 11, 14
  α-carotene 23, 294, 316
  β-carotene 23, 294, 297–8, 316–18, 324–5
carotenoids
  anticarcinogenic 313–17
  antioxidant 297–9
  biosynthesis 16–19, 272
  colour 15–25
  phytochemistry 354
  storage 22
carrot
  colour 15
  nitrates 345
  phenolic 52
  physical damage 205–6, 210, 218
  phytoalexins 223–6, 229–32, 235–6
cassava 7
catechin 302, 319
cauliflower 322–3, 338–9, 344
celery 232, 344, 347
Ceratocystis fimbriata 227, 237
Cercospora nicotianae 237
chemical identification 30
chemistry
  attractive 354–6
  protective 356–66
Chenobium amaranticolor 211
cherry, *see Prunus avium*
chestnut 96, 101
chicory, *see Cichorium intybus*
chive 138
chlorogenic acid 44, 56, 65
  absorbence 65
chlorophyll 3–4, 13, 29
chloroplast 12, 14, 56, 272
chromophore 35
chromoplast 15–16, 272
Cichorium intybus 362
cider 96
cinnamic acid 120–1
Citrus 15, 52, 207, 223, 225, 231, 271
  aurantifolia 211, 279
  limon 225, 229, 234, 279, 355
  paradisi 180
  reticulata 271, 279
  sinensis 15, 144, 271, 279
cloudberry 110, 113, 120
clove 109, 113
coconut 132
coffee 109, 146
Colletotrichum 221
  gloeosporioides 231–2
  lindemuthianum 228
  musae 228
colour
  browning 64–70
  carotenoids 15–25
  phenolic compounds 29–46
  see also enzymatic browning
co-pigment 33–46
coronary heart disease 74
coumaric acid 121
coumarins 173–93

Coumarouna odorata 173, 180
cress, *see Lapidium sativum*
Cruciferae 140, 174
cryptoxanthin 316
  β-cryptoxanthin 23
cucumber, *see Cucumis sativus*
Cucumis melo 244
  sativus 279
Cucurbita maxima 271–2, 274
currant 279
cyclodextrins 43–4, 70–1
cysteine 73
Cytisus laburnum 270

Dactylium dendroides 73
date 51
degradation 33, 181–3
delphinidin 39, 42
Desmodium 252
deterioration 22
detoxification 181
diene 231–2, 234
dihydroflavonol reductose (DFR) 257–8
Dioscoreophyllum cumminsii 365
Diplodia natalensis 232
disease resistance 221–38
DNA 217, 243
  carcinogenesis 314
  coumarin effect 184, 186
  PTOM, 13, 243–4
Duria zibethium 355
durian, *see Duria zibethium*

eggplant, *see Solanum melongena*
Eichhornia crassipes 39
Elaeocarpus angustifolius 355
Empetrum nigrum 254, 354
  rubrum 354
environmental factors 21
enzymatic browning
  absorbence measurements 64
  control 66, 70, 72
  inhibition 66, 73, 348
  natural inhibitors 73
  pH 68
  phenolic compounds 51–78
enzymatic hydrolysis 111–13
epicatechin 65
Erwinia caratovora 226, 228, 232, 234
erythorbic acid 72
ester formation 146–9
ethylene 207–9, 213, 216–18, 232, 234, 280
etioplast 13
Evolvulus pilosis 39

fat intake 162
Ferula 189
ficin 69
Ficus 51, 69, 279, 357–8
fig, *see Ficus*
flavonoids 29, 52, 300–5, 313, 318–21
flavour 5–6, 162, 170, 206
flavylium 31–2, 41–2
fluorescence emission 42
  auto 191
  toxicity effect 341
food choice 157–70
  sensory attributes 158
  theory of planned behaviour (TPB) 159–61, 163–70
  theory of reasoned action (TRA) 160
Fortunella margarita 225
Fulvia fulva 227
fungus 73
furanocoumarins 347
furanone biosynthesis 146
Fusarium oxysporum 227

gallic acid 37
gamma radiation 4, 56, 218, 236
garlic 138, 306
genetic engineering 6
  ethical issues 166–9
geranylgeranyl diphosphate (GGDP) 272
gibberellins (GAs) 263–80
Ginkgo biloba 320, 355
Glomerella cingulata 225, 229
glucosinolates 140–3, 313–14, 322–4 (*table*), 337–9, 363–5
glutathione 73
glycoalkaloids 340–3, 361
glycoproteins (PRG) 88–9, 91–2
grape, *see Vitis vinifera*
grapefruit 146

guar, *see Psophocarpus tetragonolobus*
guarana, *see Paullinia cupana*

health 23, 25, 74
heart disease 287–306
Helicoverpa zea 353
Heracleum 180
  lanatum 181
honey 73
hydraulic signal 210–13
hypoglycin 332–3

isoflavones 336–7
isomerization 22
isoprenoid 17
  pathway 143–6
isothiocyanates 338, 363

kale 146, 339
kojic acid 69
kumquat 229, 236

laccase 55, 63
lactones 132
Lactuca sativa 12–13, 205, 213–14, 216, 218, 223, 226, 344
lactucaxanthin 12
Lapidium sativum 141
lectins 334
lemon, *see Citrus limon*
lettuce, *see Lactuca sativa*
Levistium officinale 181
light absorption 29
lime, *see Citrus aurantifolia*
lipid oxidation 54
lipoxygenases 22
Lotus corniculatus 251, 255–60
low density lipoprotein (LDL) 288–306
  oxidation chemistry 293
LOX pathway 128–32
lutein 12–13, 17, 294, 298, 316
lycopene 16, 18–19, 21–2, 294, 298, 316, 318
Lycopersicon esculentum, see tomato
  pimpinellifolium 271

macrophages 287, 289–92, 300, 303
maize 15, 211
Malus domestica, see apple
malvidin 45, 302
malvin 44
Mammea africana 180–1
mandarin, *see Citrus reticulata*
mango 52, 56–7, 109–10, 112–13, 116–17, 132, 146, 231–2, 237, 359
meat 162–3
Melilotus 181
Melissa officinialis 116
melon, cantaloupe 247–9
membrane depolarization 215
Menolinia fructicola 228
metallic complexation 45
molecular genetics 21–2
monoene 231
mouth 88
mucins 91–2
Mucuna deeringiana 213
mushroom 69, 74, 131
mustard 76
myocardial infarction 288, 291

Nectria galligena 224
neoxanthin 12
Neurospora 148
Nicotiana benthamiana 211
  tabacum 182
nitrate 343–5
nitrogen compounds, *see* amino acid
nuclear magnetic resonance (NMR) 118
nutrition 5, 23

oat 76
odour 109–21; *see also* aroma
olive oil 306
olives 56, 76, 130–1
onion 76, 138, 228, 301, 319
orange, *see Citrus sinensis*
oregano 76
oxalates 345–7
oxalic acid 69, 348
oxidation 51, 54, 63, 70, 126, 129–30
  see also enzymatic browning
oxygen 217

cell wounding 207–9
effect 58, 70, 234
oxidized LDL 289–94
uptake 189

parsnip 228
parthenocarpic fruits 271, 279
patulin 333–4, 355 (*table*)
Passiflora edulis 109–21
passion fruit, *see Passiflora edulis*
Paullinia cupana 356
PCR (polymerase chain reaction) 244–5
pea, *see Pisum sativum*
peach, *see Prunus persica*
peanut 76
pear, *see Pyrus communis*
pelargonidin 37
Penicillium spp. 221
digitatum 225, 229, 232
expansum 224, 333
pepper, *see Capsicum annuum*
peroxidase (POD) 51, 55, 210, 217
persimmon 52, 95, 101
Petroselinum sativum 180
Peucedanum 180
Pezicula malicorticis 224
pharbitins 38–9
Pharbitus nil 31, 37
Phareolus coccineus 270, 274
phenol 55, 58, 60, 67–78, 99–100, 120–1, 302
phenolic 110, 112–16, 206, 210, 313
compounds 51–78

phenylalanine ammonia–lyase (PAL) 210, 213, 217–18
Phoma complanata 228
exiguea 236
phospholipids 57
phytoalexins 190, 221–38, 359
phytochemistry 353–66
phytoene 16, 18, 22, 272
phytohormone, *see* ethylene
Phytophthora infestans 227
pigment bleaching 34
pineapple 51, 132, 146
Pisum sativum 57–8, 211, 213, 215, 265, 270, 273–8
plastids 16, 21, 59, 272
plums, *see Prunus domestica*
polyphenoloxidases (PPO) 54–60
polyphenols 29, 35, 41, 43
polysaccharides 99
pomegranate 95, 101
Ponciras trifoliata 180
pork 168–9
postharvest 2–3. 207, 221–2, 231, 238
disease resistance 223 (*table*), 234–6
potato
alkaloids 361–2
browning 69
nitrate 340–44
phenolic 52, 56–7, 59
phytoalexins 223, 226, 231–7
pigment 36
Prangos 181
processed fruits and vegetables
effects 210–17
mechanical shock 207
packaging 209
wounding 206–7, 210–18
proline rich proteins (PRP) 89–90, 103–4
proteins 99, 217
Prunus 57
amygdalus 109, 111, 146, 278
armeniaca 109–10, 112, 118, 132, 278, 332
avium 116, 211, 278
domestica 51, 278
persica 132, 278
Psophocarpus tetragonolobus 334
Psoralea 187
subcaulis 173
psoralen 187–90, 347
Pueraria lobata 213
Pyrus communis 65, 126, 279, 332

quercetin 75, 301–2, 314, 319–21, 325
quinones 60, 62, 71–2, 323

rape 76, 142, 252, 339, 365
raspberry 109, 120, 135, 146
respiration 216–17
retinol 316–17
Rheum spp. 70, 345
rhubarb, *see Rheum* spp.

rice 76
RNA 184, 217, 244
Rosaceae 96, 301
rosemary 76
Ruta graveolens 180, 191

Saccharomyces cerevisiae 148, 243, 245
saliva 92–4, 96–102
  anatomy 88
  changes in menstruation 93–5
  glands 88
  pH 93, 102
  protein 89–91, 100, 103
Sclerotinia sclerotiorum 225, 228–9, 236, 347
Sechium edule 270
senescence, peel 280
sesame seeds 76
sesquiterpene lactones 362–3
shelf-life 218
shellfish chitin 71
Sium 187
S-methylcysteine sulfoxide (SMCO) 339–40
Solanum dulcamara 357
  nigrum 357
  melongena 227, 279
  tuberosum, see potato
sorghum 252
soyabean 21, 52, 76–7, 211
  isoflavones 336
spectrometric techniques 46
spices 76
Spinacia oleracea 14, 57, 70, 272
  nitrate 344–6
  physical damage 211, 216
spinach, *see Spinacia oleracea*
Sporix® 70
Sporobolomyces odorus 132
storage 22, 210, 221, 233–5
strawberries
  aroma 132, 1135, 146, 148–9
  gibberellins 277, 279
  phenolic 32–3
  phytoalexin 233
stroma 16
sugar residues 31, 45
sulphur 5, 54, 61
  compounds, *see* amino acid
sunflower 211
superoxide dismutase 74
surfactants, *see* phospholipids
swede 15
sweet potato 15, 76, 223, 227, 237

tannins 7, 89, 95–100, 103
taste 87–104, 206
  bitter 92
  buds 88
  choice 170
  pH 101
tea 96 102, 260, 301
  antioxidant 76–7
  green 319
terpenes 143–5, 313
theory of planned behaviour (TPB), *see* food choice
theory of reasoned action (TRA), *see* food choice
Thuja occidentalis 356
thyme 76
α-tocopherol 294–8, 301
tomato
  aroma 129–30, 1137, 44, 146
  genetics 21–2, 244, 247
  genetic engineering 1167–9
  gibberellins 271, 279
  phenolic 56, 59–60, 121
  physical damage 207, 211–12, 215–17
  phytoalexins 231–2
tongue 87–8
tonka bean tree, *see Coumarouna odorata*
toxic compounds 331–49

ultraviolet
  colour inducement 4
  flavonoid 319
  furanocoumarins, absorbence 181–91, 347
  phytoalexin 226–7, 229–30, 236
  spectroscopoy 45–6

Vaccinium angustifolium 278
vacuum 70
vanilla 109–10, 112
violaxanthin 12

vitamin A 23, 313, 316–17
vitamin C 296, 298–301, 313, 333
vitamin D 346
vitamin E 295–7, 298, 318
Vitis vinifera
  aroma 120
  gibberellins 277, 279
  oxidative mechanisms 52, 57, 59, 70, 73, 76
  phenolic 29, 34, 42, 70, 73

walnut 52
wheat, durum 211
white cedar, *see Thuja occidentalis*
wine (red)
  anticarcinogen 319
  antioxidant 302–4
  aroma 112–13, 120
  astringency 96, 101–2
  oxidative mechanisms 74
  tannins 260
xanthophylls 11–13, 24
xanthotoxin 187–90, 228, 347
Xenopus 244

yam tuber 34, 36

Zea mays 258, 272–3
zeaxanthin 12, 17
Zebrina pendula 37, 39

The manufacturer's authorised representative in the EU for product safety is Oxford University Press España S.A. of el Parque Empresarial San Fernando de Henares, Avenida de Castilla, 2 – 28830 Madrid (www.oup.es/en or product.safety@oup.com). OUP España S.A. also acts as importer into Spain of products made by the manufacturer.

www.ingramcontent.com/pod-product-compliance
Ingram Content Group UK Ltd.
Pitfield, Milton Keynes, MK11 3LW, UK
UKHW041131071026
15587UKWH00009B/427
* 9 7 8 0 1 9 8 5 7 7 9 0 4 *